Kim,

I wish you all the best in everything you do.

Aloha,

The Small Manufacturer's Toolkit

Series on Resource Management

Titles in the Series

The Small Manufacturer's Toolkit

A Guide to Selecting the Techniques and Systems to Help You Win

Steve Novak

with a foreword by Senator Daniel K. Inouye

Auerbach Publications
Taylor & Francis Group
Boca Raton   New York

Published in 2006 by
CRC Press
Taylor & Francis Group
6000 Broken Sound Parkway NW, Suite 300
Boca Raton, FL 33487-2742

International Standard Book Number-10: 0-8493-2883-7 (Hardcover)
International Standard Book Number-13: 978-0-8493-2883-1 (Hardcover)
Library of Congress Card Number 2005045210

Library of Congress Cataloging-in-Publication Data

Novak, Stephen.
    The small manufacturer's toolkit : a guide to selecting the techniques and systems to help you win / Stephen Novak.
        p. cm. — (St Lucie Press series on resource management)
    Includes bibliographical references and index.
    ISBN 0-8493-2883-7 (alk. paper)
    1. Production management. 2. Manufacturing processes 3. Small business—Management
I. Title. II. Series.

TS155.N657   2005
658.5--dc22

                                      2005045210

Dedication

This book is dedicated to my wife, Arlene,
without whom nothing would be possible.

Contents

Foreword

Small businesses are the lifeblood of our economy, and the spirit of America's entrepreneurs revitalizes our communities. Today's marketplace offers unprecedented opportunities — a small business can extend its reach globally to find partners all over the world and develop as a member of a multinational production chain. Through the Internet and advanced telecommunications, consumers can seek out products and knowledge tailored to meet their individual needs instead of settling for what is locally available.

Seizing these opportunities challenges the ingenuity of an entrepreneur. Now more than ever, the business strategy, accounting practices, and attention to the regulatory environment must temper a small businessperson's passion and expertise. Without attention to the business side of creating products and services, a small business may face difficulty and perhaps even failure.

Mentors and technical assistance from government programs can help, but ultimately the responsibility for managing a business falls on its owner — and that means attention to detail and much hard work. Fortunately, small business owners possess both of these characteristics in abundance, and with the valuable guidance of *The Small Manufacturer's Toolkit*, success is within reach.

Senator Daniel K. Inouye

Acknowledgments

Many people helped to make this book possible, although some of them don't know it. Many people have either taught me or encouraged me through the years, but I have to especially thank Cecily Barnes, who got me involved with APICS and started me down this path. I also must thank Mel and Sue Nelson and Bill Latham, who have taught me many things through the years and have shown me what dedication and profession-alism can do. Special thanks to Ken Nixon for his friendship and encour-agement. Thanks to Dennis Lord, Lee Wallace, Becky Morgan, Walt Williams, and the rest of the APICS SM SIG committee for allowing me to work with them and learn from them. Special thanks to Gavan Daws for the valuable advice that made this book possible. And thanks to all my friends and colleagues, who are too numerous to mention, but who keep me going through good and bad. Aloha.

Introduction

The purpose of this book is to help guide the reader through the maze of management philosophies and process improvement techniques that are constantly being promoted. Trying to sort through all the tools available and deciding which one, if any, you should adopt can be overwhelming. Moreover, small manufacturers often lack the resources and expertise available to larger organizations, making it even more difficult to choose. This book attempts to help you learn enough about the tools available so that you can make an informed choice. Many books and other resources detail the individual tools discussed in this book (see Appendix A), but few present the tools together in a format that allows you to compare and contrast them. Also, a large number of current texts are implementation or how-to guides. This book purposely avoids that format because it is not intended to advocate any one tool over another. After reading this book, and using the diagnostic tool, you should be able to make an informed choice on which tool or tools you plan to adopt. Once you decide on the tools you will use, you should find the resources to help you implement them correctly. You will gain the most benefit from this book by reading through the chapters in order, because some of the topics in later chapters build on ideas discussed in earlier chapters. However, the intent was to write the book so that each chapter can be read independently, or you can skip around as the topics interest you.

Every tool presented in this book has some relevance to a manufacturing organization. Many of the tools work well together or share some characteristics. There is no natural progression of all the tools, but I have attempted to order the chapters in a logical manner. Chapter 1 introduces you to the ideas behind this book and will help you think about your strategy through the rest of the book. Chapters 2 and 3 relate to planning.

Chapters 4 through 7 relate more to execution and control, and Chapters 8 through 11 discuss more ancillary tools, or tools that fit on top of the other tools. Chapter 12 helps you work through the decision of which tool or tools you should adopt or consider. Appendix A provides a partial list of resources to find out more about each of the individual tools.

Chapter 2 presents some of the most fundamental inventory management and planning tools that have been proven over many years. This chapter will help you assess your current performance, which will give you a starting point for where to go next. A thorough understanding of the basics and flawless performance in the use of these tools is a necessary first step for any organization that wants to improve. If you cannot perform the basics well, the chance of success with more advanced techniques is less likely.

Chapter 3 discusses a powerful yet underutilized high-level planning tool. Planning is done in all organizations, but not always in a structured and formal way. You can execute your daily tasks flawlessly, but if these tasks do not mesh with the company's plans and strategy, it's all a waste. The sales and operations planning stage occurs between the high-level strategic plan and the midrange master production schedule.

Lean Manufacturing is a popular topic and a frequently implemented method of operation. Chapter 4 discusses many aspects of Lean Manufacturing systems. Popular does not mean it is right for everyone and in every situation. This chapter will help you decide if Lean is right for you, or which aspects of Lean might be appropriate for your company.

Six Sigma is often so intertwined with Lean Manufacturing that many people think they are, in effect, one system. Chapter 5 follows the chapter on Lean for this very reason. Many companies implement Lean Manufacturing without Six Sigma and vice versa. However, a growing number of organizations are implementing systems they call Lean Sigma, Lean Six Sigma, or some other variation, which formally incorporates both Lean Manufacturing and Six Sigma.

Chapter 6 covers several aspects of the management and scheduling systems developed by Eli Goldratt and presented in his near-famous book, *The Goal*. Many companies spend time reading and discussing *The Goal*, but the implementation of the theory of constraints (TOC) is not as widespread as the book. This chapter explains some of the aspects of TOC that cause confusion, which hinders implementation.

Chapter 7 reviews several of the most popular quality tools and quality management systems. Their names are widely known, but the details of how they work are not as widespread. Quality is, or should be, a foundation of any manufacturing company (any organization, really). Quality is often given lip service without the systems to back it up. This chapter will help you gain a better understanding of some of the more popular quality management systems.

Implementing any of the tools in this book would probably be considered a project. Many of the tasks you undertake in the course of your work are projects. Because projects are so prevalent, a thorough understanding of project management is necessary to help ensure success. Chapter 8 explores project management, to help you identify and manage projects more effectively. Not all tasks require formal project designations or management as projects, but you need to know when they do and when they don't. This chapter should help you determine when to work a task as a project and how to manage it when it is.

If your company does not already use an enterprise resource planning (ERP) system, it probably will soon, or you may be planning upgrades or enhancements. The use of computerized planning and execution systems is almost required, even in the smallest organizations. Chapter 9 examines the requirements surrounding the use and implementation of ERP systems. This chapter will discuss some of the many things you must consider when deciding on an ERP system. The actual choice of systems and the implementation comes late in the decision and selection process.

Chapter 10 examines tools that reach beyond the walls of your company. No company stands alone, but instead fits into a chain that stretches from the mining or processing of raw minerals and compounds all the way to the final consumer or user of a product, and beyond, all the way to the waste stream. First, companies must recognize that they are part of a supply chain; then they must utilize tools that allow them to be successful members of the chain.

We are living in tomorrow. Technological advances are made daily across the globe. One visible reminder of this is the popularity of the Internet. Chapter 11 explores some of the aspects of doing business in the modern, electronic world. E-commerce is multifaceted, and the company Web site is just one element. This chapter will focus on some of the most common aspects of modern business that fall under the heading of e-commerce.

Finally, after gaining an understanding of the various tools available to the small manufacturer, what comes next? Chances are good that you are considering implementing one of these tools or you simply know you need to undertake some sort of improvement activity. Chapter 12 helps you work through the process of deciding which tools, if any, you should consider more closely. Every organization is different. You have different products, different customers, different delivery channels, and different resources. The tools presented in this book are appropriate in a wide variety of situations and organizations, and many of them have aspects that complement or overlap others. However, you will not be able to implement all of these tools, at least not all at once. You may want, and be able, to implement some aspects of several of the tools, but you will probably have to choose which parts to implement first. This chapter, and the diagnostic tool included, will help you work through these decisions.

Author

Steve Novak is the founder and president of PPR Management Services, LLC. He has held a variety of management positions in various industries with responsibility for operations, purchasing, inventory management, planning, production scheduling, warehousing, safety, sanitation, and facilities. Steve has performed process improvement activities throughout his career, has worked to develop quality systems, and has held key positions with multiple enterprise resource planning (ERP) system implementations. He has earned the prestigious certified in production and inventory management (CPIM), certified in integrated resources management (CIRM), and certified quality manager designations. He earned his bachelor's degree in business administration with accounting concentration from the University of New Mexico, Albuquerque, New Mexico. He is currently working on his MBA with industrial management concentration from Baker College, Flint, Michigan.

Steve has been an officer of the Hawaii chapter of APICS (the Association for Operations Management) for several years, holding various positions including president. As the chapter's vice president of education he developed the chapter's education program and has served as an instructor since its inception. He has taught CPIM review courses and fundamentals of materials and operations management courses. He served on the APICS small manufacturing special interest group (SM SIG) steering committee and the APICS 2004 nominating committee. He authored, with Lee Wallace, the inventory essentials for small manufacturers course for the SM SIG.

In 2004 Steve served as an examiner for the Hawaii Award of Excellence, which is based on the Malcolm Baldrige National Quality Award. He is a member of the following organizations:

- APICS — The Association for Operations Management
- ASQ — American Society for Quality
- Hawaii Food Manufacturers Association
- Chamber of Commerce of Hawaii
- Hong Kong China Hawaii Chamber of Commerce
- ISM — Institute for Supply Management

Chapter 1

Assembling Your Toolkit

There are many tools, techniques, and theories on how to operate your business. There are many process improvement methodologies. There are innumerable consultants ready to sell you their brand of services that will transform your business. There are numerous societies and organizations devoted to each one of these tools, both nonprofit and for-profit organizations. How do you decide what is best for you?

Too often, businesses make the decision of what tools to use based on incomplete information. It is a hectic and competitive world. Who has the time to study all of the tools and techniques available, compare them, and make an objective decision on which direction to take? The answer, really, is no one. Add to this the small manufacturer. Small manufacturers face many of the same challenges as their large counterparts, but small manufacturers do not have the same resources available. Trying to sort through this maze is especially difficult and challenging for the small manufacturer.

Many people make a career of studying, in depth, only one, or a small number, of these tools and techniques. There are organizations and consultants dedicated to only one particular methodology. It can become mind boggling and frustrating. Small manufacturers, even more than larger organizations, rely on outside experts to help guide them through the decisions of selecting a business process or management philosophy to adopt. Small manufacturers simply do not have the level of expertise (usually) that can be found in a large company. Outside experts can be beneficial, but reliance on them to make the incredibly important decision on how to operate the business can be damaging. Outside experts do not know the details of your business as you do. Although they can learn a great deal in a short amount of time, they cannot fully understand your culture, know your history, nor be totally aware of your financial condition. Outside experts and consultants should be used when appropriate, but the important task of making an informed decision about the management systems and process improvement tools that you will use should be made within your company. Once this decision has been made, you can, and probably should, bring in outside experts to help implement the method you have chosen.

The problem is, you may not have the knowledge to make those informed decisions. Thus, you may make decisions based on luck ("I heard a great presentation on Lean Manufacturing"), hype ("Everybody's doing Six Sigma. We'd better get on the ball"), and marketing ("Increase your bottom line using theory of constraints"), rather than on objective comparison, need, and appropriateness to the organization. The good news is that all of these tools are useful and will provide some benefit if applied correctly. The bad news is that you can spend a lot of time and money implementing tools that are not appropriate to your unique situation. You want to maximize every dollar spent. Implementing the right tools at the right time will help ensure the greatest return on your investment. The goal of this book is to provide enough information on the various tools to help you make a better decision on which ones to use. This book is targeted specifically to the small manufacturer. Consider it an investment in knowledge that will contribute to your success.

1.1 Knowledge Is Power

As a small manufacturer, you operate in a competitive environment, not just with companies that make the same or similar products, but even with companies whose products or services are very different from yours. Consider the following example. An office furniture manufacturer that makes bookshelves is in competition with an Internet search engine. If people stop buying books because they can get all their information on the Internet, sales of books, and then bookshelves, will plummet. Think that may be a bit of a stretch? I don't think so. Competition is everywhere. It's across the street, it's across the country, and it's across the globe. Advances in technology, improvements in materials and product design, trade agreements, the availability and cost of labor, and other factors contribute to competition. Small manufacturers, with their fewer resources, are especially vulnerable to competition.

Competition is not a death knell for the small manufacturer. Competition is good. Competition drives innovation and improvement and often facilitates grand achievement. However, you must be prepared for competition, and this preparation must take a multifaceted approach. You will not have the expertise or resources available to cover every facet right away. Some areas where you need to prepare for competition are the general business climate; the political and regulatory environment; your supply chain; the logistics involved with your industry and your products; your plant, equipment, and other capital investments; your internal systems; your financial situation; your workforce; and your culture.

Although you have little control over some of these areas, such as the general business climate or current economic conditions, you can exert some influence. By remaining a vital, competitive company you are contributing to the economy and helping to maintain or improve the business climate. Overall, small manufacturers are struggling, but thousands of remain successful and are an important part of the economy. Stay competitive and you will be a contributor to the success of the entire manufacturing sector. As a small manufacturer you may find it difficult to have an impact on the laws and government regulations that affect you — if you try to act alone, that is. However, as a group you have a strong voice. You can become involved in any number of trade groups and advocacy organizations. Many of these organizations devote a large part of their time and resources to lobbying local and national lawmakers on behalf of themselves and their members and constituents. Many also provide education and training to their members. Your local chambers of commerce, industry trade associations, and national organizations, such as the National Association of Manufacturers, are examples of such important groups.

One of the areas over which you have the most control is your internal systems. This book focuses primarily on the internal systems and the choices of tools that you have available. Education in the various tools is how this book will help you. You must constantly improve your internal systems to remain competitive and successful. Your internal systems cover all areas of your organization, from the accounting and human resources functions, to the operations and manufacturing areas, to maintenance and customer service. Your internal systems also encompass those functions that interact with your customers and suppliers. Purchasing, logistics, and customer support functions are internal systems, even though they have a heavy emphasis on looking outside. If any area is not keeping up or not operating at peak productivity, the entire organization will suffer. This book deals primarily with the operations and manufacturing areas and the tools available to help you operate these areas more effectively. Many of these tools are also applicable and appropriate in the support functions and other areas of the organization.

Another area over which you have great control is the education and training of your employees, and maybe even your suppliers and customers.

You wouldn't go to a doctor who hasn't had any training beyond medical school. You wouldn't allow an accountant who hasn't kept up with the latest tax laws to complete your tax return. Why would you let the managers and employees of your company make vital, daily decisions without ongoing education and training in the latest tools and techniques in business and manufacturing management? Many small manufacturers have little or no formal education in materials management, operations, logistics, supply chain management, or any of the areas that are important to the success of the company. Although many are very successful and have employees who are well educated in general business and have many years of experience, it is difficult to make the improvements necessary to stay competitive without ongoing, formal education. Education (classes, seminars, conferences, books) is often viewed as an expense or a luxury. Education, ongoing education, is a necessity and an investment.

Education and training should be treated and evaluated as an investment. You wouldn't buy any old piece of equipment you found, just because the price was reasonable. And you shouldn't send your employees to educational offerings just because they are available and reasonably priced. You buy equipment that you need to replace outdated or obsolete machinery or to support the strategic goals of the company. The same should be done with education. You should educate or reeducate your employees to replace outdated ideas or techniques, and the education should support the strategic goals of your organization. Just as you would perform a return on investment analysis for equipment purchases, you can do the same for education. What is the investment in the educational offering? Consider the price, time, and other expenses. What should you expect in return for that investment? Will you be able to schedule more effectively so that throughput increases? Will you be able to increase quality, thus reducing scrap and rework and, therefore, costs? This is the analysis you should undertake when addressing your education needs.

When you can't find the equipment you need locally, you search nationally or globally. When you can't find exactly the equipment you want, you work with the manufacturer or distributor to have it made to your specifications. If you can't find the education you want locally, you should search nationally and globally. If you can't find exactly the education you need, you should work with the education suppliers to tailor the education to your needs. Where do you find education? Many options and opportunities exist. I advocate using professional organizations, some of which you will find listed in Appendix A. Besides providing educational opportunities, usually of very high quality in a variety of formats, they present exceptional networking opportunities. Chances are good that other companies have experienced many of the same challenges that you have, and their employees are probably willing to talk to you and help you.

Other avenues for education are industry and trade organizations, for-profit education providers, and universities and community colleges. Education opportunities are increasingly available over the Internet, from a variety of sources. Given the great availability of education, the only hurdle is a lack of desire and commitment.

I firmly believe that the small manufacturer can compete in today's global marketplace. It may not be easy and there are many obstacles, but you can compete. You may not be able to compete as the low-cost producer of mass-produced products, but you can compete in smaller or specialized niches or in markets where cost is not the only or primary consideration. You can compete and you can succeed. To compete, you need to have systems in place that allow you to produce your products at the lowest cost and the highest quality and ensure that they are delivered to your customers when they want them. You need a well-educated and well-trained workforce. You may need to make some changes to your operations. If you do not already have the right systems in place, you need to change the current systems. And that is the $64,000 question — how do you know what tools to use and what to change to?

That is the purpose of this book: to give you enough information on the various tools at your disposal, so that you can make an informed decision on which ones to use. As you narrow down your choices, you will need to attain more in-depth knowledge of the tools you choose. A great many resources are available (see Appendix A). This book is not an implementation guide. When you get to the point of implementation you will probably need to find outside help. The last chapter of this book provides information on how to find such outside expertise. This book is only one piece of your education. You need more, and you need an ongoing program. As they say at APICS — the Association for Operations Management — "lifelong learning for lifetime success." This applies to you as well as your organization.

1.2 Assembling Your Tools

This book is intended to add to the tools in your toolkit and to explain, or clarify, some of the tools that may already be in it (Figure 1.1). These tools can help you become more successful; in other words, more profitable. You increase your chances of being successful and profitable when you are better able to manage your assets, improve your processes, increase your productivity, and enhance your quality. Although success and profitability cannot be guaranteed, the odds can be increased by adding tools to your toolkit and knowing how to use them, when to use them, and why to use them. Just implementing some new technique or

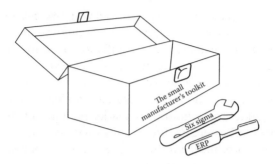

Figure 1.1   Toolkit.

system is not enough. In fact, it can be the wrong thing to do. You need to know the strengths of each tool as well as the costs and resources needed to use them. You need to know as much as possible about a wide variety of tools, so that you know when it is appropriate to use each one and so you are able to pick the right one at the right time. This book attempts to help you by making you aware of and more familiar with the variety of tools available to you as a small manufacturer.

You may notice that I did not mention lowering costs in the preceding paragraph. Cost has many components, some of which you have more control over than others. However, there are usually tradeoffs involved when trying to control the different components of cost. For example, although you have some control over your labor costs, wages are largely determined by market conditions; the overall economy, the availability of labor, prevailing wages, and whether the workforce is unionized. You can choose to pay less than market wages, but you have to consider whether productivity and quality will suffer. You may choose to use overtime regularly rather than hire additional workers, but again, you need to determine whether productivity and quality may suffer in the long run. And you need to determine if you might see increases in injury and health-related costs when workers are on the job for long hours over extended periods. This book does not address cost reduction directly, but by choosing and using the tools that are most appropriate to you, you will gain more control over your costs.

All of the tools in your toolkit are valuable when applied correctly. However, just like using a hammer when a screwdriver is called for, misapplication of the tools can be costly. These costs include dollars spent, time lost, and a decline in morale and productivity. Many of these tools, when applied completely and correctly, require multi-year implementations. Losing interest halfway through an implementation can be costly and devastating. All of the tools require commitment; most require sustained executive-level commitment. Before committing valuable resources, it is best to learn as much as you can about each of the tools to increase the chances of success.

Some of the tools contain many similarities, causing confusion and making choices between them difficult. Some of the tools complement each other, making them even more powerful. Some do not require full implementation but can be very valuable if only parts of them are used. This probably makes the whole idea of deciding which tool or tools to use even more confusing and frustrating, but this book will attempt to clear up the confusion and give you enough information to set you down the road to success.

The final chapter of this book provides a diagnostic tool that will help you work through the choices and selection of tools. Your company is unique. Your products may be unique or may be widely available, but the combination of your products, processes, location, management, work-force, level of education and training, and corporate culture makes your company unique. All of these factors affect the choice of tools that you will or should adopt and implement. The performance level of your existing systems will affect your choices. If you are planning well but not executing as well as you would like, you will choose different tools than if you have flawless execution but no clear plan or direction. The regulatory environment, customer requirements, complexity of your product, and any storage and delivery requirements will affect your choices. The diagnostic tool provided in Chapter 12 should help you work through the complexities of deciding which tools are appropriate to your unique situation.

My hope is that you will use this book as a guide and reference to help you make any necessary changes to your operations that will allow you to maximize your profits, realize your potential, and compete in the global marketplace. Let's go!

1.3 Challenges of the Small Manufacturer

Small manufacturers face many challenges (Figure 1.2). Although you share many of the challenges of your larger cousins, you also face challenges unique to a small business. Small manufacturers have fewer resources available. They have fewer employees, which means they have fewer specialists, or fewer people who are experts in any one particular area. Managers at small manufacturing companies tend to oversee multiple functional areas; that is, they wear many hats. They are often responsible for several departments, some of which they have little experience with. Much of what they know about the functions they manage they have learned on the job or "the hard way." Whereas a larger company might

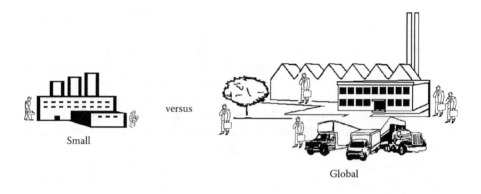

Figure 1.2   Small Co. versus Global, Inc.

have a separate warehouse manager, logistics manager, and supply chain manager, a small company may have one person responsible for all of operations. In the larger company, each manager may have a college degree, specialized training, and many years of experience in their particular specialty. The manager at the small manufacturing company may have worked his way up from a forklift operator, to supervisor, to manager with only experience and on-the-job training. These are generalizations; the small manufacturer may be very successful and the manager highly competent, but size and numbers limit the small manufacturer's potential for growth and improvement. To grow and improve, you need to be exposed to new ideas and other ways of doing things. Managers at small companies may have inventive and innovative ideas about how to change and improve their operations, but why reinvent the wheel? There are proven techniques that can be adopted and known pitfalls to avoid. Through education and networking, you can gain the needed expertise in these techniques and learn about the pitfalls. If you do not have the expertise in-house, you will have to find it outside your company. Although outside expertise is valuable, and often desirable, it limits your company's growth because you are dependent on outsiders who do not know your business as well as you do and aren't always available when you want them. If you don't even know that there may be better ways to do things, you may not know that there are experts available to help you in those areas.

Managers and staff that wear multiple hats cannot focus all their energy on one area. They may also have conflicting priorities between the various tasks for which they are responsible. If you are responsible for both production and warehousing, one part of you may want to boost production to increase inventory levels, whereas another part of you may want to lower production so inventory in the warehouse decreases. In some cases, these conflicts and decisions may be easier for the small manufac-turer, but in other cases they may be more difficult. A larger company, with separate managers, may have more incentive to develop a solution that benefits both production and the warehouse. The manager at the small manufacturing company may simply balance the needs of each department on a daily basis, rather than coming up with a long-term, systems approach. Add to this that the manager at the small manufacturing company may not have a high level of education and expertise in each of the areas, so the solution may not be as effective as possible. Remember, the small manufacturer may be very successful and the staff highly effective, but to remain successful, to grow, and to compete in the global marketplace, you need to move to the next level. You need to improve your operations, streamline your systems, increase your quality, lower your costs, and improve your service. To do all that you need to expand your knowledge.

The lack of formal education in production and inventory management, operations, and quality is not easily overcome for the small manufacturer. Employees at small manufacturers often cannot take the time off for education courses because they do not have the level of backup for managers and staff that exists in larger companies. Work hours and the employees' other commitments limit the after-hours time available for education. Small manufacturers often have small, or nonexistent, budgets for education and training. There may be state or federal funding available for workforce education and training, but the small manufacturer may not have human resources personnel with the knowledge and experience needed to get this funding. Much of the training at a small manufacturer is on-the-job training. Formal training programs may not exist. Managers and staff are often expected to join the company with the necessary education and experience. Even if this is true, education becomes outdated and must be kept current. There also seems to be a lack of material specifically targeted to small manufacturers, which compounds the prob-lem. This book attempts to fill part of the gap of available material.

The small manufacturer may have fewer cash reserves and more cash flow concerns than a larger company. Small manufacturers have less investment in plant and equipment than larger manufacturers. This lower ability to generate cash and lower investment in capital equipment works to limit the small manufacturer's ability to raise additional capital and increases the reluctance of investors to make investments in the company. This means that the small manufacturer may need to utilize their equipment longer than they would like, rather than purchasing new equipment. This then limits their ability to innovate and grow. Productivity gains that might be achieved with new equipment must be attained through other means and might not achieve the same level of improvement that would be made with new equipment.

A related issue is space — production, warehouse, and office space. Additional equipment, if it can be obtained, may not be possible if there is not sufficient space available to operate it safely and effectively. Even if you use your warehouse space as efficiently and effectively as possible, there may be a point when the space is no longer sufficient. Even if you are moving toward Lean operations that would allow you to produce the same volume in a smaller amount of space, maybe you have made changes in the way you do business, or maybe your volume has increased enough so that you now need additional warehouse space. If you do not have the space readily available, or you cannot easily obtain it, you will be limited in the changes you can make or your ability to service your customers. Office space may not seem as important to a manufacturing company, but your office and administrative personnel have a tremendous impact on the success of your organization. Just as your production employees need sufficient space to work effectively and safely, so does your administrative staff. If you cannot add staff because you have no place to put them, or you pack them tightly into cubicles, the performance of your entire organization will suffer.

Small manufacturers do not have the leverage with suppliers that larger manufacturers have, because they cannot purchase in the same volumes. You may not receive the same level of service as your larger counterparts. You will receive fewer quantity discounts, and the discounts you do get may be lower. The lot sizes you are required to purchase may be determined by the supplier rather than by your needs. Your shipping costs per unit may be higher because of the low shipping volume and the type of carrier used to ship smaller quantities. You may also lack the logistics expertise in-house that could help you lower your shipping costs. Your payment terms may not be favorable. Your supplier base may not be as extensive as that of a larger company. Some suppliers may not be willing to deal in the smaller quantities that you require, which limits your choices of suppliers. Suppliers may not be fighting for your business, because of the volume.

A serious challenge for many small manufacturers is a small customer base. You may rely on only a few customers for the majority of your business, and in some cases you may have only one customer that accounts for almost all of your business. In this situation, if one of your primary customers suddenly decreases their orders or completely stops buying from you, you face the possibility of going out of business fast. Until you can find new customers, you face a loss of income while still having the costs of keeping your company alive.

The good news is that all of these challenges and constraints can be overcome. It may not be easy, but you can do it. This book presents some of the tools available to you to overcome these challenges and to remain profitable and successful. If you are not already profitable and successful, the tools discussed in this book can help you get there. They won't be all that you need. You need to have a variety of tools at your disposal, and you need to know which tool to use for each situation. Add these tools to your toolkit and work your way to increased profitability, growth, and success.

Being a small manufacturer has its benefits as well. Because price may not be the deciding factor in your customer's decision to buy from you, you may be able to compete more effectively on quality and service than larger competitors. The multiple hats that can be a disadvantage in some situations may be a benefit in others; staff may be more flexible and innovative because they have experience in multiple areas of the company. And sometimes a lack of formal education in a particular discipline reduces barriers to change; if you don't know that you can't do something, you might just go ahead and do it. However, I advise getting the education while keeping an open mind to new ideas and innovations.

MATERIALS
MANAGEMENT

Chapter 2

The Basics: Are You at the Starting Line?

Many organizations are not achieving the results that they would like. They are not as productive or efficient as they think they can be; they ship more orders late than they feel is acceptable; they are out of stock of materials when they are needed; and they cannot rely on their data and records. At some point, these companies will begin to look for a fix for their problems. They will send their managers to presentations and seminars looking for the newest management program; they will hire consultants; or they will simply pressure everyone to "get better, or else."

Many of these organizations will fail to take the most basic step of evaluating their current processes and procedures. There are two things that you can do right now, before going out and looking for the Holy Grail of manufacturing perfection. First, review your current operations as compared with your established policies and procedures. Are you following your procedures as they are documented? (They are documented, aren't they?) Presumably, you established procedures that were found to be appropriate and effective for you. If you have established policies and procedures but are not following them, this is an immediate cause for concern. Review the procedures to determine if they are still appropriate, then either enforce them or change them and enforce the new procedures. Policies and procedures are worthless if ignored or not enforced. If you find that your current policies and procedures are no longer appropriate or need to be changed or modified, go ahead and change them. Determine what the process should be, document it, and enforce it.

The second thing to do is determine if you are performing the basics well. By the basics, I am referring to the most fundamental processes of planning and execution related to manufacturing. These basics, or fundamentals, are well-established and proven techniques that have been used successfully for decades by thousands of organizations. However, many companies do not perform these basics well. In fact, many companies do not even use them at all, and many managers have no experience with and may not even have heard of some of them. I can attest to this from personal experience, so if you or your company falls in this category, don't be embarrassed. You should feel embarrassed only if you know what you could be doing but aren't doing it. But be heartened — in this chapter we will review some of the basic tools that every manufacturer should at least be familiar with, if not performing well.

Excellence in performing the basics in manufacturing is as critical as the flawless performance of the fundamentals in sports. I have played rugby for many years and I have seen firsthand the difference between teams that execute the fundamentals well and those that do not. Anyone familiar with athletics knows it's the same in every other sport. For me, watching the 2003 Rugby World Cup highlighted the difference too. Complex plays and tricky passing invariably fell to well-controlled balls and solid tackling. Before entering the field for a match, a good team will have a strategy. You will know at least a little bit about your opponent; what their style of play is, how big they are, how fast. This will, in part, determine your strategy. If your opponent is big, but relatively slow and likes to control the ball in the forwards, you may want to try to open up the field and run at them with your faster back line. Once you are on the field, you have to execute your strategy. If you want to win, you need to execute flawlessly, especially the basics, and you need to work as a team. Very often, the team with the best athletes will lose because everyone is trying to be the star instead of working together. Ball control, tackling, and avoiding penalties are probably the most fundamental aspects of rugby. These must be practiced every day. Ball control includes keeping hold of it in various situations (it gets pretty rough out there) and passing. Behind-the-back passes, overhead passes, and 30-meter passes look impressive, but they are prone to failure. One bad pass can lose the game in a hurry. Short, quick, controlled passes are preferred, and they need to be practiced. And your teammates need to be right where they are supposed to be. One-armed tackles and bone crunching hits may impress the fans, but clean, solid tackles will stop your opponent, break their momentum, and win games. A fast back-line runner, dodging tackles, and running back and forth across the field looks great on television. But these bursts are few and far between and often lead to mayhem, confusion, and turnovers. It is better to run straight up the field, use your support, and control the ball if you want to win the game. Avoiding penalties is important, because the opposing team can add points as a direct result of your penalty. If the opposing team is within kicking range, they can kick for points. More than one rugby game has been won with only points from penalty kicks.

The rugby game can be used to illustrate challenges in the business environment. The opponent is your competition. The basics of ball control, tackling, and avoiding penalties are the same as performing the basics of planning, executing the plan, and conforming to applicable regulations and laws in the manufacturing plant. Before trying to implement the hottest new management system, perform the basics well. You need to practice every day. This means follow your procedures and policies, establish relations with your suppliers, and service your customers. Review your current operations, processes, and procedures and make sure you are performing them flawlessly before trying something new.

This chapter will help you see if you are even at the starting line with the competition. Are you performing the basics well? We will go over some of the basics, and you will be able to assess how you are doing with them. However, even if you are not doing the basics well, that doesn't mean you can't use some of the tools in the toolkit. But it does mean that you should assess whether your organization is ready to move on to something new. It takes discipline to perform the basics well on a continuing basis. If you don't have the discipline to do this every day, you should assess whether you have the discipline to move on to more advanced or complex systems.

2.1 Inventory Basics

Let's review some fundamental things about inventory before going any further. You are manufacturing something, so you will have and will be responsible for inventory at some point. Some understanding of inventory is essential to establishing a firm foundation upon which to build. If you do not understand your inventory, are not managing it well, and are not keeping good control of it, it may mean that your organization is not ready to use more sophisticated or more complex tools.

Inventory is increasingly being outsourced, or the responsibility for managing inventory is being pushed back on to the supplier. Vendor-managed inventory is an option for many organizations. However, even if you do not own and manage some portion of your inventory, you need to understand inventory management and the techniques associated with it. Even retailers and distributors will benefit from knowing more about inventory management as it relates to the manufacturing process. The more knowledgeable you are about your suppliers' and customers' systems and techniques, the better able you are to work with them to your mutual benefit.

Start with the basics and do them well, then evaluate your needs. Remember, just as in sports, if you can't execute the basics well, you're not ready for the complex plays.

2.1.1 Why Do You Have Inventory?

Too many people do not really understand why they have inventory. It is so ubiquitous that most people never even question it. Occasionally, someone, often an executive, will question the amount of inventory. Or, more typically, they will just come out and say that inventory needs to be reduced. This is a blanket statement that lacks any specifics and is left to someone else to interpret. However, after the command to reduce inventory is given, an inventory reduction program usually starts. Maybe this is a good idea, and maybe it's not.

I would agree that many companies hold too much inventory, but I wouldn't agree that you should just make a generic statement to reduce inventory. Besides, many organizations that have "too much" inventory also have a lot of out-of-stock situations or cannot deliver customers' orders on time; so something is out of whack. You must understand why you have inventory before you can determine if you have too much or too little.

As a manufacturer you are building a finished product from raw materials or purchased components. No matter how fast the process, you will have inventory at some point. It is possible that you could pre-sell the finished product before it is made and even before you buy the raw materials, but you will have inventory to manage at some point. You need to have an inventory of materials and components so that you can build the finished product. Whether this inventory is officially owned by you and listed as an asset on your books is irrelevant. You need to have it, so you need to take the ultimate responsibility for managing it. If you are outsourcing the management of the inventory, you still need to take responsibility for ensuring that you have what you need when you need it. You may also need to have an inventory of finished products if your customers are not willing to wait until you build them. During the production process you will have an inventory of partially completed items (work in process, or WIP). If you cannot complete the production of the finished items in one workday, you will have WIP inventory until you can complete the production. In short, you need to have inventory

Table 2.1 Roles of Inventory

Fluctuation inventory
Anticipation inventory
Transportation inventory
Hedge inventory
Lot size inventory

to support production and sales. You need to have raw materials and components available when they are needed for production, and you need to have finished goods when the customer wants them.

One important point is that you need to plan your inventory. "Inventory happens" is not a good slogan for your company. Without good inventory management, total inventory can increase or decrease, or certain pockets of inventory can increase or decrease, all to the detriment of the company. Too often, inventory results as a by-product of other policies and procedures. The sales department may call for increased inventories for one reason or another, the production department may want to increase the efficiency of their schedule, and accounting may think that inventory is just too high. Although valid reasons may exist for increasing or decreasing inventory levels, inventory cannot be managed in the aggregate. Customers buy individual products, and these products are made from individual components. Each of these individual products and components must be managed separately. This does not mean that you can't group them together or set policies for a range of items, but it is difficult to manage your inventory effectively as one aggregate group.

Why else do you have inventory? There are several reasons, and some of them are good reasons. Keep in mind that these inventories should be planned; don't just let them happen. The five roles that inventory plays (Table 2.1) can result in your company having inventory on hand. The following is a brief explanation of each type:

1. *Fluctuation inventory* — This is also referred to as buffer stock or safety stock. It is inventory that is held as protection. It protects you from forecasting errors or other statistical fluctuations. Forecasts are never 100 percent accurate. You need to protect yourself from those times when customer orders are greater than the quantity that was forecast. This means that you need to have more inventory on hand than the forecast says you need. The level of inventory you keep as protection against these forecast errors will depend on how big or how consistent the errors are over time. If the forecast is off by large amounts on a regular basis, you will need to keep more inventory on hand than if the forecast is off

only a little bit. Statistical fluctuations are variations that are a part of any process. Customers will not order the same quantities each time, and the time between orders will not always be the same. These variations may be small, but they will exist, and you need to account for them. (See Chapters 5 and 7 for more information on statistical fluctuations.) Safety stock also protects you from uncertainty in supply and demand, which are affected by many factors. This is especially true if you are buying from foreign suppliers or selling to foreign markets. Suppliers' and customers' lead times are often not fixed or consistent. A supplier's lead time may be affected by the size of your order, their order backlog, changing regulatory requirements, and other factors. A customer's needs may vary depending on their customers, unidentified seasonality, economic conditions, and even the weather. You need to buffer against these uncertainties.

2. *Anticipation inventory* — This refers to the inventory you build up because of an expected event, such as a sales promotion, holiday season, or plant shutdown. You build up this inventory prior to the event if you do not have the capacity to respond to the event when it occurs. Most make-to-stock inventory would fall into the anticipation category. In a make-to-stock environment you build inventory prior to receiving customer orders. You anticipate, or expect, that customer orders will follow the production of the items. Forecasts are the basis for the expected quantity of the customer orders. When the customer wants your product in less time than it takes you to make it, you will build inventory before you receive the customer's order. If a customer walks in your door and wants something immediately (think retail), you do not have time to make the product after they order it. So you build it ahead of time, expecting that because customers wanted it before, they will want it again. Many small manufacturers shut down their plants at a normally slow time of the year, to allow for rest, recovery, and maintenance. In anticipation of these events, inventory is built up ahead of the scheduled shutdown. Raw materials are considered anticipation inventory in both make-to-stock and make-to-order environments. You anticipate, or expect, to use them and you purchase and receive them before you actually do use them.

3. *Transportation inventory* — Also known as pipeline inventory, this is an often overlooked component of inventory. Transportation inventory is inventory that is in transit between distribution points. Materials shipped from suppliers but not yet received and product shipped from the manufacturing plant but not yet received by the customer fall in this category. Materials and products at any point

in the distribution system are transportation inventory. Transportation inventory is often overlooked, or forgotten, because even though the shipper or receiver owns it, it is often in the possession of a third party during transportation. If you do not have the systems in place to account for in-transit inventory, you may order more before you should. In a large or complex distribution network, transportation inventory may be quite significant. The lead time for transportation will affect the amount of inventory in the pipeline. The longer it takes for transportation, the more inventory will be in transit. If the transportation time is long enough, you may have to ship more inventory before the last shipment has reached its destination. From a supply chain perspective, transportation inventory can be significant, even though the inventory may be owned and controlled by a variety of suppliers, producers, distributors, and sellers. To manage this supply chain, individually or as a team, you must consider all the inventory in the pipeline.

4. *Hedge inventory* — Hedge inventory is similar to anticipation inventory, except hedge inventory is acquired prior to a possible or uncertain event. Currency fluctuations, import tariffs, supplier shortages, and other events that are uncertain but possible are reasons for holding hedge inventory. Hedge inventory can be an asset, or it can be a liability. If the event that causes you to hold hedge inventory happens, the inventory is an asset; you either saved money or protected yourself with the inventory you were holding. If the event that caused you to hold the inventory does not occur, you are left holding inventory that was not necessary, is in excess of your current needs, or is at a higher cost. With hedge inventory, you are basically gambling. Your risk tolerance will influence if, or how much, hedge inventory you hold.

5. *Lot size inventory* — Lot size inventory occurs when more inventory is purchased or produced than is immediately needed. For small manufacturers, suppliers often dictate the lot size of purchased items. These lot sizes are often more than is needed for immediate production, and the excess is held in inventory. Because, as a small manufacturer, you may not have a great amount of influence with your suppliers, you may hold lot size inventory regularly. This may be your normal inventory procedure. Production batch size policy often results in producing more finished goods than are needed for immediate sales. If you do not have the processes in place to produce exactly to customer demand, you probably build more than your immediate needs. The alternative would be to produce less than your immediate needs, then not be able to service your customers. Transportation costs may influence your

lot size quantities. You may ship, incoming or outgoing, in truck-load quantities to reduce transportation costs. This needs to be balanced with the holding costs of the resulting lot size inventory.

Inventory helps a company achieve its strategic goals. It is used to help level production and stabilize the workforce. It can help keep plant and equipment utilization at optimal levels. Again, inventory should be planned and should fit in with the strategy of the company. For more discussion on using inventory in an equipment utilization role, see Chapter 6.

2.1.2 Part Numbering

Part numbers and part numbering systems have been known to create a bit of controversy (as I can attest from personal experience). Although I will not be able to end all the controversy here, I will offer some explanation on part numbering that should help you evaluate your part numbering system.

There are only two basic types of part numbering systems: significant and nonsignificant, also known as random. In a significant part numbering system, each position, or group of positions, in the part number has a specific meaning. For example, the first position may stand for the color of the item; the next three positions may stand for the size; and so on. If you use significant numbering, you can describe the item from the part number if you know the coding system. The significant system has several distinct disadvantages, however: (1) it is difficult for new people to learn the coding system; (2) new items do not always fit easily within the existing numbers; and (3) part numbers are prone to data entry errors because they usually contain a combination of numbers, letters, and symbols.

The other type of part numbering system is the nonsignificant, or random, system. In this system, the part number has no meaning, does not represent any code, and is assigned randomly. In the random system, part numbers are often assigned sequentially, with new items simply being assigned the next available number. Random numbering systems often utilize all numeric part numbers, greatly reducing data entry errors. Coding systems do not have to be learned and new items are easily added. The only real disadvantage is that similar items may have part numbers with no relation to each other. However, with the almost universal use of computerized inventory systems, this disadvantage is increasingly of little concern.

Some companies try to compromise by using a random part numbering system but blocking out certain blocks of numbers for certain categories of items. Perhaps five-digit part numbers are purchased raw materials, six-digit part numbers are manufactured components, and seven-digit part numbers are finished goods. Or perhaps 6-series numbers — those part numbers beginning with a 6 — are reserved for one type of metal and 7-series numbers are reserved for another type of metal, and so on. Although this may sound like a reasonable compromise, it usually fails quickly in practice. Depending on the number of digits in your part numbers and the number of items in inventory, you may run out of available numbers in a particular series. Then there's the problem of those items that don't fall neatly into one category or another. What about an item with two different metal types incorporated in it? Into which series of numbers would it fall? Would the metal with the greater percentage determine the appropriate series? As you can see, this system will tend to become meaningless over time. Some people may continue to defend this system as long as someone can memorize every item in the system, but we're trying to improve our processes and systems, not just make them work. Figures 2.1 and 2.2 show examples of significant and random part numbers, respectively.

The purpose of a part number, or stock number, is to uniquely identify an item. Every unique item should have a unique part number. Companies sometimes make the mistake of assigning the same part number to more than one unique item because the items can be used interchangeably. Although this may sound good at the time, it will almost certainly cause problems eventually. A 2-inch bolt and a $2\frac{1}{2}$-inch bolt are two unique items, even if you can use them interchangeably in the production of your product. Another common mistake is to assign two different part numbers to the same item. Sometimes this is done inadvertently, but sometimes it is done on purpose. One common reason is that different

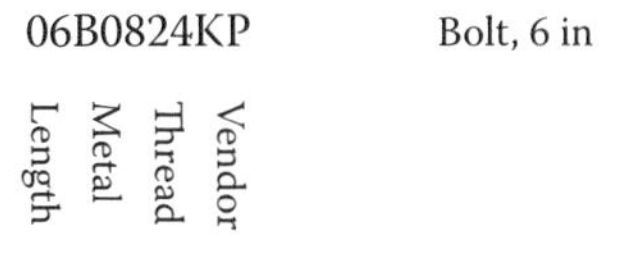

06B0824KP   Bolt, 6 in

Length Metal Thread Vendor

Figure 2.1 Significant part number.

40015	Paint, Red
5904859	Lid, 6 inch
224967	Wrench, Crescent
39549392	Flour, Whole wheat, 25 lbs

Figure 2.2 Random part numbers.

suppliers or customers use different part numbers for the item. This may sound valid, but it can cause havoc in planning, managing, and controlling the inventory. Most computerized inventory or entreprise resource planning (ERP) systems offer better alternatives, usually providing one or more fields for alternate part numbers, suppliers' part numbers, and customers' part numbers.

Many people use the terms part number and stock keeping unit (SKU) interchangeably, but they have different meanings. A part number identifies a unique item. An SKU refers to a unique item in a specific location. If you have an item stocked in two locations, you have only one part number but two SKUs. Most electronic inventory systems allow for location codes for inventory locations or warehouses. You can think of the SKU as a combination of the part number and the location code. In most organizations, using the term *SKU* when you actually mean part number will not cause any problems, because the meaning will be understood. It is important, however, to understand the difference.

The primary purpose for part numbers is for use in data transactions. This means for ordering purposes, invoicing, planning, and all the other transactions that are processed in manufacturing and inventory systems. In computerized systems, and even in manual systems, the part number is a reference number to facilitate the transactions. If any other information is needed, such as a description of the item, the unit of measure, or other specifications, the item master file can be reviewed. In computerized systems, the item master file will be accessed automatically during most transactions to get the needed information. Many organizations try to use the part number for more than is necessary. Some people memorize the part numbers for all of their products and materials. This is totally unnecessary and seems to me to be a complete waste of energy. Of course, if you work with some items and their part numbers every day, you'll naturally memorize some of them. But it seems your effort would be much better spent on other activities. Especially with the wide use of computerized inventory systems, let the computer do the work it was designed to do. Computers are fantastic at storing data and calling up information when referenced from some sort of code, such as a part number. All the information you need is at your fingertips. With the increasing use of mobile computing devices, this information is available to you wherever you go.

2.1.3 Item Descriptions

Item descriptions are another area in which many organizations could stand to improve. Sometimes item descriptions seem to be quite random, with similar items having descriptions that have no resemblance to each other. Developing a standardized format for item descriptions would make life easier in many organizations. A standardized format will help tremendously when new items are created in the system. Standardization will also help when analyzing inventory. Most computer databases, such as an inventory or ERP system, allow searches and queries by any field, including the item description field. Having standardized descriptions will make searching and sorting items much easier and more understandable.

In developing a standardized format, you should develop a company dictionary. This company dictionary can be based on standardized terminology for your industry or can be customized by using your organization's terminology. This dictionary defines the words that will be used in the item descriptions that relate to the characteristics of the actual items. Examples might include: bolt — threaded fastening device, cable — multistrand wire, and flour — ground grain products. These keyword nouns should then begin the description, so that anyone reading the description will be able to immediately identify the item. The keyword can then be followed by more descriptive information, such as the size, color, or other characteristics. Again, the format should be standardized and consistent. If your format is keyword–size–color, use that format for all your items. Resist the urge to suddenly start using a new format, such as color–keyword–size. Don't laugh — you wouldn't be the first to do something like this. Figure 2.3 lists examples of part descriptions.

Standard abbreviations should also be developed, instead of allowing the people entering or creating the descriptions to use their own abbreviations. The item descriptions need to be understood by anyone reading them or using them (within your organization, that is). Item descriptions provide a lot of information about the items, and the descriptions are useful in analyzing inventory. Standardized and consistent item descriptions will help you find information and perform analysis. Be sure that everyone who is allowed to add or change item descriptions is trained in the policies, and ensure that they are followed.

Consistent

Bolt, $\frac{1}{2}$ *inch, Steel*
Bolt, $\frac{1}{2}$ *inch, Aluminum*

Not Consistent

Bolt, $\frac{1}{2}$ *inch, Steel*
Aluminum Bolt, $\frac{1}{2}$ inch

Figure 2.3   Item descriptions.

2.1.4 Bills of Material

An often misunderstood tool of inventory management and planning is the bill of materials (BOM). Complete and accurate BOMs are vital to the proper planning and management of materials. The BOM may be the most used document in a company, and it forms the foundation of the manufacturing and cost systems. It is used to plan materials purchases and production schedules and to determine product costs. It is used to define an item and to determine how to manufacture the item.

The BOM allows the master production schedule (MPS) to be broken down, or exploded, into all of the raw materials and components that need to be purchased or produced. The purchasing department uses the BOM and the MPS to determine the materials needed and to plan the purchases. Production and assembly schedules are based, in part, on the manufactured components needed per the BOM and MPS; therefore, all materials needed to produce an item should be listed in the BOM. Some companies do not include small-value items that they always keep in stock (unless they run out). This can have serious consequences. If any item needed to produce a product is not included in the BOM, production delays and missed deliveries may occur because the item has not been properly planned. Even in a Lean Manufacturing environment, where materials are controlled by the use of kanban, BOMs need to be accurate for proper planning. It can be both embarrassing and costly to disrupt production and shipping because of parts shortages, especially small-value parts.

The BOM is a document that lists all of the materials needed to produce an item. It is also known as the recipe, ingredients list, formula, parts list, or other similar names. The BOM defines a parent–child relationship between items. The children are materials that are needed to produce the higher level item, the parent. A parent can be a manufactured component or a finished good. A child can be a raw material or a manufactured component. A parent may have one child or many children. That is, a manufactured component or finished good may have only one raw material or subassembly that is used to produce it, or the manufactured component or finished good may have several raw materials or subassemblies that are used to produce it. For example, a gear component may have a steel disk as its only component. This disk is cut and machined to the proper size and shape. A gear assembly may have multiple gear components, some pins, and some bolts making up the completed assembly.

The BOM can be represented by several different formats, but they all share similarities. Figure 2.4 shows a table that will be used to illustrate different BOM formats. Figure 2.5 shows the bill in a graphical format, whereas Figure 2.6 simply lists the materials. Some formats reflect only a

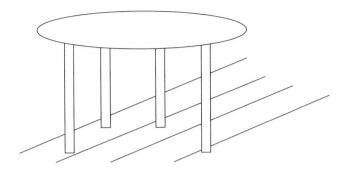

Figure 2.4 A table.

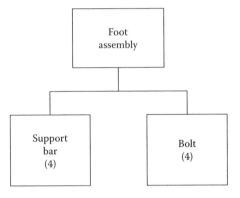

Figure 2.5 Single-level bill of materials (BOM), foot assembly.

Level	Qty per	Unit of measure
0......Table	1	Each
1.......Top assembly	1	Each
2.......Top, Round	1	Each
2.......Bracket, Top	1	Each
2.......Bolt	4	Each
1......Support	1	Each
1......Foot assembly	1	Each
2.......Support bar	4	Each
2.......Bolt	4	Each

Figure 2.6 Indented BOM.

single level (Figure 2.5), whereas others show multiple levels or the entire structure (Figure 2.7). A single-level bill shows one parent–child(ren) relationship, whereas a multi-level bill shows more than one. No matter the format, the BOM lists all of the materials needed to produce the parent

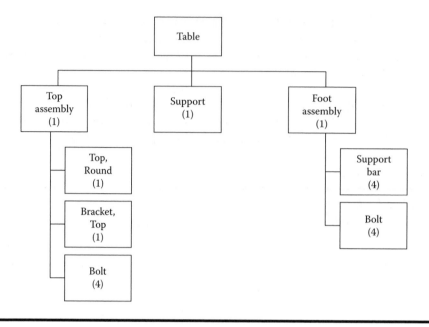

Figure 2.7 Multi-level BOM, table.

item. The BOM defines an item by showing the relationships between all the materials and components, but it does not give detailed assembly instructions. The BOM also helps to manage any engineering changes. Any changes in materials used or in the design of the product must be reflected in the BOM. By listing all of the materials and components needed, the BOM allows you to determine the direct materials costs of a product. An inaccurate or incomplete BOM can cause many problems. Purchasing, planning, scheduling, and accounting all rely on accurate BOMs. Inaccuracies can cause parts shortages or overages. Parts shortages can result in production delays, even entire plant shutdowns, and missed deliveries. Excess inventory can cause other problems besides tying up cash and space. Incorrect items listed in a BOM can cause many quality-related problems, including rework, scrap, customer returns, and safety concerns. Purchasing plans and production schedules will not be valid if there are errors in the BOM, and the materials costs of the products will not be accurate.

It is important to understand the proper use and function of the BOM. The use of different structures may be appropriate in different situations and different environments. The widespread use of the BOM throughout the organization needs to be understood, as well as the potential effects of any errors or omissions in the bill. The importance of ongoing maintenance of the bills cannot be understated. BOMs can be difficult to maintain without proper procedures in place. Continued accuracy and completeness of the bills cannot be emphasized enough. It is easy to become complacent regarding BOMs, with all the other day-to-day activities and concerns. Small manufacturers usually don't have anyone whose sole function is to maintain them and monitor engineering changes. This work may fall to several people. Maybe no one is specifically assigned to this; it is handled by whoever is most conscientious or would be most affected by any errors. The maintenance of BOMs and engineering changes is an important job; it should be assigned to a specific individual or individuals as a part of their duties.

2.1.5 Inventory Accuracy

Accurate inventory records are vital for an organization to reach its highest level of performance. Many important decisions are made based on reported inventory balances. Long-term strategic decisions as well as daily operational decisions are made based on inventory balances. Many of the same problems that result from errors in the BOM occur when there are errors in the reported inventory balances or inventory records.

Inaccurate inventory balances can result in inventory shortages, which can cause production delays or missed deliveries. Inventory overages, resulting from errors in the inventory balances, can use space, resources, and cash that could better be used for something else. Depending on whether the inventory balances are overstated or understated, purchases may be delayed or they may be made sooner than necessary. Errors in inventory records can result in production schedules that do not meet customer demand or your company's needs. Promises to customers are made based on inventory balances — current and expected balances. If the reported inventory balances are wrong, the promises to the customers might not be kept. This can result in lost sales as well as lost customers, which means lost profits.

Inventory is often the most valuable asset listed on a company's balance sheet. What that means is that more cash is tied up in inventory than in any other particular asset. Banks, investors, and other people are very interested in the financial reports of a company, and they are very concerned with the accuracy of those reports. Because inventory is often a large item on financial reports, the accuracy of the reported inventory is very important. As important as accurate inventory is, some companies do little to achieve or maintain accurate inventory records.

One all too common method companies use to try to accurately record their inventory is by conducting physical inventory counts. Some companies do this on a monthly basis. These companies convince themselves that their records are accurate after the counts are completed. Even when the counts for individual items fluctuate wildly from one count to the next or errors are revealed every time the physical counts are performed, they believe the counts are right "this time." Physical counts rarely result in accurate inventory records. Many of the individual item counts may be accurate immediately after the record adjustments are made for the results of the physical counts, but they may not remain that way for long. It may be close enough for the finance people for reporting purposes, but it is nowhere near good enough for the smooth operation of the plant or for optimal performance.

A proven method of achieving and maintaining inventory record accuracy is through cycle counting. Cycle counting is often misunderstood as just a method of counting inventory more often. Although cycle counting does use frequent, often daily, counting of items, the counting itself, and any subsequent adjustment of the records, is not the purpose. The intent of cycle counting is to find the causes of errors in the inventory records, eliminate these causes, and prevent them from recurring. Why are the records not accurate? What are the reasons for the differences between the reported inventory balances and the physical counts? The inventory records are adjusted as a result of the counts, but only after the underlying cause of the error is discovered. But it is not good enough to just identify the causes of errors. Part of the cycle counting program is to make whatever changes need to be made to prevent the errors from recurring. These could be changes to procedures, transaction processing methods, training of employees, or whatever else is necessary. Some changes may be quick and easy to make, but others may require substantial improvements or upgrades to existing policies, procedures, or equipment.

Cycle counting does require frequent counting of inventory items, but there is simple logic behind the counting. One of the principles behind cycle counting is that some items are more important, or more valuable, than other items. Items are classified as either A, B, or C items, where A items are the most important, or most valuable; B items are moderately important; and C items are the least important or least valuable. The classification is usually based on the annual value (unit cost × annual volume) of the items, but other factors are also considered. Hard-to-get items or items with long lead times may be placed in a higher category than their annual value would indicate. The idea is that the more important items need to be controlled more than the less important items. This results in the A items being counted more frequently than the B items, which are counted more frequently than the C items. Think of it this way: which do you want to keep a closer eye on — a $10,000 specialty part or a $.05 screw that you could get at any hardware store?

ABC analysis is a widely used method for classifying inventory items in cycle counting. ABC analysis is based on the 80/20 rule, or Pareto's law. The premise is that 80 percent of the total inventory value is in only 20 percent of the total number of inventory items. That is, if you have 100 separate inventory items (100 part numbers) with a total value of $100,000, then only 20 of those items account for $80,000 of the inventory. The remaining $20,000 of inventory value is spread over the remaining 80 inventory items. A general guideline is the top 20 percent of items are A items, the next 30 percent of items are B items, and the final 50 percent of items are C items. This holds true in many cases and is a useful tool, not only when analyzing inventory. Tables 2.2 and 2.3 show an example of an ABC analysis.

The steps for conducting the ABC analysis are

1. Determine the annual usage quantity for each item.
2. Determine the annual usage in dollars (quantity × unit cost).
3. Rank items in descending annual dollar value.
4. Calculate the cumulative value, cumulative percentage of value, and cumulative percentage of items for the ranked items.
5. Classify the items as either A, B, or C, based on the values.

The annual quantity usage for each item can be based on historical values or from expected, or forecast, figures. This analysis and classification by value is a good basis from which to start. You may want to change the classification of some items, however, based on other criteria. Items that are hard to get, have a long lead time, or are critical components may be classified as A items regardless of their annual dollar value. The initial analysis is a starting point, not a hard requirement. Be careful, though, when deviating from the annual dollar value method. You can

Table 2.2   Annual Usage

Item No.	Annual Usage (Qty.)	Unit Cost	Annual Dollar Value
1	12,000	$0.16	$1,920.00
2	5,000	$1.20	$6,000.00
3	3,800	$9.95	$37,810.00
4	24,000	$0.12	$2,880.00
5	1,000	$2.00	$2,000.00
6	4,500	$1.22	$5,490.00
7	6,500	$9.85	$64,025.00
8	4,800	$0.30	$1,440.00
9	5,200	$1.20	$6,240.00
10	3,600	$0.30	$1,080.00

Table 2.3   ABC Analysis

Rank	Item No.	Annual Dollar Value	Cumulative Value	Cumulative % of Total		ABC Classification
				Total Dollars	Total Items	
1	7	$64,025.00	$64,025.00	49.7%	10%	A
2	3	$37,810.00	$101,835.00	79.0%	20%	A
3	9	$6,240.00	$108,075.00	83.9%	30%	B
4	2	$6,000.00	$114,075.00	88.5%	40%	B
5	6	$5,490.00	$119,565.00	92.8%	50%	B
6	4	$2,880.00	$122,445.00	95.0%	60%	C
7	5	$2,000.00	$124,445.00	96.6%	70%	C
8	1	$1,920.00	$126,365.00	98.0%	80%	C
9	8	$1,440.00	$127,805.00	99.2%	90%	C
10	10	$1,080.00	$128,885.00	100.0%	100%	C

easily waste time and energy worrying over whether an item should be an A or B item. The idea is that you are setting levels of control, but all your items will have some level of control. The extra benefit of controlling a borderline item may not be worth the time and effort it took to classify it.

When adopting a cycle counting program, you need to constantly monitor your record accuracy. This gives you an indication of the performance of the program. You can measure the accuracy at any point in time and track it over time. It is important to know not only how you are doing today, but also if your accuracy is trending up (improving over time) or down (getting worse over time). The ultimate measure of performance, however, is whether there are any disruptions to the plant or to deliveries to customers because of inventory errors. Also, because your inventory accuracy should improve with a cycle counting program, your employees can see the improvement. Most employees like to know that their work is worthwhile, and that gives them more motivation. Measuring inventory accuracy under the cycle counting program may be a little different than you're used to. In cycle counting, the record is either accurate or it is not. If the records say you have 100, but you can find only 99, the record is not accurate. It's not mostly right, or 99 percent right; it is not accurate.* If a customer wants 100, and you promise them that you can deliver 100, but you discover you have only 99, you have

---

* Some organizations use tolerances when determining if an item is considered accurate. For example, C items may have a tolerance of 5 percent, and if the count is within 5 percent of the reported balance, that item is considered accurate. This topic is not explored further here.

Table 2.4 Inventory Record Accuracy

Part No.	Class	Inventory Record	Count	Hit	Miss
1	B	100	100	X	
2	B	100	98		X
3	C	100	105		X
4	A	100	100	X	
5	C	100	94		X
6	B	100	99		X
7	A	100	100	X	
8	C	100	96		X
9	C	100	110		X
10	C	100	100	X	
Total hits			4		
Total items counted			10		
Inventory record accuracy			40%		

a problem. And this is not a problem you want to happen very often. Table 2.4 shows an example of an inventory record accuracy calculation.

You may think your inventory is in pretty good shape, but if only 40 percent of the items are reported accurately, you have the potential for a lot of problems. Cycle counting is a good method to improve the accuracy of your inventory records, but it requires discipline and commitment. You have to ask yourself, would you rather invest in a program to improve your inventory accuracy, or invest in more inventory to protect yourself, more equipment to handle that inventory, more space to hold the inventory, more people to manage the inventory, and more people to explain to your customers why you couldn't ship their order? You may find that if you can keep good control of your inventory, you will need less of it.

2.2 Master Production Scheduling

2.2.1 Planning Hierarchy

The planning process is a vital component of your business. Many companies do not have formal processes for each level of planning, so they don't know they are doing it. Unless you're throwing darts at a dartboard and just ordering and making whatever number you hit, you are planning. It may be rough, it may be some quick calculations in your head, but you are planning. You need to plan, you are already planning, and you should formalize the planning process to get the most benefit from it.

Generally four levels of planning precede the execution phase. Execution is where you do the actual purchasing, production, shipping, and so forth. Before you get to this stage, you need to know what you are going to buy or make, how many, and when. The planning phases are where you determine the what, how many, and when. Formalized systems for planning are better than informal, or back-of-the-envelope, systems. Formal systems allow you to move through the various stages of planning with enough time allowed for each stage. Formal systems include capacity checks at each stage. This is important, because a plan is useless if it is impossible to execute due to a lack of capacity. This is often overlooked or not completely understood. Of course, all this planning and execution is ongoing and many activities overlap. The different levels of planning have planning cycles of varying lengths, and the execution consists of a series of continuous, daily activities.

The planning process starts with long-term strategic plans broadly outlined in product lines and market segments. Strategic plans are developed at the highest level of the company and set the direction for the company. Gross volumes, or market share, may be discussed at the strategic planning level. Everything else the company does should support the strategic plan. Companies often undertake new initiatives that seem to make sense but do not support the strategic plan. These initiatives may be beneficial to the company in the short term but may be detrimental in the long term. Any new initiative or major project should be evaluated for its support of the strategic plan. If it supports the strategic plan, it can be pursued; if it does not, it should not be undertaken at this time. If warranted, the strategic plan can be modified and the project undertaken. However, strategic planning is not an activity that is taken lightly, so more than likely the initiative or project should be set aside for the time being.

More detailed, but still long range, is sales and operations planning. Sales and operations planning covers anywhere from 12 to 36 months, depending on your needs. In sales and operations planning, plans are developed for aggregate volumes, usually in monthly time buckets. (See Chapter 3 for a more detailed discussion.) Master production scheduling is the intermediate level of planning and usually covers about 3 to 12 months. The MPS converts the aggregate plan into a plan for producing specific items. Master production scheduling is followed by materials requirements planning, which is the short-term level of planning (see Section 2.3). Figure 2.8 shows the planning hierarchy.

2.2.2 Volume versus Mix

Master production scheduling is often overlooked by small manufacturers, but it is an important part of the planning system. Master production

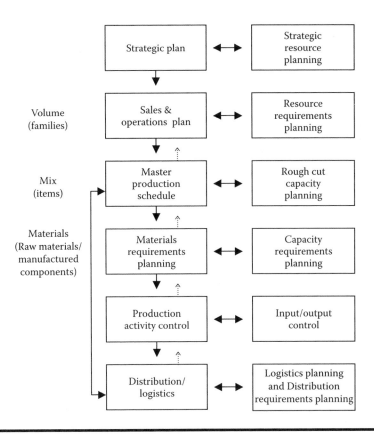

Figure 2.8   Planning hierarchy.

scheduling is concerned with balancing demand and supply. Demand is what is needed, what is demanded. This comes primarily from customer orders, but includes service parts requirements, engineering and testing needs, and any other inventory requirements. Supply is what you are going to make or what you can make. Some people think that balancing demand and supply is not that big a deal; figure out how much you need, then make it. If only it were that simple. There are many challenges on both the demand and supply sides, but we won't get too much into detail here. We will discuss the basics of master production scheduling and how it works to balance demand and supply. (See Appendix A for a list of several good references to learn more about master production scheduling and managing demand and supply.)

The MPS ties to the S&OP and is derived from the S&OP. The S&OP is stated in aggregate volumes, along product lines. The MPS is stated in specific end items, or a mix of products. The total quantity of the individual items in the MPS must equal the aggregate quantities in the S&OP. The

Volume–product families

Family		Jan	Feb	Mar	Apr	May	Jun	Jul	Aug	Sep	Oct	Nov	Dec
Amps		1,000	1,000	1,000	1,500	1,500	1,000	1,000	1,000	1,500	2,000	2,000	2,500
Disks		2,500	2,500	2,500	2,500	2,500	2,500	2,500	2,500	2,750	2,750	2,750	3,000
Speakers		500	500	500	750	750	500	500	500	750	1,000	1,000	1,250

Mix–End items (Amps family)

Model		Wk1	Wk2	Wk3	Wk4	Wk5	Wk6	Wk7	Wk8
XL 10		70	70	70	70	70	70	70	70
RG 7		50	50	50	50	50	50	50	50
RG 9		110	110	110	110	110	110	110	110
TZ 123		20	20	20	20	20	20	20	20
Total		250	250	250	250	250	250	250	250

Figure 2.9   Volume versus mix.

MPS translates the aggregate quantities from the S&OP into the specific mix of items, or product configurations, that are to be built. See Figure 2.9.

2.2.3 Disaggregation

So how do you get from the product family aggregate quantities in the S&OP to the individual item quantities in the MPS? Different companies may use different methods, but both the product family quantities and the individual item quantities are forecast amounts. In the sales and operations planning process, forecasts for the product families are developed. (This will be discussed in more detail in Chapter 3.) In the master production scheduling process, forecasts for the individual items are developed. One method of developing the MPS quantities is to disaggregate the product family forecast. This won't be discussed in detail here, but it involves using some method, usually based on past history, to break down the aggregate quantity (volume) into the individual items that make up that aggregate (mix). One method is to calculate the historical percentage (of sales) of each individual item in the family, then apply the percentages to the aggregate quantity.

In the example from Figure 2.9, there are four individual items in the Amps family. The historical percentage of each model is shown in Table 2.5. These percentages are then applied to the aggregate quantity for the Amps family for each month, which results in a forecast quantity per month for each individual item. This can then be broken down further into the quantity per week, because we are using weekly time buckets in the MPS. We will assume four weeks per month for each month. Table 2.6 shows the results for January.

Table 2.5   Amp Family Historical Percentages

Model	% of Sales
XL 10	28%
RG 7	20%
RG 9	44%
TZ 123	8%
Total	100%

Table 2.6   Amps Family Item Forecast for January (1000)

Model	% of Sales	Quantity	Qty. per Week
XL 10	28%	280	70
RG 7	20%	200	50
RG 9	44%	440	110
TZ 123	8%	80	20
Total	100%	1,000	250

2.2.4 Forecasting

Disaggregating the forecast for the product families is one method of forecasting the quantities for the items in the families. You can also forecast the individual items directly, through whatever forecasting method you choose. In this case, you will have two separate forecasts. You will have one forecast for the product family and one forecast for each individual item in the family. These two forecasts must then be reconciled, if they do not agree (or are not very close). This is then one validation of the S&OP. If the forecast for the aggregate quantity matches the totals of the individual item forecasts, that validates the quantities.

There are some basic rules of forecasts. One is that forecasts have errors.* If you forecast twice, you will introduce error twice (not necessarily twice as much error). The disaggregation process is a form of forecasting, so it seems that you can't get away from introducing error more than once in these planning processes. Another rule of forecasting is that

---

* Some people believe forecasts are always wrong. I don't like to think that way. Forecasts are never 100 percent accurate, but that doesn't mean they are wrong. Forecasting is a process, and all processes have some variation. Ideally, you want to reduce the variation (the forecast error), but because of the nature of forecasting, you will never eliminate it.

forecasts are more accurate in the aggregate. That is, a forecast for the total quantity of a product family will be more accurate than forecasts for each individual item in that family. Keep these rules in mind when you are developing the forecasts for the product families (S&OP) and for the individual items (MPS).

2.2.5 Planning Horizon

The MPS is the planned build schedule for a certain period of time, known as the time horizon. The time horizon generally covers the longest cumulative lead time for the products in the MPS. The cumulative lead time is the length of time it would take to acquire all the materials, build any subassemblies, assemble the final product, and ship the product to the customer. The reason the MPS time horizon covers the longest cumulative lead time is because you want to plan far enough in advance to handle any problems that may occur. You want that visibility into the future to avoid any foreseeable problems. If your cumulative lead time is six months, but you are planning ahead for only three months, you may not be able to react to issues or problems that arise.

The MPS shows specific configurations of finished products, in specific quantities, in specific time periods. Time phasing refers to breaking down the planning horizon into smaller time periods and planning the quantities to be produced in each of those periods. (Time phasing will be discussed more in Section 2.3.) These time periods are referred to as time buckets. In master production scheduling, time buckets may be days or weeks, or something else, depending on your needs. The time periods you choose will depend on various factors, including company policies, the products you produce, and customer requirements. Specific configurations refers to the variations of the finished product that will be built. This means that if there are several options, or choices, available, the MPS must reflect the quantity of each final configuration that is planned. The MPS will show all of the specific finished items to build, the quantity of each of these items, and the time period in which they will be completed. Again, the total of all these individual items will add up to the aggregate amount shown in the S&OP. Figure 2.10 shows an example of an MPS. Let's look at the MPS and discuss its components; then we'll work through the calculations.

Let's look first at the gross demand row. The gross demand is the total quantity that is needed, or required, in each time period. Notice that the gross demand is equal to the quantity that we derived from the S&OP (see Figure 2.9). Remember, we started with the total quantity for each product family (volume), then broke that down into the quantity for each individual item in each family (mix). The other option would be to forecast

Model XL 10, Lead time = 1 week, Lot size = 100

Week		Week1	Week2	Week3	Week4	Week5	Week6	Week7	Week8
Forecast		40	50	55	65	70	70	70	70
Customer orders		30	20	15	5				
Gross demand		70	70	70	70	70	70	70	70
Projected available balance	50	80	10	40	70	0	30	60	90
MPS quantity		100		100	100		100	100	100
Planned order release			100	100		100	100	100	
Available to promise		100		80	100		100	100	100

Figure 2.10   Master production schedule, model XL 10.

the individual items, then reconcile any differences in the total for the items with the aggregate forecast for the family. With this option, the individual item forecast would be the gross demand.

Notice that the gross demand equals the forecast plus customer orders. Now, you may ask, if the gross demand is derived from the S&OP, which begins with a forecast, what, then, is the forecast? The forecast row may show different quantities depending on your company's policies. In this example, the forecast quantity is a net quantity. It shows the original forecast quantity (derived from the S&OP) less any customer orders. In Week 1, the original forecast quantity is 70. With customer orders for 30 booked in Week 1, the forecast row shows the unbooked balance of the forecast. We expect to receive customer orders totaling 70 in that week, but so far we have only received 30. The remaining forecast quantity is 40, which is shown in the forecast row. This process of reducing the forecast quantity by the amount of firm customer orders is called consuming the forecast.

Some organizations may choose to do this differently. For some portion of the schedule, they may feel that the forecast is no longer relevant and that only customer orders should be taken into account. The schedule is fixed for some period of time out from the present, so only the customer orders are accounted for. Any remaining portion of the original forecast is dropped. See Figure 2.11. In this example, the schedule is fixed for the first two weeks, so take into account only the customer orders in those weeks. (Section 2.2.6 will explain this further.)The projected available balance is the expected amount of inventory that will be on hand at the end of the period. This is calculated from the available balance at the beginning of the period, plus any receipts during the period, less any demand during the period.

The MPS quantity is the quantity that will be completed in the period. This quantity will be affected by any lot size requirements that are in place.

Model XL 10, Lead time = 1 week, Lot size = 100

Week		Week1	Week2	Week3	Week4	Week5	Week6	Week7	Week8
Forecast				55	65	70	70	70	70
Customer orders		30	20	15	5				
Gross demand		30	20	70	70	70	70	70	70
Projected available balance	50	120	100	30	60	90	20	50	80
MPS quantity		100			100	100		100	100
Planned order release				100	100		100	100	
Available to promise		85			95	100		100	100

Figure 2.11   Master production schedule, model XL 10, alternate.

The planned order release shows when production needs to begin so that the order will be completed in the proper time period. This is the same quantity as shown in the MPS line, but it is in the time period when the order is released rather than when it is completed. This is important for linking the MPS to the materials requirements plan (MRP).

Available to promise (ATP) causes great confusion for many people, because they think projected available balance and ATP are the same. They are not! The projected available balance is the quantity you expect to have in inventory at the end of each period. But having this quantity in inventory does not mean it is available for customers. A portion of that inventory may already be committed to customers who have placed orders for delivery in some future period. Available to promise is the portion of your current inventory and future production that is not yet committed to customer orders. Available to promise is the quantity you are able to commit to new customer orders.

Now let's work through the calculations that bring us to the numbers we see in our example. Figure 2.12 shows our original example. Let's see where we're starting from. Start with the gross demand. We derived these quantities by disaggregating the S&OP forecast, so they are more or less given to us here. Next, notice that our projected available balance shows a quantity of 50 prior to Week 1. This is our current on-hand inventory balance. The convention for reading the MPS grid is that Week 1 (the leftmost time period that is showing) is the current period. Because the projected available balance is the expected balance at the end of the period, we show the current (or beginning) inventory balance to the left of the current period. We also see that we are expecting a receipt of 100 in the current period (Week 1). This order was placed prior to the current period, so the release of the order has already occurred and will not show on our grid. The final item at our starting point (not yet having calculated anything) is the customer orders. Customers have placed orders for our products, and we haven't delivered them yet, so they are still open and

Model XL 10, Lead time = 1 week, Lot size = 100

Week		Week1	Week2	Week3	Week4	Week5	Week6	Week7	Week8
Forecast		40	50	55	65	70	70	70	70
Customer orders		30	20	15	5				
Gross demand		70	70	70	70	70	70	70	70
Projected available balance	50	80	10	40	70	0	30	60	90
MPS quantity		100		100	80		100	100	100
Planned order release			100	100		100	100	100	
Available to promise		100		80	100		100	100	100

Figure 2.12   Master production schedule, model XL 10.

we need to see them. They are shown in the period in which the customer expects to receive them or we expect to ship them.

Now that we've established where we are starting from, let's walk through the rest of the grid. First is the forecast. We said that the forecast is the gross demand less any customer orders. So in Week 1 we have gross demand of 70 minus customer orders of 30, so the remaining forecast is 40. The calculations for the forecast row are as follows:

The customer orders row is easy. This is the firm, or booked, orders

Period	Gross Demand – Customer Orders = Forecast
Week 1	70 – 30 = 40
Week 2	70 – 20 = 50
Week 3	70 – 15 = 55
Week 4	70 – 5 = 65
Week 5 to Week 8	70 – 0 = 70

from customers, shown in the period in which they will be completed or are required. This is the total quantity required for all customer orders in each period. Here, they are simply quantities plugged into the grid as an example.

Let's walk through the projected available balance. The beginning, or on-hand, quantity is 50. This is just given in this example. To calculate the quantity for Week 1, which is the quantity that is expected to be on hand at the end of the period:

Projected available balance = beginning balance
+ MPS quantity (receipts) – gross demand

So, for Week 1 we have a projected available balance = 50 + 100 – 70 = 80. For Week 2 we have 80 + 0 – 70 = 10. The beginning balance for Week 2 is the ending balance for Week 1, and we have no MPS (no receipts) for the period. The same calculations carry through the remainder of the planning horizon.

Next is the MPS quantity. This is the quantity we are planning to complete in each period. We will add a quantity in each period if the projected available balance were to fall below zero. As you work through the calculations from week to week, you will first calculate the projected available balance. If the projected available balance falls below zero, you will add a quantity in the MPS row. The quantity will depend on your lot size policy and will be a multiple of your lot size. The quantity will be at least enough for the projected available balance to be equal to or greater than zero. If you have a safety stock quantity for an item, you will add an MPS quantity that will keep the projected available balance from falling below the safety stock quantity. In this example, Week 1 shows an MPS quantity of 100. The order for this quantity was placed prior to the current date, so it was added to the MPS during a prior scheduling run. If you look at Week 3, you see an MPS quantity of 100. As the calculations for this item were performed, the projected available balance for Week 3 would have worked out to be–60 (beginning balance of 10 minus gross demand of 70 equals –60). Because we don't want the balance to fall below zero, we add an MPS quantity in that period. The lot size is 100, so we add one lot. The calculation for the projected available balance then becomes 10 + 100 – 70 = 40. Perform this logic through the remainder of the schedule.

The planned order release is simply the amount of the MPS, placed in the period the order must be released so that it is completed in the proper period. In our example, the lead time is one week, so the order for the MPS quantity must be released one week prior to the period it is needed. The MPS for Week 3 must be released in Week 2 so that the order is completed in Week 3. Again, follow through for the remainder of the schedule.

Finally, we have the ATP. Although the calculation itself is simple, the concept of ATP tends to cause confusion. We will discuss here only one of the many ways to show ATP. It is a single-period ATP. Many people also calculate a cumulative ATP. We won't discuss that here, but the logic follows from the single-period calculation. Many people calculate an ATP for each period in the schedule. We just need to understand the logic and calculations, so we will calculate an ATP only for each period that we have an MPS quantity.

Calculating an ATP for the first period is slightly different than for the remaining periods. In the first period, we know the actual on-hand quantity of inventory, so we will take that into account. In the remaining periods, the inventory balance is a calculation, not an actual quantity, so we do not take it into account. This next thing is very important and probably causes the most confusion. The customer orders that are included in the ATP calculation are only those customer orders in the periods up until the next MPS receipt. The reason for this is that you don't want to commit customer orders beyond the next MPS receipt, because they will be committed to the next MPS receipt. The calculation is:

Available to promise = on-hand balance (first period only)
+ MPS receipt – total customer orders in periods up to next MPS receipt

For our example, the ATP for Week 1 is:

$$50 + 100 - (30 + 20) = 100$$

Try working through the remainder of the schedule yourself. Start at Week 1 and work through the forecast, projected available balance, MPS, planned order release, and ATP. You'll find that it is not as difficult or daunting as it may first appear. After working through the calculations, step back and think about the concepts behind them, and master production scheduling as a whole. As I mentioned, master production scheduling is an important but often underutilized tool for the small manufacturer.

2.2.6 Time Fences

Time fences are an important concept in master production scheduling. Some people find the concept difficult to accept and difficult to adhere to. The MPS is usually separated into three sections, by time fences. The first section is the frozen, or firm, zone and is the time period in the immediate future. The second section is the slushy, or trading, zone. The frozen zone and the slushy zone are separated by the demand time fence. The final section is the liquid, or open, zone. The liquid zone is separated from the slushy zone by the planning time fence. The liquid zone is the time period that is farthest into the future. The idea behind the time fences is similar to the ABC classifications of inventory. They have to do with levels of control and the costs associated with changing the plan or schedule. In the frozen zone, changes to the plan or schedule can be very costly or difficult to manage, because resources have already been committed. Work orders and purchase orders have been issued, manufactured components may already be in process, inventory is in transit, workers have been scheduled, and so on. Because changes within this zone can be costly and disruptive, they need to be closely controlled. Authorization for changes within this time period must be made by someone with the authority to incur the costs of the changes, which can be substantial. In the slushy zone, changes can still be costly and disruptive, but not as much as in the frozen zone. Some resources may already be committed, but there is still some flexibility. Changes can be authorized at a lower level. In the liquid zone, no resources have been committed and there is virtually no cost to any changes. No authorization is needed to make changes. A computerized system will automatically make any changes to the schedule in this zone. In the frozen zone, within the demand time fence, the schedule is derived from firm customer orders only. Between the demand time fence and the planning time fence, within the slushy zone, the schedule is derived from the greater of forecast or customer orders. Beyond the planning time fence, in the liquid zone, only the forecast is used. See Figure 2.13.

One reason the concept of time fences and the policies regarding changes in the different zones is so difficult for many is that the costs associated with the changes are not always clear and direct. Many managers do not fully understand that so-called "hidden" costs are just as real as direct costs. Disruptions to the schedule may not mean immediate cash outlays, but the costs of stopping one job and laying it aside while expediting another results in very real costs to the company. The extra setups that may be required for this schedule change have a direct cost associated with them. Extra setups mean that you are paying people to set up or change over equipment instead of paying them to produce products that will generate income. Putting aside one production lot while expediting another directly results in increased inventory in the system. Extra inventory means extra costs. Materials are purchased sooner than needed, meaning cash is paid out or tied up sooner than needed. Extra inventory means extra monitoring and control of that inventory and the possibilities for loss and damage of that inventory. Managers need to realize that these and other hidden costs are very real and result in higher

Model XL 10, Lead time = 1 week, Lot size = 100

Week		Week1	Week2	Week3	Week4	Week5	Week6	Week7	Week8
Forecast		40	50	55	65	70	70	70	70
Customer orders		30	20	15	5				
Gross demand		70	70	70	70	70	70	70	70
Projected available balance	50	80	10	40	70	0	30	60	90
MPS quantity		100		100	100		100	100	100
Planned order release			100	100		100	100	100	
Available to promise		100		80	100		100	100	100

◄— Firm zone —►◄———— Slushy zone ————►◄— Liquid zone —►

Figure 2.13   Time fences.

product costs. Many of these hidden costs end up in the overhead, or burden, component of the product cost. This overhead component is often a mystery to many managers, but one they should know more about, especially how it is determined and the components of it.

2.2.7 Capacity

Just because you have done all this planning doesn't mean you're finished. Remember that at each level of planning we have to make a capacity check. When we develop the plan of what is to be produced, we need to check whether that plan is actually doable. It is a waste of time and money to develop a plan that you cannot possibly accomplish. The MPS is not a wish list. Do you have the ability, the capacity, to build what you plan on building? If your plan calls for building 100 units a week for the next six weeks, can you actually build 100 units a week? If all you've managed in the past is 80 units per week, how would you suddenly be able to build 100 per week? You may think that you should be able to build 100 per week; so if the plan calls for 100 per week, you think you can do it. But again, if the best you've ever been able to do is produce 80 per week, why would you believe that 100 is doable? If the schedule cannot be met, either the plan must be reworked or more capacity must be added. Adding capacity is usually not a short-term solution. It takes time and money to add, or reduce, capacity; so, for the short term, you usually need to change your plan. This may mean that you also have to rework the S&OP. But this is better than coming up with a plan that nobody pays any attention to because it cannot be done.

At the MPS level of planning, the capacity check is known as rough-cut capacity planning. Every item that is produced uses up a certain amount of available resources, or available production capacity. If you make only one item, determining the amount of capacity that is used up for each unit made or planned is pretty straightforward. If you make more than one item or if you have more than one workcenter, things are more complicated. But at the MPS level, we are not looking for an exact match when comparing the quantity we are planning to make with the resources available to make them. That is why it is called rough-cut capacity planning; we just need to know if it looks like we have enough capacity available for the plan to be achievable. With rough-cut capacity planning, we are concerned with only those critical resources, or workcenters, that are capacity constrained (bottlenecks). A more detailed capacity check will be done at the next level of planning, materials requirements planning.

When talking about capacity you need to compare the capacity available and the capacity required. That is, how much do you have and how much do you need? When we talk about available capacity, we'll talk

Table 2.7 Capacity in Hours

Model	Units/Hr	Hr/Unit	MPS (Units)	Capacity Required
XL 10	25	.04	70	2.8
RG 7	20	.05	50	0
RG 9	20	.05	110	5.5
TZ 123	4	.25	20	5.0
Total			250	15.8

about rated capacity and demonstrated capacity. Rated capacity is also known as calculated capacity. And we need to have a standard measure of capacity so that all of the items we make can be compared with each other. Before we talk about available, required, rated, and demonstrated capacity, we need to talk about how to measure capacity.

Units of product is not always a good standard of measure to use, because the resources needed to produce a unit of one item may be vastly different than the resources needed to produce a unit of another item. For example, Workcenter A may be able to produce 10 units of Item A in an hour, 50 units of Item B in and hour, or 100 units of Item C in an hour. These units are not a comparable measure. You cannot say that you have available capacity of X amount of units in Workcenter A, because it would depend on what you are making. You can't use an average either (53 units per hour in this example) because the number of products you can produce in an hour depends on the mix of items you are planning to make. So you need to come up with a standard measure that allows you to compare all of the items you make. Remember, the measure must be common to all items you make. A good standard measurement for capacity planning in industries that produce a variety of products is hours.* Every item you make takes a certain amount of time, so a unit of time is a good standard of measure.

Let's look at an example using the four models of amps for Workcenter A (Table 2.7). Model XL 10 requires .04 hr/unit (1 hr per 25 units); model RG 7 requires .05 hr/unit (1 hr per 20 units); model RG 9 also requires .05 hr/unit; and model TZ 123 requires .25 hr/unit (1 hr per 4 units). Now you can compare the capacity available (or the amount of time available) with the capacity required (or the amount of time needed) to make the mix of products you are planning to make in Week 1.

This example shows the number of units per hour and the time per unit. Many people think, or do rough calculations, in terms of the number

---

* If your company uses a different measurement, such as gallons or tons, the same principles apply. The measure used must be common to all your products.

of units per hour that can be processed. The processing time per unit is a better figure to use, though. Also, inherent in these figures is the setup time. It is important to account for setup time when performing capacity calculations.

2.2.8 Rated Capacity

Theoretical capacity is the amount of capacity that is theoretically available. This is not very useful in itself, but it is your starting point. Using our standard measure of hours, a workcenter has a theoretical capacity of 24 hours per day. If you have two machines in that workcenter, the workcenter has 2 × 24, or 48 hours available per day. However, the machinery or people at that workcenter most likely don't work continuously 24 hours a day, even if you are running three shifts. There are several reasons for having less than 24 hours available, which brings us to rated capacity. First, start with the number of shifts the workcenter is used and the length of each shift. If you operate one 8-hour shift, start with 8 hours available. If you shut down the workcenter for a half-hour lunch break, you're down to 7.5 hours. Then you may have equipment shutdowns for maintenance (planned), equipment breakdowns, and changeovers. This brings down your available capacity. Let's say that over time you have tracked your shutdowns to an average of 2 hours per day. This brings your available capacity down to 5.5 hours per day. Then you have other factors that decrease the amount of capacity you actually have available. Most companies take these data that they have collected over time to calculate efficiency and utilization factors for each workcenter. These efficiency and utilization factors are then used to calculate rated capacity.

Efficiency is a measurement of actual output as compared to a standard. For capacity planning, the efficiency of a workcenter is the standard amount of output expected divided by the actual amount of output produced. If you expect to produce 1,000 tons in an hour, but you end up producing only 900 tons, you're not 100 percent efficient. In using hours as the measurement, it is the standard amount of hours that should have been needed divided by the actual amount of hours that were used. And as you are aware, different people work at different paces, and no one is 100 percent efficient all the time.

$$\text{Efficiency} = \frac{\text{Standard Output Expected}}{\text{Actual Output}} \times 100\% \textbf{ or}$$

$$\frac{\text{Standard Hours Expected}}{\text{Actual Hours Worked}} \times 100\%$$

From Table 2.7, the standard hours expected are the hours required as shown in the table for each of the items in that workcenter (Capacity Required). This is how long we expect it take to produce these items. If we have a schedule that calls for 70 units of model XL 10, 50 units of model RG 7, 110 units of model RG 9, and 20 units of model TZ 123, the standard hours of output for Workcenter A is 15.8 hours. After production is completed we may find that it actually took 18 hours to complete the schedule. This gives us an efficiency factor of 88 percent (15.8 ÷ 18 × 100 percent). A single event, or a short period of time, is not sufficient to determine efficiency, so this data is collected over time to determine the efficiency factor. This efficiency factor should be reevaluated every so often, because it may change due to a variety of factors, such as worker training, improved processes, and so forth.

Utilization is a measure of the amount of time that something is used as compared to the amount of time it could be used. In our case, we are concerned with the utilization of the workcenter. Utilization is calculated by dividing the amount of time the workcenter is actually producing product by the amount of time the workcenter is available.

$$\text{Utilization} = \frac{\text{Amount of Time Actually Used}}{\text{Amount of Time Available}} \times 100\%$$

From the example above, we see that Workcenter A is producing products an average of 5.5 hours per day, out of 8 hours available in our one-shift, one-machine operation. The utilization factor of Workcenter A is 69 percent (5.5 hours ÷ 8 hours × 100 percent).

This brings us to the calculation of rated capacity. Rated capacity is one way of determining the available capacity. Rated capacity is calculated by multiplying the time available, the efficiency factor, and the utilization factor.

$$\text{Rated capacity} = \text{time available} \times \text{efficiency} \times \text{utilization}$$

For our Workcenter A, the rated capacity is 4.9 hours per day (8 hours available × 88 percent efficiency × 69 percent utilization), or 24.5 hours per week (4.9 hours per day × 5 days per week).

Another way to determine the available capacity is to simply keep records of the output of the workcenter and take an average. This is called demonstrated capacity. Demonstrated capacity is similar to rated capacity. The difference is that in rated capacity, the efficiency and utilization factors are calculated once and then are used to calculate the rated capacity, or available capacity. In demonstrated capacity, the efficiency and utilization are inherent in the average output that has been recorded, rather than calculated separately and used in a further calculation. The question, then, is why would you use one over the other — rated or demonstrated capacity?

One reason to use rated capacity is because the efficiency and utilization data are very informative and useful. If either efficiency or utilization is not at the level expected, they can be questioned or examined independently. They can also be tracked over time, to reveal if improvements have been made, or if they are declining. Another reason is that the efficiency and utilization are usually calculated once, then used until they are recalculated. With demonstrated capacity the efficiency and utilization are not calculated separately, so no information can be obtained from them. And the average output is tracked on a more continual basis, which requires the effort to continuously track the output and calculate the average.

This brings us back to comparing available capacity and required capacity as the capacity check for master production scheduling. The MPS is a plan of what to produce. If we know how much capacity is required for this plan, we can then compare this with how much capacity we have available. Let's look at an example. First, we've developed the MPS (Figure 2.14). Then we check the capacity required in each critical workcenter for each time period. In this example, we're looking at Workcenter A for Week 1 (Table 2.8). We see that we need, or expect to use, 15.8 hours

MPS–Amps

Model		Wk1	Wk2	Wk3	Wk4	Wk5	Wk6	Wk7	Wk8
XL 10		70	70	70	70	70	70	70	70
RG 7		50	50	50	50	50	50	50	50
RG 9		110	110	110	110	110	110	110	110
TZ 123		20	20	20	20	20	20	20	20
Total		250	250	250	250	250	250	250	250

Figure 2.14   Master production schedule — amps.

Table 2.8   Capacity Required — Workcenter A

Model	Units/Hr	Hr/Unit	MPS Week 1	Capacity Required
XL 10	25	.04	70	2.8
RG 7	20	.05	50	2.5
RG 9	20	.05	110	5.5
TZ 123	4	.25	20	5.0
Total			250	15.8

Table 2.9   Disk Family Capacity Requirements

Model	Units/Hr	Hr/Unit	MPS Week 1	Capacity Required
C5	25	0.04	300	12
C100	10	0.10	125	12.5
D1	25	0.04	125	5
D5	25	0.04	75	3
Total			625	32.5

Table 2.10   Speaker Family Capacity Requirements

Model	Units/Hr	Hr/Unit	MPS Week 1	Capacity Required
T75	20	0.05	75	3.75
T100	20	0.05	50	2.50
Total			125	6.25

of capacity in Workcenter A in Week 1 for the schedule that we've developed. We know how much capacity we need and we can then check that against how much capacity we have (for this workcenter). We calculated our rated capacity at 24.5 hours per week (8 hours per day × 88 percent efficiency × 69 percent utilization × 5 days per week = 24.5 hours per week). So now we simply compare the available capacity of 24.5 hours with the required capacity of 15.8 hours, and we see that our schedule for Week 1 should easily be achievable, with regard to capacity.

Capacity Available > Capacity Required   Valid Schedule
24.5 hours   >   **15**.8 hours   Valid schedule

This is a fairly simple example. We are looking at one product family consisting of four different items. If Workcenter A also processes the items in other product families, we need to add the capacity required to produce those items to the available capacity in the workcenter. Let's add the disk and speaker families to our calculations. Tables 2.9 and 2.10 show the capacity requirements for Workcenter A in Week 1 for the disk and speaker families, respectively.

If we total the capacity required for all of the items to be processed in Workcenter A in Week 1, we find that we need 54.55 hours (15.8 + 32.5 + 6.25 = 54.55). Now we have a problem:

Capacity Available < Capacity Required   Invalid Schedule
24.5 hours   < 54.55 hours   Invalid schedule

The schedule we have developed is not valid; it is not possible to produce the scheduled items in the planned time period. We have only two options. We must either add capacity or change the plan so that we will not require more capacity than we have available. You may be able to add capacity by using overtime, adding people or equipment, outsourcing, offloading some of the work to an alternate workcenter, or some other method. Changing the schedule is certainly an option, but you need to look closely at how the schedule was developed in the first place. The MPS includes both forecast and customer demand components. For the customer demand component, you will have to either negotiate with your customers to change the delivery date or change the ship date without consulting your customers. Neither option is good. Either way your customers will lose some level of confidence in your ability to deliver as promised.

The point is that you need to check the MPS for validity by performing a capacity check. If the available capacity and the required capacity are not in balance, the plan is not valid. Many people will wish for it to be in balance, won't believe that it is out of balance, or will think it can be accomplished by "putting some effort into it." These same people will then wonder why delivery promises weren't met, the expediters aren't doing their jobs, customers aren't happy, and profits are down. Don't be one of these people! If there is an imbalance, do something about it. Add capacity or change the schedule. Don't work with a plan that isn't doable; you're wasting your time.

2.3 Materials Requirements Planning

The MPS is concerned with finished goods. It tells you what items you need to make, how many to make, and when they are needed. Materials requirements planning is used to break down the MPS and determine what materials you need to purchase and components you need to make, the quantity of these materials and components, and when they are needed. It also uses time phasing, but the time buckets are shorter time periods than the ones used in the MPS. Time buckets in an MRP may be weeks, days, or even hours, depending on your needs.

In addition to the MPS, the MRP uses previously planned or released work orders and inventory status as inputs to the process. It also uses the BOMs and information from the item master file (e.g., purchasing and production lead times, safety stock, and lot size data) to calculate requirements. Raw materials and components may be used in multiple finished goods. The MRP accounts for all the needs for all items. Quantities are consolidated and placed in the appropriate time buckets. The MRP calculates when raw materials and components are needed, then calculates when purchases should be made and production of components started so that they are available when they are needed. The MRP is not magic, as some people seem to think. It is basically a big calculator. It takes inputs from the MPS, inventory and work order status, and item master file to come up with a plan for purchasing materials and building components.

Materials requirements planning has been in use for more than 40 years, yet many small manufacturers either do not use it or do not use it properly. At times another inventory replenishment system may be appropriate, but for most manufacturing operations, materials requirements planning is probably the best choice.* One reason companies do not use materials requirements planning is because they do not understand it, the concepts behind it, and the development of it. One point of confusion is dependent versus independent demand. The raw materials and manufactured components that are needed to produce finished goods are dependant demand items. This means that the need for these materials and components depends on the need for the finished goods. If you need a finished good, then you need the materials and components that are required to produce that finished good. Finished goods, on the other hand, are independent demand items; meaning that the need for them does not depend on the need for any other items. Customers order, or demand, finished goods individually, without regard to any other item. They are independent of each other.

Inventory replenishment systems other than the MRP, such as order point or periodic review systems, mostly react to current conditions. Some sort of forecast need is taken into account, but current inventory levels trigger a reaction for ordering. Many times, materials needs are forecast as independent demand, which is not appropriate and introduces more error into the requirements.** Materials requirements planning is forward looking and time phased. The MRP takes the expected future demand for finished goods from the MPS and the expected future inventory levels, then calculates the needs based on these. Expected future inventory levels are calculated from current inventory levels and expected future usage. Time phased means that the date, or time period, when the materials are

---

* Even in a Lean Manufacturing environment, planning must be performed. Some modifications to a traditional planning system may be warranted in a Lean plant, but planning is still required.

** Forecasts have errors. If you forecast both finished goods and raw materials you will have more error in the system than if you forecast just finished goods then calculate the materials needs for those finished goods.

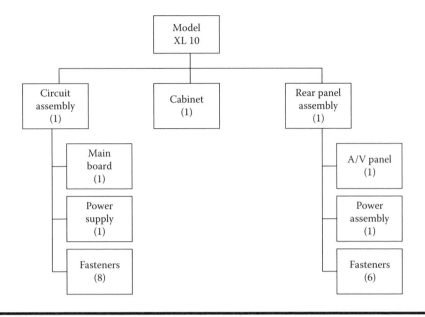

Figure 2.15   Multi-level BOM, model XL 10.

needed is calculated and the quantities needed in each time period are reported.

Materials requirements planning is best shown with an example. Figure 2.15 shows the BOM for the model XL 10 that will be used in the example. Figure 2.16 shows a portion of the MPS for the XL 10 and the MRP grid for the circuit assembly and fasteners. Before we proceed with the calculations, let us define the row headings in the MRP grid:

- *Gross requirements* — This is the total quantity needed. It is determined by the need for the parent item and is derived from the MPS and the BOM. The MPS shows the quantity needed of the finished good and the BOM shows the quantity of material needed for each finished good. The gross requirements are shown in the time period when the materials are needed, which is the time period in which production of the parent item is started.
- *Scheduled receipts* — This is the quantity expected from orders that have already been released. These are shown in the period in which the receipt is expected.
- *Projected on-hand* — The projected on-hand is the calculated on-hand quantity at the end of the period. This is also the calculated, or expected, quantity that will be on hand at the beginning of the following period. The current on-hand quantity is shown to the left of, or prior to, the first period, which is the current period.

Master production schedule, model XL10

Item	Week1	Week2	Week3	Week4	Week5	Week6	Week7	Week8
XL 10		100	100		100	100	100	

Materials requirements plan
Circuit assembly (1 per), Lead time = 1 week, Lot size = 50

Week		1	2	3	4	5	6	7	8
Gross requirements			100	100		100	100	100	
Scheduled receipts		50							
Projected on-hand	20	70	20	20	20	20	20	20	20
Planned receipts			50	100		100	100	100	
Planned order release		50	100		100	100	100		

Fasteners (8 per), Lead time = 1 week, Lot size = 2,500

Week		1	2	3	4	5	6	7	8
Gross requirements		400	800		800	800	800		
Scheduled receipts									
Projected on-hand	1,200	800	0	0	1700	900	100	100	100
Planned receipts					2500				
Planned order release				2500					

Figure 2.16   A portion of the master production schedule for model XL 10 and the materials requirements plan (MRP) grid for the current assembly and fasteners.

- *Planned receipts* — This is the quantity that is expected to be received from planned orders, or orders that have not yet been released. As with the scheduled receipts, they are shown in the period in which the receipt is expected. The difference between scheduled receipts and planned receipts is whether the receipt is from an order that has been released or is still only in the planned stage.
- *Planned order release* — This is the quantity that is to be ordered. The time period in which this is shown is the period in which the order is to be released. Release in this time period allows the quantity to be ready when it is needed. The period in which the order is released is offset from the period that it is needed by the amount of the lead time. This is referred to as lead time offsetting.

Now let's walk through the MRP calculations for the circuit assembly and the fasteners. We already know where the gross requirements for the circuit assembly come from (from the need for the XL 10). The scheduled receipts row tells us that we have 50 circuit assemblies due to be completed in Week 1, and the projected on-hand figure tells us that we are starting with 20 completed circuit assemblies on hand.

From this we calculate that we will have 70 circuit assemblies on hand at the end of Week 1; we have 20, we will receive 50, and we won't use any because there are no gross requirements in Week 1. So, 20 + 50 – 0 = 70. Then in Week 2, we start with 70, receive 50, and use 100, leaving 20 on hand at the end of the week. Because we will only have 20 on hand at the beginning of Week 3, but we need 100, we have to order 100 in Week 2. The lot size is 50 (× 2) and the lead time is one week, so the 100 we order in Week 2 will be available in Week 3. These calculations continue through the balance of the schedule. The calculations for the projected on-hand balance are

Projected on-hand = beginning balance + receipts – gross requirements

Week	Calculation
Week 1	70 = 20 + 50 – 0
Week 2	20 = 70 + 50 – 100
Week 3	20 = 20 + 100 – 100
Week 4	20 = 20 + 0 – 0
Week 5	20 = 20 + 100 – 100
Week 6	20 = 20 + 100 – 100
Week 7	20 = 20 + 100 – 100
Week 8	20 = 20 + 0 – 0

The same calculations are used with the fasteners, except that the gross requirements for the fasteners come from the planned order release for the circuit assembly. If we are planning to release an order for 100 circuit assemblies in Week 2, then we need 800 fasteners in that same week (eight fasteners per circuit assembly) so that we can begin the production for the circuit assembly. Note that the fasteners needed for the circuit assembly would be added to the fasteners needed for the rear panel assembly to arrive at the total number of fasteners needed per time period.

The planned receipts are the quantity needed to prevent the projected on-hand from falling below zero, or below any safety stock quantity that may be required. The planned receipts quantity will be a multiple of the designated lot size.

The planned order release is simply the planned receipt quantity, offset into the period that allows the order to be received in the period it is needed. In our example, the release is one period prior to the receipt, because we have a one-week lead time.

As you can see, the basic process of MRP is fairly easy. It gets more complicated and requires the use of computers when you have more than a few finished goods with several components. To get useful information out of the materials requirements planning system, you need to have good data in the system. This means that all lead times, inventory records, lot sizes, and all other data used in the calculations needs to be accurate. Making sure all the data is accurate when the system is set up is a big task, but keeping it accurate may be an even bigger job. Lead times may change, or may even be variable. Lot size requirements may change, and engineering changes may be made that change the BOMs. It is a constant battle to maintain the system, but is an absolute requirement.

2.3.1 Capacity

Remember that with each level of planning we need to perform a capacity check. It is important for the plans to be valid and achievable. If they are not, they are worthless and will be ignored. At the materials requirements planning level, the capacity check is known as capacity requirements planning. Capacity requirements planning is a detailed capacity check. It differs from rough-cut capacity planning in a couple of ways and may arrive at different results. It should not be surprising that the detailed capacity plan may arrive at a different result, or conclusion, than the less detailed plan. As they say, the devil is in the details.

One difference between rough-cut capacity planning and capacity requirements planning is the frequency of planning. The MPS is generally run monthly, so rough-cut capacity planning is also performed monthly. The MRP is run more frequently — weekly or even daily — so capacity requirements planning is also performed more frequently. Another difference is that rough-cut capacity planning is performed only for critical resources, or capacity-constrained resources. Capacity requirements planning is performed for all machines and workcenters to arrive at a detailed expected load schedule for all resources.

In capacity requirements planning, the capacity required for the setup time and the run time are calculated separately. By combining the capacity requirements for setups, the capacity requirements for processing (run time), and the time-phased order quantities, we arrive at a detailed capacity plan for the building of all manufactured components. If you do not have a separate final assembly schedule,* the detailed capacity plan will include the building of all finished goods.

---

* Some companies break down the master production scheduling process into two components: the MPS and the final assembly schedule. This depends on the nature of your products and your production strategy.

Let's look at an example. Figure 2.17 shows the MRP for circuit assemblies and rear panel assemblies. Figure 2.18 shows the setup and run times for the circuit and rear panel assemblies for Workcenter A. The run time is equal to the run time per unit times the number of units. The number of units is taken from the planned order releases in the MRP.

The total of the setup and run times is the capacity required in Workcenter A in each time period for the circuit assemblies and rear panel assemblies. The same procedure would be performed for all manufactured items in all workcenters. These detailed capacity calculations are performed for both released orders and planned orders. The capacity required would then be compared with the capacity available to determine if the

Circuit assembly (1 per), Lead time = 1 week, Lot size = 50

Week		1	2	3	4	5	6	7	8
Gross requirements			100	100		100	100	100	
Scheduled receipts		50							
Projected on-hand	20	70	20	20	20	20	20	20	20
Planned Receipts			50	100		100	100	100	
Planned order release		50	100		100	100	100		

Rear panel assembly (1 per), Lead time = 1 week, Lot size = 50

Week		1	2	3	4	5	6	7	8
Gross requirements			100	100		100	100	100	
Scheduled receipts									
Projected on-hand	200	200	100	0	0	100	100	100	0
Planned Receipts						100	100	100	
Planned order release					100	100	100		

Figure 2.17   MRP for circuit assemblies and rear panel assemblies.

	Week1	Week2	Week3	Week4	Week5	Week6	Week7	Week8
Setup								
Circuit	0.25	0.25	0	0.25	0.25	0.25	0	0
Rear panel	0	0	0	0.17	0.17	0.17	0	0
Run			0				0	0
Circuit	1.25	2.5	0	2.5	2.5	2.5	0	0
Rear panel	0	0	0	2.0	2.0	2.0	0	0
Total	1.5	2.75	0	4.92	4.92	4.92	0	0

Figure 2.18   Setup and run times (hr) for the circuit and rear-panel assemblies for Workcenter A.

schedule is valid, or feasible. As you can see, with many items and more than a few workcenters, this can become very complex. But you can also see that the basic principles and logic behind materials requirements planning are pretty straightforward.

Materials requirements planning is an important part of the planning process. It is very detailed and depends entirely on valid and accurate data. BOMs, item master data, inventory status, and a valid MPS are required for a valid and useful MRP. However, the rewards are great. A valid and feasible schedule of production and a recommended schedule for purchases are at your fingertips. And with the power and flexibility of modern electronic systems, these schedules can be kept up to date and current with considerable ease.

2.4 Warehousing Basics

Small manufacturers have limited resources. Employees' responsibilities often cross multiple functions, and there is usually little time to specialize in any one particular area. The core business is the production of the company's products. This often means that the warehousing function does not receive the attention it deserves or an equal share of resources. As long as products are getting out the door, little attention is paid to the warehousing functions and processes. This is a mistake, because the operations of the warehouse have a huge impact on the success and profitability of the company. This section will discuss some of the fundamentals of warehousing that every small manufacturer needs to know.

2.4.1 Layout, Picking, and Putaway

Successful warehouse operations depend greatly on the layout of the warehouse and the picking and putaway processes. Unless you are building a new facility, the layout of the warehouse is often not given much thought. Even if you recognize that the current layout is not the best, the time, effort, and money needed to change it usually acts to prevent any change. But the layout you have today may not meet your needs today. Maybe it met your needs initially, or several years ago, but that doesn't mean it still fits you now. Your product mix may have changed, your products may have changed entirely, your volume may be different, customer ordering patterns may be different, and you may be using different equipment and technology than when the warehouse was first set up.

Your first consideration when designing the warehouse layout should be how much overall space you need. This will depend on a number of factors. Obviously, the amount of inventory normally kept in the warehouse is one factor. However, that is not always as straightforward as you might think. You have an average amount of inventory that is stored in the warehouse, but you also have spikes, or peaks, in the inventory levels. This is generally due to seasonal demand. You need to know the average storage requirements and the peak requirements. You can then calculate the ratio of peak requirements to average requirements to determine how much of the peak requirements you should accommodate in the warehouse. If the ratio is high (the peak is much greater than the average) and the time period of the peak is short, you may not want to plan for the warehouse to hold all the inventory. Instead, you may want to use temporary outside storage to accommodate the spikes. If the ratio is low (there is not much difference between the average and the peak), you may want to plan the warehouse so it can hold the peak amount. See Table 2.11 for an example of the inventory calculations and Figure 2.19 for a graph of the planned inventory levels.

Of course, the inventory levels are only part of the equation. The space required for the inventory is another factor. That is, how much space does this inventory take up? Well, that depends on how the inventory is stored. Are individual cases stored in some sort of case-dispensing racks, are they palletized and stored in pallet racks, or are the items small and held in some type of bin storage? The unit inventory needs to be converted into space requirements. This is relatively straightforward. If you know the quantity of inventory and the space requirements based on how they are stored, you can determine your space requirements. For example, if you pack items into cases then palletize the cases, you can calculate the pallet space required. Let's convert the unit inventory, from Table 2.11, into space requirements (Table 2.12). Although the actual amount of floor space required for this inventory will depend on the type of rack system you have, you can easily approximate how much will be needed from these calculations of the number of pallets you expect to store.

The warehouse space required will also depend on the utilization rate of the storage spaces. It is not realistic to plan for a warehouse where every storage space is used all the time (100 percent utilization.) How often will you immediately fill a space after you empty it? It probably won't be all the time. You also need to consider the amount of space needed for receiving, shipping, full case picking, piece picking, packing or palletizing, and other assorted functions. And don't forget the space needed for offices and administrative functions, break rooms, lockers, restrooms, and any other accommodations. Of course, any special handling needs will have to be considered in the calculations, such as needs for temperature-controlled storage, oversize items, and so forth.

Table 2.11 Average and Peak Inventory (Units)

	Jan	Feb	Mar	Apr	May	Jun	Jul	Aug	Sep	Oct	Nov	Dec	Avg.
Amps	1,400	1,400	1,400	1,900	1,900	1,400	1,400	1,400	1,900	2,600	2,600	2,800	1,842
Disks	3,200	3,200	3,200	3,200	3,200	3,200	3,200	3,200	3,500	3,500	3,500	4,200	3,358
Speakers	550	550	550	820	820	550	550	550	820	1400	1,400	1,600	847
Total	5,150	5,150	5,150	5,920	5,920	5,150	5,150	5,150	6,220	7,500	7,500	8,600	6,047
Average	6,047	6,047	6,047	6,047	6,047	6,047	6,047	6,047	6,047	6,047	6,047	6,047	—
% of avg.	85%	85%	85%	98%	98%	85%	85%	85%	103%	124%	124%	142%	—

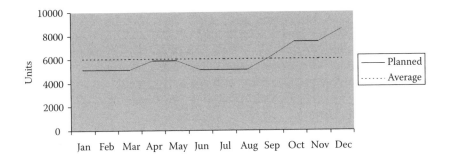

Figure 2.19 Planned inventory levels.

Table 2.12 Pallet Requirements

	Units/Pallet	Avg. Units	Peak Units	Avg. Pallets	Peak Pallets
Amps	24	1,842	2,800	77	117
Disks	24	3,358	4,200	140	175
Speakers	6	847	1,600	141	267

The usage rate of the items in the warehouse affects the layout as well as the picking and putaway systems. High-use items should be stored closer to the doors or docks so that the distance covered when picking and putting them away is minimized, thus saving time. The picking and putaway systems will depend on how the items are stored, the equipment used, and the technology used. Items stored on pallets will be picked differently than items shelved in full cases, and items stored in broken cases or bins will be picked differently than items stored in full cases.

Picking and putaway can be performed by zones, in waves, by single order, or by order batches. In zone picking, a person (the picker) is assigned to an area of the warehouse, and that picker picks all items located in their zone. In wave picking, the pickers travel through the warehouse, ideally in one direction, picking as they go. In single-order picking, the picker picks all the items on an order before moving on to the next order. Picking by batches means that multiple orders are consolidated for picking purposes, and the total quantity for each item is picked before moving on to the next item. See Figure 2.20.

Depending on the size, shape, weight, and quantity of items picked, the picker may sort and pack the order or there may be a separate designated area for sorting and packing. These are just some of the considerations in designing and planning your warehouse and warehouse processes. As your products and customer requirements change, you need

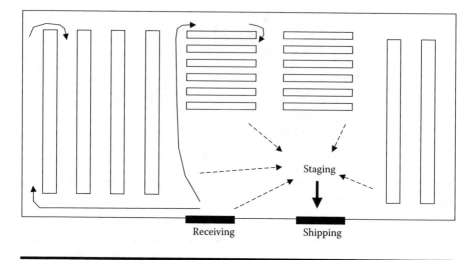

Figure 2.20 Warehouse layout.

to review and evaluate your warehouse layout and your picking and putaway processes. The proper layout and design of your processes will have a major impact on the success of your warehouse.

2.4.2 The Warehouse as a Strategic Weapon

I have mentioned just a few requirements for success in the warehouse, but how do you define success for the warehouse? For a distributor or an independent warehouse, success closely matches profitability. For a small manufacturer, the warehouse has a less direct impact on profitability, so it is not as straightforward in determining success. However, the warehouse should have well-defined goals and performance measurements. If these goals and measures are achieved, the warehouse is considered successful.

Financial performance of the warehouse can be evaluated by comparing operating costs with prior time periods and with benchmark data for third-party facilities. Various costs, such as the cost per transaction for picking, packing, putaway, and delivery, can be evaluated. You can than compare these transaction costs over time or compare them to industry standards. If your transaction costs are decreasing and they are below the industry standard, you are doing well. Productivity measures usually consist of some sort of output per labor hour. These include the number of line items picked per labor hour, items put away per labor hour, and line items shipped per labor hour. Space utilization is a performance measure. Quality should also be measured. The percentage of line items picked accurately and shipped accurately are quality measures. Inventory accuracy is another important quality measure.

Success for the warehouse can also be defined in terms of how well it supports the strategic plan of the company. The strategic plan will include things such as the markets the company is targeting, the amount of market share desired, and the growth and growth rate being sought. The warehouse is a vital tool in achieving these strategic goals. The warehouse needs to develop strategies that will help the company reach the markets it is targeting. The warehouse can improve their processes so that the company can gain a greater market share. The warehouse is closer to the customer than the core business of production. This means that the warehouse plays an equal, if not greater, role in customer satisfaction. The warehouse plays a large role in the quality of the products that reach the customer. Obvious damage to a product can be spotted by warehouse personnel, who can then prevent the item from being shipped. The warehouse monitors the shelf life of products. Items with little shelf life remaining can be flagged for high-volume customers or customers that are located closer to the warehouse.

Whatever strategic plans the company has should be supported by the warehouse. And the warehouse should be evaluated in relation to these plans. The warehouse is not just a place to store inventory until it moves out the door. The warehouse plays a vital role and should be given the resources needed to perform at its best. The warehouse interacts closely with logistics providers, distributors, and customers and is important to the long-term profitability of the company.

2.4.3 Technology in the Warehouse

This ain't your daddy's warehouse. The modern warehouse is technology driven and increasingly automated. Bar coding is nearly universal, is more sophisticated than ever, and is starting to become outdated. Bar codes are being replaced by radio frequency identification (RFID) tags. Radio frequency tags have been common in distribution centers and shipping ports, but they are getting smaller and cheaper and are now being used on individual items. With them, individual items can be tracked and monitored in real time throughout the supply chain.

Picking has evolved from paper-based pick tickets. New picking methods include pick-to-light and pick-to-voice systems. These systems allow hands-free picking and improvements in pick accuracy. Bar coding, RFID, and visual systems allow automated sorting and flow through the warehouse. Vehicle-mounted computer terminals or other devices speed picking, putaway, and inventory monitoring.

Warehouse management systems are commonplace and continue to increase in functionality. For small manufacturers, ERP systems may include some warehouse management functions. (See Chapter 9 for more on ERP.) It may be necessary, however, for the small manufacturer to explore a separate warehouse management system in order to meet the strategic goals of the warehouse. If a separate warehouse management system is purchased, it is important for it to integrate with the ERP system.

In the warehouse, as in all other areas, it is a mistake to use technology just for technology's sake, or because some salesperson gives you a slick presentation. The benefits of any technology need to be evaluated in relation to its cost and applicability. In today's globally competitive marketplace it is a requirement to use some level of technology. However, improper selection, implementation, and use of technology can be costly or devastating to your company.

Chapter 3

Sales and Operations Planning: Know Where You're Going

The first question we need to answer is, what exactly is sales and operations planning? Sales and operations planning is a well-defined, ongoing planning process that ties together the sales planning and operations planning processes, to better satisfy customers and meet the goals of the company. Sales and operations planning helps you balance the demands from your customers (sales planning) with the resources at your disposal (operations planning). Financial planning is also integrated into the process, because this is an important resource component. The subtitle for this chapter is, "Know Where You're Going," because sales and operations planning sets the direction for the entire company. If you are not performing sales and operations planning, your company is probably trying to head in several different directions at the same time. Sales and operations planning develops one plan, one set of numbers that everyone will use. This sets everyone on the same path, for better results. You are doing sales and operations planning, aren't you?

The good news is that you are already doing it to some extent, even if you don't know you are and aren't calling it by this name. What sales and operations planning does is formalize the processes and tie them together into one coherent plan that becomes a valuable tool to help you run your business. Many companies that begin the sales and operations planning process wonder how they ever managed to run their business without it.

Sales and operations planning helps you plan at a macro level without getting into too much detail. It works at the product family level and is concerned with volume; that is, the overall levels of product rather than the mix of individual items. However, detailed plans must be developed from the sales and operations plan (S&OP), and then they must be executed. The S&OP is an executive-level plan that allows you to plan, monitor, and make corrections to the overall course of your business on a regular basis. Let's review the planning hierarchy (Figure 3.1) and see where the S&OP fits.

As you can see from the planning hierarchy, sales and operations planning is a high-level planning process. At this level, we are concerned with balancing demand and supply at aggregate levels, and we are looking more long term. When we speak of demand, we are talking about all of the requirements, or demand, for our products. Many people think that sales and demand are the same thing, but they are not. Sales are a subset of demand. Demand includes any and all needs that must be planned for — not just customer sales. Demand includes any service parts requirements, safety stock requirements, marketing supplies, engineering needs, and any other needs for the products. In sales and operations planning, we also time phase the demand, meaning that the demand is placed into separate, discreet time periods that reflect when the products are required. Once all the demand has been identified, product supply can be planned and managed.

Balancing supply and demand is not always as simple and straightforward as we would like. Identifying, and predicting, demand is full of challenges. But planning and managing the supply to meet that demand has its own set of challenges. The ability to satisfy the demand depends greatly on the resources available and the ability to manage those resources effectively. Sales and operations planning is a valuable tool that helps balance demand and supply and identify and manage the resources needed to meet the demand.

3.1 Volume versus Mix

Although sales and operations planning is not common among small manufacturers, it is a very important part of the planning system and a valuable tool. Some people think that balancing demand and supply is not that big a deal — figure out how much you need, then make it. If it were that simple, your customers would always be satisfied, you would

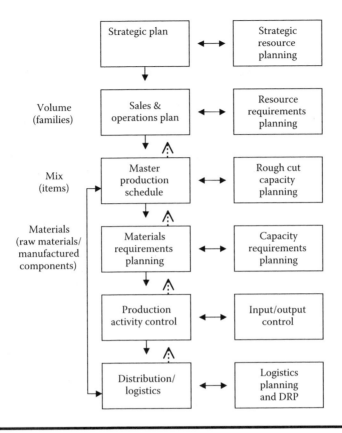

Figure 3.1 Planning hierarchy.

never be out of stock, and all of your deliveries would be on time. Because it is not as simple as we would like, we need to develop some plans, then perform some capacity checks to validate those plans. And we need to perform this planning and validation at several levels. Many small manufacturers are missing out on the full benefits of planning by not planning at all levels.

Many small manufacturers develop good item-level production plans and schedules, with the necessary capacity checks performed. And many small manufacturers execute their plans very well, procuring materials, building products, and shipping to customers. But many of these same companies are not reaching the level of satisfaction with their performance that they would like. One of the missing links to greater satisfaction could be sales and operations planning.

The master production schedule (MPS), the item-level plan, should be derived from a higher level plan, the S&OP. The S&OP is stated in

Volume–Product families

Family	Jan	Feb	Mar	Apr	May	Jun	Jul	Aug	Sep	Oct	Nov	Dec
Amps	1,000	1,000	1,000	1,500	1,500	1,000	1,000	1,000	1,500	2,000	2,000	2,500
Disks	2,500	2,500	2,500	2,500	2,500	2,500	2,500	2,500	2,750	2,750	2,750	3,000
Speakers	500	500	500	750	750	500	500	500	750	1,000	1,000	1,250

Mix–End items (Amps family)

Model	Wk1	Wk2	Wk3	Wk4	Wk5	Wk6	Wk7	Wk8
XL 10	70	70	70	70	70	70	70	70
RG 7	54	54	54	54	54	54	54	54
RG 9	106	106	106	106	106	106	106	106
TZ 123	20	20	20	20	20	20	20	20
Total	250	250	250	250	250	250	250	250

Figure 3.2   Volume versus mix.

aggregate volumes, along product lines, and the MPS is stated in specific end items, or a mix of products. The total quantity of the individual items in the MPS must equal the aggregate quantities in the S&OP. The MPS translates the aggregate quantities from the S&OP into the specific mix of items, or product configurations, that are to be built. If a company is developing an MPS without performing sales and operations planning, they are missing an important verification of the MPS. See Figure 3.2 for an example of the link of volume and mix.

The MPS can be developed by forecasting all of the individual items that are produced, but this does not always provide a complete picture. As will be discussed in more detail in Section 3.2.1, one rule of forecasting is that forecasts are more accurate in the aggregate than in detail. This means that you have a better chance of correctly forecasting the total volume of a product family than correctly forecasting each of the individual items that make up that family. The volume forecasts developed in the sales and operations planning process are a verification of the individual item forecasts developed during the master production scheduling process. If the total of all the individual item forecasts does not match the volume forecast, within a specified range, the difference must be investigated and reconciled. If the volume forecast for a product family is 1,000 units, but the individual item forecasts for the items in that family total 2,000 units, there is obviously a discrepancy. This discrepancy needs to be investigated and the cause determined. Sometimes, each individual item is forecast optimistically, but then the items are not summed together and given a sniff test to see if it makes sense as a whole.

3.2 The Sales and Operations Planning Process

The size of your organization will, in part, determine the complexity of your sales and operations planning process. The smaller your company, the more integrated you are likely to be. This is, of course, a generalization, but it is natural that in smaller companies people's duties tend to cross functions more than in larger companies, and people tend to be physically closer, which facilitates communication. By communication, I mean that people actually talk with each other and discuss problems they are having. The larger the company, the more compartmentalized it tends to become. This is precisely what sales and operations planning is designed to overcome. Sales and operations planning brings the functional managers together to develop a common plan.

The sales and operations planning process is a bit circular — a chicken and egg sort of thing. Which comes first — the detailed plans that are rolled up into the overall plan or the overall plan that is then used to develop the detailed plans? Actually, it's a little of both. If you are going to start doing sales and operations planning, you have to start somewhere, so don't get caught up worrying about where to start. Just start; see where you are and go from there.

Once the S&OP is developed, managers in the functional areas (sales, operations, finance) develop more detailed plans that are derived from the S&OP. The planning process usually works on a monthly cycle, which culminates in a monthly S&OP meeting. This is an executive-level meeting. During this meeting the plan will be reviewed, questions will be asked, assumptions will be challenged, and the plan will be updated and approved. The actual meeting should be relatively short. Much of the work has already been done when putting the plan together each month. The meeting ensures that everyone is in sync and formally reviews and approves the plan.

The S&OP works on a monthly cycle, with a planning horizon of about 12 to 18 months. This means that each month, the plan is reviewed and updated, with the past month "falling off" and a new month added on. The past month or months do not just disappear from the plan. It is a good idea to have past performance included in the review, but the S&OP is a rolling plan, so that you are always looking out at a fixed planning horizon.

Once the plan is approved, the functional managers develop detailed plans derived from the S&OP. These detailed plans are used to drive the actions in their areas. Next comes execution of the day-to-day actions of the various functional areas. Either ongoing during the month or at the end of each month (probably both), actual results are reviewed. These results are compared to the plans, any variances are noted, reasons for the variances are stated, and the plans are updated accordingly.

The updated detailed plans are rolled up and tied together into the S&OP. As you have probably guessed, if any changes were made to the plans from the previous month, there must be some communication among the functional areas before the plans are rolled up into the S&OP. For example, if production (operations) produced fewer units than planned in the month, but they intend to make up the production in the next month, they need to update their plan. This affects the finance department, because the increased production in the next month may mean an increase in expenditures in that month (meaning spending more than was initially planned). Therefore, the finance department needs to update its plan or the operations and financial plans won't match when you try to roll them up into the S&OP. Sales may need to update its plan also. If the reduced production caused sales to be shifted from one period to another, the sales plan must reflect that change.

If this sounds complex and time consuming, remember, you are already doing many parts of this process. The S&OP formalizes the process and ties together the plans of the functional areas in a way that probably has not been done before. Yes, when you first get started with sales and operations planning it will take some time and effort to get going. But it will quickly become a regular part of your monthly activities and prove its worth almost immediately.

The first thing that needs to be made clear is that the S&OP should be a unit- or quantity-based plan, not a dollar-based plan. This is a major change in thinking for some companies. But don't worry. You can (and probably will want to) easily convert the units into dollars. There are two primary reasons for basing the plan in units. First, you make and sell units, whether they are discreet units, pounds of product, gallons, or whatever. You plan production in units, purchase materials in units, stock units, move units, and sell units of product; therefore, you should plan in units. Second, units are equivalent over time, whereas dollars are not. Your prices change, your costs change, and the value of the dollar changes over time. Comparing dollars from one time period to another is not always easy. Of course, you will review your performance in financial terms, but dollars should not be the basis for your planning. It is easy to convert your units into dollars to use for the necessary review.

Another thing that needs to be clear is that the S&OP is developed at a product group or family level, not at an individual item level. The individual item level is too detailed for the S&OP. Individual items are planned at the MPS level (this is derived from the S&OP). In sales and operations planning you plan overall production by product groups. Some companies will have distinct product groups that make it easy to identify the groups. Other companies may have to spend a little more time identifying the product groups. Sometimes the question of which items should be grouped and which should be separated is not clear. The product groups need to be few enough in number to be easily reviewed, but there needs to be enough groups to provide useful information. Each product group will be reviewed separately during the monthly S&OP meeting. Keep this in mind when assigning the product groups.

One starting point to identify product groups is by the production process. Identifying products that share production processes or facilities may be a good way to separate the groups. Just to add a wrinkle, if you subcontract some of your products or some assembly, you need to take that into account in the sales and operations planning process, and it may or may not affect the grouping of your products.

The S&OP documents will consist of one page for each product group, with various supporting documents to back it up. The plan is easily reviewed when all the information is summarized on one page. The supporting documentation is needed only for answering questions during the meeting. Several key people from the various departments may attend the monthly meeting to provide additional information or to answer questions. Some of these people will attend only that part of the meeting that affects their department or product groups. Remember, this is an executive-level meeting, and as such it should be high level and short.

I will use a sample S&OP spreadsheet to illustrate the process during the following discussion. Getting started with sales and operations planning requires only a simple spreadsheet. As you get more sophisticated, you may add multiple users entering data, the detail plans rolling up into the final document, or data importing from other programs. Some software vendors are beginning to develop and market sales and operations planning programs as add-ons to their enterprise systems. However, always remember that the goal is to use technology as a tool to increase your profitability, not as way to impress anybody or make things more complex.

3.3 The Sales Plan

As a concept, the sales plan is very simple, as is the operations plan. However, the actual planning can be a little difficult and requires resources and expertise. But I'm going to say it again — you're already doing it, whether you know it or not, so why not formalize and improve it? The sales plan is simply how many units you plan or expect to sell during each time period over the planning horizon. Okay, it's not really as simple as that. For one, when we talk about sales we really mean demand. Demand includes all need for the products. This includes any items to be used by marketing for promotions and advertising, any items needed by engineering for testing or other purposes, any requirements to replenish safety stock, any service parts requirements, and any other needs. Also, you probably really mean shipments, or units that are leaving your possession and are moving to your customers, rather than only those items that are technically considered sales. It is important to understand these differences and to define the terms as you use them in your company. You always want to make sure that everyone understands the terms and that they are well defined, so there is no confusion or misunderstanding when using them.

3.3.1 Forecasting

In discussing the sales plan, we have to talk a little about forecasting. Again, don't worry. I guarantee that you are already forecasting. Someone, somewhere in your company is forecasting. Maybe it's not the right person, maybe several people are doing it and not telling each other, and maybe you're doing it very poorly. But someone in your company is forecasting.

Ideally, your sales department is doing the forecasting. Ideally, the person or people who are doing the forecasting have the training and experience to forecast well. Ideally, these people are communicating well with your customers. In reality, forecasting is often done by someone outside the sales department. In reality, the people doing the forecasting often have no training and little experience in forecasting. In reality, the people doing the forecasting often do not communicate with your customers whatsoever.

I've seen decent-size companies that did no formal forecasting and where the production department decided what to build and when. The sales department complained loudly whenever an item was out of stock, the warehouse questioned why they had so many of certain items, and finance was constantly worrying about the money tied up in inventory and the epidemic inventory variances. But I've also seen very small companies that had, and continued to develop, very good forecasting systems. The sales department develops the forecast, in units, based on past performance and dialogues with customers. The forecast is updated regularly and incorporates feedback from various departments. Production, shipping, marketing, and sales meet and communicate regularly to discuss the forecast, the production needed to support the forecast, and the marketing and shipping requirements. What this should show you is that, yes, someone is doing forecasting, and that the process can be performed well and can be improved no matter the size of your organization.

Many methodologies have been developed to forecast sales, or more accurately, demand. Many are complex and required highly educated and talented mathematicians to develop. Numerous software packages utilize many different forecasting algorithms. This means that you have many forecasting methods and tools to choose from. But for you, the small manufacturer, there are some basic components of forecasting that you should be familiar with, regardless of the methods or tools that you use. It is important to understand some of the basic principles behind forecasting and the nature of demand. Too often people rely on technology or the expertise of others without even a basic understanding of the principles behind them. Although it isn't possible to be an expert in everything, and you should rely on the expertise of others sometimes, you should also take the time to understand some of the principles behind many of the tools that your company uses on a daily basis.

Two important factors in the development of forecasts are past history and your current customers. First, history: How have your sales performed in the past, and is this history a good indicator of future performance? If your sales history is a good indicator of future performance, use it as a component in the development of your forecast. If it is not a good indicator, ignore it, and use other information. Many people use sales history to develop their forecasts without questioning it. But history is not always a good indicator of future performance. Sometimes it is pretty obvious that the history should not be taken into account, but sometimes it is not so obvious. You should question the validity of your historical data whenever using it to develop a forecast. If it is valid, use it. If not, use other data. Besides sales history, you need to consider your customers. What are your customers' plans? Are they planning to increase purchases from you, decrease them, or keep them at the same level? Do they see their markets changing, their customer base changing, or any trends? You need to have good relations with your customers so that they are willing to share this information with you, because you need to incorporate this information into your forecasts. Question your customers. Just asking them for their expected need, or their forecasts, is not enough. You need to understand where their demand is coming from, and the likelihood that their projections are realistic. If you think they are being overly optimistic or you think they are doing more speculating than forecasting, you need to take that into account in your forecasts.

Historical data and information from your current customers are only two factors that you need consider. You also need to account for any expected new customers — either several small-volume customers or one large-volume customer — that would impact future demand. You need to consider the impact of any new product introductions. Will the new products enhance sales of your current products or take away from them? Or will they increase the total market for all your products? Any promotions that you have planned, and any price changes, will also affect your demand and therefore your forecasts.

Other terms you will probably come across when learning more about forecasting are intrinsic, extrinsic, qualitative, and quantitative. Intrinsic and extrinsic refer to the sources of the data you use when developing the forecasts. Intrinsic refers to data that is internal to your organization, such as historical sales, planned promotions or price changes, and demand other than sales. Extrinsic refers to factors that are external to your company. This includes your customers (both current and new customers), new competitors, and the general economic climate. Qualitative and quantitative refer to the type of data. Information other than hard data, or that is not easily verifiable, is qualitative. Qualitative forecasts use this type of information, such as a manager's experience or intuition, when other data is not available. Verifiable, reliable data is quantitative. Quantitative forecasts use both intrinsic and extrinsic data.

3.3.2 Characteristics of Forecasts

You should be familiar with some general characteristics of forecasts. The first is that forecasts will not be 100 percent accurate. Many people like to say that forecasts are always wrong, but I don't agree with that view. It is impossible to be 100 percent accurate with forecasts because there are too many variables, and these variables do not obey any strict natural laws. Customer desires, changes in technology, economic conditions, and many other factors affect customer demand. These factors, and simple random variation (discussed in more detail in Chapters 5 and 7), result in forecasts that will have a certain level of error.

However, when developing forecasts, past errors can be tracked over time and can be incorporated into future forecasts. The knowledge that there will be some level of error should not prevent us from developing useful forecasts. Actually, all forecasts should include the expected level of error, so that a forecast range can be determined. This means that instead of a forecast that says you expect demand of 500 units in January, your forecast should say something like you expect demand of 500 units, plus or minus 20 units, in January. This provides more information to the users of the forecast and aids in planning. For example, the following three forecasts are considerably different:

- A forecast of 500 units in January
- A forecast of 500 units, plus or minus 20 units, in January
- A forecast of 500 units, plus or minus 100 units, in January

The first example does not provide enough information. The forecast is 500 units, but there is no range or expected variation from the forecast quantity. The second and third examples both provide the forecast and the range, but the second example shows a forecast with a much lower level of error. The planning will be considerably different under these three forecasts.

Forecasts are more accurate for larger groups of items and less accurate for individual items. This means that aggregate forecasts are more accurate than detailed item forecasts. A forecast for the total unit sales of a product line will be more accurate than the forecasts for each individual item or configuration within that product line. Forecasts are more accurate for nearer time periods and less accurate for time periods further into the future. The forecast for the next three weeks will be more accurate than the forecast for nine months from now. This does not mean that we should develop forecasts only for a short period into the future; it only means that the accuracy will decline over the time horizon of the forecast. It is important that the forecast cover a time frame that is long enough for the company to plan for any potential problems.

Before we move on, there is one more thing of vital importance: the responsibility for the forecast. This is often a contentious issue, but there is only one answer. The sales and marketing* department is responsible for the forecast. Sales and marketing will be responsible and accountable for the sales results, so they must be responsible and accountable for the forecast. It is totally unreasonable, and unrealistic, for one person to be responsible for the forecast and another to be responsible for the sales results. The company develops many plans based on the forecast, so there must be accountability for it. If someone other than sales and marketing is responsible for the forecast, there will be a disconnect between responsibility and accountability, between planning and execution. Sales and marketing is responsible and accountable for sales. They are held accountable if sales are less than forecast and are often rewarded if sales exceed the plan. Sales and marketing executes the sales plan. They use the plan, or forecast, as a goal and a guide, and they monitor their results and measure themselves against it.

Because sales and marketing is responsible and accountable for the sales results, they should be responsible for developing the forecast and be accountable for the final forecast. Some might argue that if sales and marketing is rewarded for beating the forecast (selling more than forecast), they will purposely or consistently underforecast. If your company allows this to happen, sure, some people will take advantage of it. However, rewards can be tied to both sales performance *and* forecast performance.

---

* I use the term *sales and marketing* to refer to one entity or one person. Some companies may have separate sales and marketing departments, but if they develop individual forecasts, somebody must be responsible for the total forecast, or for rolling the separate forecasts into one final number.

Others may argue that in this case, some sales people will hold back on sales so that they do not exceed the forecast, thus missing out on the sales and associated revenue. This is probably not likely, and either way, rewards can be developed that will allow for maximum sales and more accurate forecasts. A lot of this depends on the corporate culture, the company environment, and the relations between management and employees and between departments. It is important that the forecast be as accurate as possible, because the performance of the entire company is directly tied to the forecast, not just the performance of the sales and marketing department. As we will see in the rest of this chapter, the forecast is the basis for the S&OP, which is the basis for the MPS, then the materials requirements plan.

3.3.3 Demand

As noted earlier, all sources of demand need to be taken into account. If you account for only part of the total demand, the forecast will not be accurate. The parts you are accounting for may display more volatility than the data as a whole, which will affect your forecast. The demand that you leave out, or account for, may include other characteristics that will affect your forecast and reduce the accuracy of the forecast. The point is, include all sources of demand in your forecasting model.

Demand over time will display one or more distinct patterns. To identify these patterns, it often helps to graph the data. Graphs tend to bring out, or highlight, patterns that you might not see if you just look at the numbers. Let's look at some examples, starting with sales, or demand, history, then looking at the associated graphs. Table 3.1 provides historical data, and Figures 3.3 and 3.4 show this data in graphical format. The two graphs show the same data, but in different formats.

Looking at the graphs of the data, it is easier to pick out some of their characteristics, such as the seasonality. Seasonality is characterized by a regular and substantial fluctuation in the demand, as compared to the average. Just looking at the numbers in Table 3.1, you might not pick out the seasonal spike that is apparent in May. But it is easily detected when viewing the graph. Most people are familiar with seasonality through the year, such as the spikes in retail sales during the "holiday season" in the United States and travel for the lunar New Year celebrations in China and other Asian countries. But seasonality occurs during shorter time periods in different industries. Restaurants often display both weekly and daily seasonal patterns. In the United States, some restaurants are closed on Mondays due to much lower than average sales, but they show definite peaks on Friday and Saturday nights. And most restaurants are busier during the lunch and dinner hours than at other times of the day. But

Table 3.1 Demand History

Family	Jan	Feb	Mar	Apr	May	Jun	Jul	Aug	Sep	Oct	Nov	Dec
Year 2												
Amps	986	1,015	1,145	1,328	1,689	878	928	995	1,487	2,213	2,067	2,399
Disks	2,671	2,588	2,543	2,376	2,218	2,409	2,472	2,617	2,890	2,913	2,598	2,947
Speakers	436	491	530	687	772	581	583	509	816	1,200	876	1,312
Total	4,093	4,094	4,218	4,391	4,679	3,868	3,983	4,121	5,193	6,326	5,541	6,658
Year 1												
Amps	905	936	1,050	1,227	1,548	812	845	911	1,368	2,033	1,899	2,211
Disks	2,457	2,380	2,358	2,177	2,038	2,221	2,268	2,401	2,663	2,672	2,402	2,718
Speakers	389	458	493	621	704	534	542	475	760	1,111	812	1,193
Total	3,751	3,774	3,901	4,025	4,290	3,567	3,655	3,787	4,791	5,816	5,113	6,122

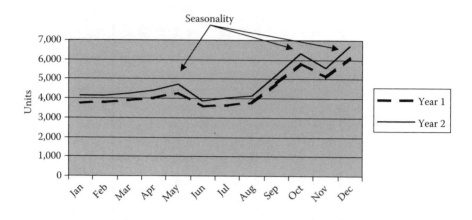

Figure 3.3   Demand history graph — year over year.

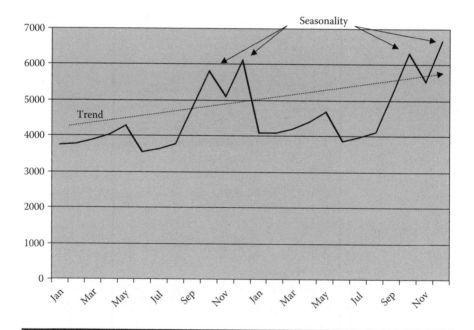

Figure 3.4   Demand history graph — multi-year.

these daily patterns vary from restaurant to restaurant. In an industrial area, lunch and dinner peaks may be several hours earlier than in a downtown, or office, area. Every company and every industry will display their own unique seasonal patterns, so you must analyze your own data carefully.

Another common demand pattern is a trend. Trend is a general upward or downward movement of the average over time. Again, reviewing the numbers in Table 3.1, you might not notice the upward trend of the data over the two years. But the graph in Figure 3.4 clearly shows this upward movement. This brings up another point — sometimes more than one graph is needed. As you can see from the graphs in Figures 3.3 and 3.4, the same data shown in different formats reveals different information. The trend can be seen in the first graph, but it is not as noticeable because of the format. The graph in Figure 3.4 better highlights the trend pattern.

A longer term pattern is a cycle. Economists often talk of business cycles. These are long-term periods — years or decades — in which overall business rises and falls as general economic conditions change. Your business may be affected by cycles tied to the general economic climate, or the cycle may be related to changes in the public's tastes or the life cycle of your products.

Other patterns may be found in the demand data, but it is not necessary to be familiar with all of them. You need to be aware that there are different patterns associated with demand and that different forecasting techniques can utilize these patterns. For example, if you have a distinct trend, you want to use a forecasting technique that incorporates this instead of ignoring it. By being aware of the different demand patterns, you will better understand the forecasts that are developed from the demand data and any patterns inherent in the data.

3.3.4 Some Forecasting Methods

This is not a lesson in how to forecast, but, as with demand patterns, it is important to understand some basic forecasting methods so that you will better understand the resulting forecasts. But first, it is important to be sure that you are forecasting the right things. As we discussed in Section 2.3, independent demand should be forecast, whereas dependent demand should be calculated. Independent demand items (finished goods and service parts) are not tied to the needs of any other items; they stand alone. Although some relationship may exist between independent demand items, you cannot calculate the need for one from the need for another. Dependent demand items, on the other hand, can be calculated directly from the need of other items. Raw materials and manufactured components are dependant demand items. This means that the need for these materials and components depends on the need for the finished goods and can be calculated from the need for the finished goods.

I mentioned above that forecasts are more accurate in the aggregate than in detail. But I also discussed tying the S&OP to the MPS by reconciling the aggregate forecast of the S&OP with the detailed forecasts of the MPS. This may seem contradictory and confusing. Well, unfortunately, things are not always as easy or as clear cut as we would like them to be. You may have to develop more than one forecast and reconcile them or tie them together somehow. As if forecasting wasn't bad enough, now you may have to develop several of them? If it were easy, everyone would be doing it, right? The effort you put into forecasting will pay off in all the areas that depend on the forecast (which is the entire company).

Sales and operations planning is concerned with aggregate demand at the product line level. Therefore, during the sales and operations planning process, the forecasts will be developed for aggregate demand. During the master production scheduling process, you may want to develop forecasts for individual items, then reconcile the total of these individual item forecasts with the aggregate forecast. You may choose other alternatives, however. One alternative is to forecast the aggregate demand for the product line, then disaggregate this into the quantities of the individual items that make up the product line. One method of disaggregation is to calculate the percentage of each individual item in the product line from historical data. These percentages are then applied to the forecast aggregate demand to arrive at the forecast demand for each individual item. See Figure 3.5 for an example of disaggregation.

An alternative to disaggregation is to forecast each individual item, then total the quantities for each product family (roll up) to arrive at the forecast for the product families. Just remember one of the rules of forecasting: forecasts are more accurate in the aggregate. Forecasting at the individual item level will have more error than forecasting at the aggregate level.

The most basic forecasting technique is the simple average. Add all of the historical data and divide by the number of periods you added (Table 3.2). With the simple average, all past history is generally used in the calculation. This can be a lot of data, depending on the length of the time periods you use (weeks or days). The advantage of using the average is its simplicity and ease of calculation. Just about everyone understands and can calculate an average. One disadvantage is that the past average may not be indicative of future demand. This can be especially true if the data being used extends far into the past. The older the data, the less likely it is to be an indicator of future demand. Another, rather big, disadvantage is that when using an average, you can effectively only forecast for one period into the future. You can use it for more than one period, but they will all be the same number, which is not realistic.

A step up from the simple average is the moving average. The moving average uses only a select amount of past data, data that emphasizes the most recent past. Because the most recent past is usually a better indicator of the future, the moving average is better than a simple average. A variation of the moving average is the weighted moving average. The

Year 2 actual demand–families

Family	Jan	Feb	Mar	Apr	May	Jun	Jul	Aug	Sep	Oct	Nov	Dec
Year 2												
Amps	986	1,015	1,145	1,328	1,689	878	928	995	1,487	2,213	2,067	2,399
Disks	2,671	2,588	2,543	2,376	2,218	2,409	2,472	2,617	2,890	2,913	2,598	2,947
Speakers	436	491	530	687	772	581	583	509	816	1,200	876	1,312
Total	4,093	4,094	4,218	4,391	4,679	3,868	3,983	4,121	5,193	6,326	5,541	6,658

Year 2 actual demand–Amp family & percent of each item in family

Amps	Jan				Feb					
Model	Wk1	Wk2	Wk3	Wk4	Wk5	Wk6	Wk7	Wk8	Total	% of Total
XL 10	73	56	82	55	87	67	63	81	564	28.2%
RG 7	45	67	64	72	42	41	36	68	435	21.7%
RG 9	93	116	83	107	98	100	126	122	845	42.2%
TZ 123	15	18	16	24	28	17	12	27	157	7.8%
Total	226	257	245	258	255	225	237	298	2,001	100.0%

Forecast year 3–product families

Family	Jan	Feb	Mar	Apr	May	Jun	Jul	Aug	Sep	Oct	Nov	Dec
Amps	1,000	1,000	1,000	1,500	1,500	1,000	1,000	1,000	1,500	2,000	2,000	2,500
Disks	2,500	2,500	2,500	2,500	2,500	2,500	2,500	2,500	2,750	2,750	2,750	3,000
Speakers	500	500	500	750	750	500	500	500	750	1,000	1,000	1,250

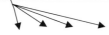

Disaggregated forecast year 3 (Amps family)

Model	% of Total	Wk1	Wk2	Wk3	Wk4	Wk5	Wk6	Wk7	Wk8
XL 10	28.2%	70	70	70	70	70	70	70	70
RG 7	21.7%	54	54	54	54	54	54	54	54
RG 9	42.2%	106	106	106	106	106	106	106	106
TZ 123	7.8%	20	20	20	20	20	20	20	20
Total	100.0%	250	250	250	250	250	250	250	250

Figure 3.5   Disaggregation.

weighted moving average assigns weights to the data, which acts to emphasize even more select periods of history. Generally, the most recent past is weighted more than further past data. See Tables 3.3 and 3.4 for examples of moving average and weighted moving average. For each of these examples, I have selected nine periods of data to use in the calculations.

You can see from these examples that the moving average represents the most recent past better than the simple average and the weighted moving average represents it even better than the moving average. The simple average, moving average, and weighted moving average are all easy to

Table 3.2 Simple Average

Family	Jan	Feb	Mar	Apr	May	Jun	Jul	Aug	Sep	Oct	Nov	Dec	Avg.
Amps	986	1,015	1,145	1,328	1,689	878	928	995	1,487	2,213	2,067	2,399	1,428

Table 3.3 Moving Average

Family	Apr	May	Jun	Jul	Aug	Sep	Oct	Nov	Dec	Avg.
Amps	1,328	1,689	878	928	995	1,487	2,213	2,067	2,399	1,554

Table 3.4 Weighted Moving Average

Family	Apr	May	Jun	Jul	Aug	Sep	Oct	Nov	Dec	Avg.
Weight[a]		.2			.3			.5		
Amps	1,328	1,689	878	928	995	1,487	2,213	2,067	2,399	1,714

[a] .5 (most recent three months), .3 (next three months), .2 (oldest three months).

calculate and easy to understand. The next step in forecasting is a little more complex and a little harder to calculate but still fairly easy to understand.

The next step up in forecast development is exponential smoothing. Exponential smoothing is similar to the moving averages but is a little more sophisticated and usually gives better results. Exponential smoothing also incorporates past errors in the forecast, meaning the difference between the forecast and the actual results. All past data is incorporated, but the most recent past data is much more heavily weighted. The older data become significantly less important, and less weighted, as new periods of data are added. The amount of the past forecast error that is incorporated into the current forecast is determined by the use of a *smoothing constant.* The smoothing constant is a number (between 0 and 1) that is included in the exponential smoothing formula. The idea behind the smoothing constant is that if there are errors in the forecast, part of the error is due to an underlying change in demand, and that needs to be added to the new forecast. For example, if you forecast 500 units, but the actual result is 550 units, part of that 50-unit difference is due to a change in the underlying demand pattern. The smoothing constant adds some amount of that change in demand to the new forecast. Not all of the error in the forecast is due to a change in demand (some is due to random variation and some is due to other factors), so not all of the error is added to the new forecast. The smoothing constant determines the amount of error to incorporate into the new forecast. Someone needs to choose the number to use for the smoothing constant. Although that seems somewhat arbitrary, some guidelines exist for selecting the smoothing constant.

The basic exponential smoothing formula can be enhanced to incorporate demand that shows the characteristics of trend and seasonality, or both together. The trend-enhanced forecast calculates a base value for the demand and a trend value. The base value is calculated by using exponential smoothing with the smoothing constant. The trend value is also calculated using exponential smoothing, but a different smoothing constant is used. This is known as the trend smoothing constant. Again, someone needs to choose the number that is used as the trend smoothing constant, but there are established guidelines that help with this decision. Like the smoothing constant, the trend smoothing constant is used to incorporate a certain amount of past trend information into the new forecast. Unless the trend is constant, there will be some fluctuations in the trend from period to period. The trend smoothing constant attempts to incorporate the actual trend component while removing the fluctuations. The number that is used for the trend smoothing constant determines how much of the total trend (net of actual trend plus random fluctuations) will be incorporated into the new forecast. To put it more simply, if your demand increases every period (say, every month), the rise in demand probably will not be constant, but will increase by different amounts every period. Part of the change is due to a change in demand, and part is due to random fluctuations. Ideally, you would like to use only that part of the increase that is due to the change in demand. However, you cannot know how much is due to the demand change and how much is due to randomness. The trend smoothing constant helps to dampen the random portion, and the choice of trend smoothing constant determines how much weight is put on the most recent past. One important consideration is that the trend enhancement should be used only if there is a definite trend pattern to your data. If there is a definite trend pattern, the trend enhancement will produce a better forecast.

A seasonally enhanced forecast incorporates seasonality into the forecast. Similar to the trend-enhanced forecast, the seasonally enhanced forecast calculates a base value, but then calculates a seasonal index. The seasonal index is a measure of the seasonal variation compared to the base value. Just as with the trend enhancement, the seasonal indices (each season has its own index) are calculated using exponential smoothing. And, of course, we need a new smoothing constant; in this case, the seasonal index smoothing constant. All of these different smoothing constants are represented by letters of the Greek alphabet, as is common in math and science formulas.

Finally, for maximum complexity, we can incorporate both trend and seasonality to produce a trend seasonal enhanced forecast. Because many products show both trends and seasonality, you want to incorporate both components into your forecasts. Although these formulas can look complex, you can understand the principles and logic behind them. Forecasting is not easy, but it is vitally important, and you should understand the basic underlying principles. There are many more forecasting models, some of which are very complex and hard to understand. However, all of them try to incorporate all of the data available, including any patterns in the demand, and they all derive in some way from these basic principles that we have just discussed.

Before we move on, look at Formulas 3.1 to 3.4, just for familiarity. These formulas come with a warning.

WARNING: Complex-looking formulas ahead! Don't be afraid! They are for illustrative purposes only. I'm not trying to teach you the mechanics of forecasting, but you should be familiar with some of the underlying principles and formulas. If you're feeling squeamish, don't look.

Formula 3.1 Exponential Smoothing Formula

New forecast = prior period forecast + α (prior period actual − prior period forecast)

where α = smoothing constant

Formula 3.2 Trend-Enhanced Formula

New base value = α (prior period actual)
+ $(1 - \alpha)$(prior period base + prior period trend)

New trend value = β (new base value − prior period base)
+ $(1 - \beta)$(prior period trend)

Forecast = new base value + x (new trend value)

$$\begin{aligned}
\text{where } \alpha &= \text{smoothing constant} \\
\beta &= \text{trend smoothing constant} \\
x &= \text{number of periods into the future to forecast}
\end{aligned}$$

Formula 3.3 Seasonally Enhanced Formula

New base value = α prior $\left(\dfrac{\text{Prior Period Actual}}{\text{Prior Indes}}\right)$ + $(1 - \alpha)$ (prior period base)

New seasonal index = $\gamma\left(\dfrac{\text{Prior Period Actual}}{\text{New Base}}\right)$ + $(1 - \gamma)$ (prior index)

Forecast = (new base value) (new seasonal index)

$$\begin{aligned}
\text{where } \alpha &= \text{smoothing constant} \\
\gamma &= \text{seasonal smoothing constant}
\end{aligned}$$

Formula 3.4 Trend Seasonal Enhanced Formula

New base value = $\alpha \left(\dfrac{\text{Prior Period Actual}}{\text{Prior Index}}\right)$

+ $(1 - \alpha)$ (prior period base + prior period trend)

New seasonal index = $\gamma\left(\dfrac{\text{Prior Period Acual}}{\text{New Base}}\right)$ + $(1 - \gamma)$ (prior index)

New trend value = β (new base value − prior period base)
+ $(1 - \beta)$ (prior period trend)

Forecast = (new base value + [x] [new trend value])(new seasonal index)

where
α = smoothing constant
β = trend smoothing constant
γ = seasonal smoothing constant
x = number of periods into the future to forecast

One of the rules of forecasts is that there will always be some error in the forecast. Some of the error is due to random variation, but some may be introduced by the forecast or the forecasting system. This means that the choice of forecasting method, choice of smoothing constant, or some other factor within the forecasting system is introducing error into each new forecast. One possible introduced error is bias. Bias means that the forecast is consistently either greater or less than the actual results. In an unbiased forecast, the actual results will sometimes be greater than the forecast and sometimes less than the forecast. But a biased forecast will almost always be greater than the actual results or less than the actual results. Bias can be measured and the results used to adjust the forecast to help eliminate the bias in the future. No matter what anyone tries to tell you, bias is bad. Some people, maybe the sales department, may try to convince you that if the actual results are always greater than the forecast, they are doing great, selling more than expected. Don't be fooled by these arguments. If actual sales are always greater than forecast, there is something wrong with the forecasting, not something great going on with sales. Remember, the forecast is the starting point for the sales and operations planning process, which is the basis for the rest of the planning process. If the forecasts are always biased, the S&OP will either have to be adjusted from the start to take the known error into account or be pretty useless right from the start because you are planning based on faulty information. Besides, if you know the actual results are always greater (or less) than the forecast, why wouldn't you adjust the forecast? Would you willingly tell everyone, "here's the forecast, but we always underforecast"? The first thing any sensible person would ask you is, "by how much do you always underforecast?" Then you would say something such as, "always by about 100 units." Then they would ask, "then why don't you just add 100 units to the forecast to make the forecast more accurate?" There is no reasonable explanation for not doing that.

Bias is a measure of the average, or mean, error over time. Simply total all the error — the actual demand minus the forecast demand — and divide by the number of periods being added. If the result is large (either positive or negative), the forecast has a lot of bias. If the result is small, there is little bias. The goal is to eliminate bias altogether, so that the only error in the forecast is due to random variation. Formula 3.5 shows the formula for calculating bias.

Formula 3.5 Bias

$$\text{Base} = \frac{\sum(\text{Actual demand - forecast demand})}{\text{Number of Periods}} \quad (\textstyle\sum \text{ is the symbol for Sum})$$

Another tool for evaluating the forecast over time is the tracking signal. The tracking signal incorporates the bias and the mean absolute deviation, which is a measure of the size of the forecast error. The tracking signal is useful for detecting whether the forecasting method is appropriate for the demand pattern. That is, if there is a trend or seasonal pattern in the data and the forecasting method does not incorporate trend or seasonality, the tracking signal will help detect this defect in the forecast.

I'd like to make a comment on random variation in the forecast. It is easy to attribute much of the error in the forecast to random variation, meaning that there is no assignable cause for the variation or error. However, some effort into searching for causes of the error may be well worthwhile, especially if the error is large. Variation that looks random at first glance may have a concrete cause that is not readily apparent but can be discovered upon investigation. Your customers may have a reason for erratic-looking purchases. Perhaps they have an underlying seasonality that they do not fully understand. Or they may have a purchasing policy that causes them to order erratically. Maybe you could work with them to eliminate this policy and smooth their ordering with you. That would probably benefit both of you. The point is, put some effort into investigating the causes of the forecast error, or demand variation, and you may be able to develop better forecasts or eliminate some of the variation. This will benefit your entire planning process.

3.3.5 From Forecast to Sales Plan

Once you have developed the forecast, then what? The forecast is the foundation, or the starting point, for the S&OP. On an ongoing basis, the forecast will incorporate information from sales (actual sales results) and operations (any problems or delays), but the forecast begins the process. The next step, then, is to convert the forecast into a sales plan. You might ask what the difference is between the forecast and the sales plan. As I said, the forecast is the starting point. The sales department reviews the forecast and makes any adjustments based on information from the salespeople in the field or any information that was not incorporated into the forecast. New products, discontinuations, and any other product line changes need to be adjusted for. Finally, once all adjustments have been made, the sales manager (or VP of sales, or whatever title the responsible person holds) approves the sales plan.

Now, remember the discussion in Section 2.2.7 about capacity checks? This is the point where the capacity check for the S&OP is done, beginning with a feasibility or reality check. The first question that needs to be answered is, can production support the level of sales you have forecast? Remember, at this level we are dealing with volume (product lines), not individual items. Can operations support the forecast volume? Also remember that the forecast and sales plan are time phased. This means that the sales are broken down into time periods of some length (weeks or months). Operations needs to be able to supply the quantities in each of these time periods. Can they do it? Can the production department produce the quantity in the week or month that it is forecast to be sold? Or, depending on your production lead time, is there enough time for production to produce the quantity needed prior to the time it is forecast to be sold? If the answer to either of these questions is no, then the sales plan must be modified to allow for production's constraints. You will have to talk to your customers to see if they can delay some receipts or if you can ship some of their orders incomplete. Or you can try to add production capacity in enough time to make the shipments. Adding production capacity cannot usually be done quickly, but it can be added for short periods of time. Usually by adding temporary workers or outsourcing some production you can add short-term capacity. The addition or shrinkage of production capacity is usually a long-term strategic decision that is not undertaken lightly.

This is an area where good communications and good relations with your customers really pay off. Let's say your customers are willing and able to accept smaller shipments spread out over a longer time, instead of a usual ordering pattern of one large shipment every six months. You would modify your sales plan to reflect these changes and show the quantities that will be shipped in the appropriate time period. For practical purposes, this is a revised forecast, although it is really the sales plan developed from the forecast and modified by capacity constraints and communications with customers.

You have completed your forecast. You have reviewed the feasibility of production being able to support the forecast. You have made needed modifications based on production capabilities and negotiations with customers. You now have a sales plan. Let's start to look at the format in which the S&OP will be presented. This is only one possible format, but it is easy to use and to develop. Some of the new enterprise resource planning packages that include S&OP functionality may look a little different, but all the same components should be there (and probably more). The format used in this book is patterned after the format presented by Thomas F. Wallace in his 1999 book, *Sales & Operations Planning: The How-To Handbook*.

Let's look at what we have so far (Table 3.5). That doesn't look like much, does it? But that's just the beginning; we're going to add to it. Note that this is the forecast presented above; no modifications were made during the capacity check process. Once you have your sales and operations planning process up and running and it becomes an ongoing part of your business, you will have actual results to compare with your plans.

Let's see what the sales plan will look like with some actual results for a couple of months (Table 3.6). Notice that we also added a row for the cumulative difference? This helps you keep track of how you are doing on a continual basis. If the cumulative difference keeps increasing, either positively or negatively, you can quickly see that you have a deepening problem that needs to be resolved. On the other hand, if the cumulative difference is small or fluctuates between positive and negative, it is an indication that you are doing well, or you have a different sort of problem. The cumulative difference is the sum of the prior period's cumulative difference plus the current period's difference between planned and actual sales. In this example, the difference between planned and actual sales for the first period (January) is negative 14 units. The difference for the second period (February) is positive 15 units. The cumulative difference, therefore, is positive 1 unit. For some reason, actual sales were less than planned in January but greater than planned by almost the same amount in February. This leaves us with a cumulative difference between planned and actual sales at the end of February of only 1 unit. This is useful information for the executives who review these summarized results each month, and they should always be provided with detailed backup information to refer to as needed. Perhaps top management is interested in knowing why there are nearly offsetting differences in these two periods and if conditions indicate that pattern will continue. Or they may not feel the differences are significant enough to investigate further and are just a normal part of the business pattern. Either way, the backup information can be referenced if wanted.

This example is of the most basic form of the S&OP spreadsheet. The intent of this book is not to teach you how to perform sales and operations planning, but to give you an overview that will allow you to evaluate sales and operations planning as one tool among the various tools in the toolkit. You may want to develop your own variations of this spreadsheet to suit your needs. One variation is to add a row for revised planned sales. Depending on actual performance or the frequency of updates to the sales plan, the original sales plan may be revised as time passes. It may be useful and informative to review the original sales plan along with the revised sales plan. There may be questions as to why the revisions were made, and the revisions may need to be shown so that the original plan is not forgotten. By showing the sales plan this way, it allows you to review the original plan, the revised plan, and the actual sales together, which can be very informative. However, be careful not to overcomplicate things. You will develop considerable backup material and a lot of data to support the plan, but the final S&OP document is a summary for

Table 3.5   Planned Sales, Family — Amps

	Jan	Feb	Mar	Apr	May	Jun	Jul	Aug	Sep	Oct	Nov	Dec
Planned sales	1,000	1,000	1,000	1,500	1,500	1,000	1,000	1,000	1,500	2,000	2,000	2,500

Table 3.6   Planned and Actual Sales, Family — Amps

	Jan	Feb	Mar	Apr	May	Jun	Jul	Aug	Sep	Oct	Nov	Dec
Planned sales	1,000	1,000	1,000	1,500	1,500	1,000	1,000	1,000	1,500	2,000	2,000	2,500
Actual sales	986	1,015										
Cumulative difference	(14)	1										

executive management to review. Top-level management does not have time to sort through all the detail used in the development of the plan. If they need further explanation or information, they will ask for it (so keep it handy).

3.4 The Operations Plan

The sales plan is concerned with the demand portion of the business. The operations plan is concerned with the supply side of the business. The S&OP attempts to balance these two sides of the business and synchronize them, so that all needs are met and profits are generated. If left alone, sales might develop a forecast, or perhaps a wish list, that they would expect to be filled without question or complaint. Conversely, if operations were left to their own devices, they might generate an optimal production and delivery plan that would maximize productivity and efficiency with little regard for the needs of the customers and the other areas of the business. Sadly, all too often, this is not too different from what is seen in manufacturing organizations.

Of course, achieving balance between sales and operations is not always easy. Operations generally faces a host of constraints that must be worked through in order to supply the needed products that satisfy the customers' demands. Operations needs to supply these products not only on time, but profitably. Larger, publicly held companies face shareholder and market pressures to constantly increase both sales and profits. News reports are regularly filled with the results of these companies' struggles or victories. Small manufacturers face the same sales and profit pressures, but from different parties. The owners need to increase profits so they can reinvest in the company for expansion and equipment upgrades. They want to increase profits so they can invest in new ventures or simply increase their standard of living (don't we all?). Lenders and investors want to see profit improvements before making new loans or investments. Employees want to see greater profits so that their wages, bonuses, or profit sharing increases. But along with these pressures come difficult to overcome constraints.

Your facility has only a limited amount of space. You have only a limited amount of time. And you have limits on all your other resources: people, equipment, cash, and so forth. You need to increase sales and profits with all these limits on your resources, all these constraints. It is a constant struggle, but sales and operations planning is one tool to help you overcome these constraints. Sales and operations planning helps by coordinating these often conflicting parts of the business through the development of one unifying plan. If sales and operations each develop their own plan without coordination with the other, they will most likely head off in different directions. This results in underutilized resources in some areas, overtaxed resources in other areas, and generally less than desired outcomes.

Let's look at the operations portion of the sample S&OP (Table 3.7). If you have been following along with these examples, you will notice that this production plan is exactly the same as the sample sales plan (Table 3.8). This is one possibility. This assumes that the items needed in each period can be produced in that same period. What happens, though, if we cannot produce what is needed in the same period? Maybe the production lead time is greater than the length of the time period, or maybe we have only enough capacity to produce a certain amount in each period. What if the capacity is less than the amount needed in one or more periods? In that case we have to adjust the production plan so that the products are available when they are needed. Let's modify our example so that there is a maximum of 1,500 units we can produce in each period (Table 3.9). This plan looks quite a bit different, doesn't it? With a capacity limit of 1,500 units per period, some of the production to support the final three months of this plan has to be moved forward. This means we have to build up inventory in periods where the planned sales are less than our production limit. We then have choices of when, exactly, to build the inventory. You will notice that in this example, I chose to wait until the last possible moment to build up the inventory. Even so, we have to start building inventory far ahead of the periods in which it is expected to be needed. Let's look a little closer and walk through this example.

The final three months of the time horizon of our plan require more units than we can produce in each of those periods. Therefore, we have to produce some of the needed items prior to the first time period where demand exceeds production capacity. In October, the planned sales are 2,000 units. Because we can produce only 1,500 units in October, we need to produce the 500 units that are needed in October sometime before October. We don't want excess inventory sitting around for very long, so we will want to produce those extra 500 units in September. Well, September requires 1,500 units. We can produce those 1,500 units in September, but we then have to bump up the production of the extra 500 units to August. We are not really specifically identifying the items being built in August for sales in October, but it helps to look at it this way for illustration. Now, November also requires 500 units more than we can produce in that period, so we have to bump those 500 units forward too. We cannot put them in September or August, because we are already producing to capacity in those periods, so we have to move it to July. Continuing with this, in December we need 1,000 units more

Table 3.7 Planned Production, Family — Amps

	Jan	Feb	Mar	Apr	May	Jun	Jul	Aug	Sep	Oct	Nov	Dec
Planned production	1,000	1,000	1,000	1,500	1,500	1,000	1,000	1,000	1,500	2,000	2,000	2,500

Table 3.8 Planned Sales & Planned Production, Family — Amps

	Jan	Feb	Mar	Apr	May	Jun	Jul	Aug	Sep	Oct	Nov	Dec
Planned sales	1,000	1,000	1,000	1,500	1,500	1,000	1,000	1,000	1,500	2,000	2,000	2,500
Planned production	1,000	1,000	1,000	1,500	1,500	1,000	1,000	1,000	1,500	2,000	2,000	2,500

Table 3.9 Revised Planned Sales and Planned Production, Family — Amps

	Jan	Feb	Mar	Apr	May	Jun	Jul	Aug	Sep	Oct	Nov	Dec
Planned sales	1,000	1,000	1,000	1,500	1,500	1,000	1,000	1,000	1,500	2,000	2,000	2,500
Planned production	1,000	1,000	1,500	1,500	1,500	1,500	1,500	1,500	1,500	1,500	1,500	1,500

than we can produce. We have to bump up these extra 1,000 units and split them between June and March. (Notice that both April and May require as many units as we can produce.) So we have to produce half of the 1,000 extra units in June and the other half in March.

We had to bump production up to March for items that are not needed until October, November, and December. This means that we are building inventory that will sit in the warehouse for at least six months. This is important information and needs to be clearly understood by everyone involved. Inventory has a big impact on many areas of the company, so it is important that inventory be included in the sales and operations planning process. But before we get to that, let's see what the operations portion of the S&OP spreadsheet looks like, with some actual production figures for the first two months of the plan (Table 3.10). This looks very much like the sales portion of the S&OP spreadsheet, doesn't it? It's supposed to. A consistent format facilitates review and analysis of the information. The same type of data is also presented, planned versus actual, so why not use the same format to make it easier to review?

The operations plan consists of two parts: the production portion and the inventory portion. As you plan for production in support of sales, you may find that you do not have the capabilities to produce all the products in the same time period in which they are expected to be sold. When production and sales do not match, you will either add inventory or use up existing inventory. You may also find that actual production does not equal the planned production. You may produce more than planned because of batch size requirements, or you may produce less than planned because of materials shortages, unplanned down time, or decreased yields. As with the sales and production portions of the spreadsheet, the inventory portion shows the planned inventory levels along with the actual inventory and the cumulative difference (Table 3.11). We've looked at the sales portion, the production portion, and the inventory portion of the S&OP; now let's look at another aspect of interest to most executives.

3.5 The Financial Plan

The financial plan serves two purposes. It reflects the financial impact of the S&OP and it acts as a mechanism to determine the feasibility of the plans. There are financial and cash flow impacts both for sales and operations and for production and inventory within the operations plan. The sales department brings in revenue, usually with delayed cash payments, but also requires financial resources and expenditures. Production uses resources and cash to pay for facilities, materials purchases, and

Table 3.10 Planned and Actual Production, Family — Amps

	Jan	Feb	Mar	Apr	May	Jun	Jul	Aug	Sep	Oct	Nov	Dec
Planned production	1,000	1,000	1,500	1,500	1,500	1,500	1,500	1,500	1,500	1,500	1,500	1,500
Actual production	1,025	1,025										
Cumulative difference	25	50										

Table 3.11 Planned Inventory, Family — Amps

	Jan	Feb	Mar	Apr	May	Jun	Jul	Aug	Sep	Oct	Nov	Dec
Planned inventory	0	0	500	500	500	1,000	1,500	2,000	2,000	1,500	1,000	0
Actual inventory	39	49										
Cumulative difference	39	49										

payroll. The levels of sales and production, and any increase or decrease in inventory, have dramatic impacts on the financial plan.

There are two elements to the financial plan as it relates to sales and operations planning. The first part is simply converting the quantity-based plans so we can view them in dollar-based figures. The second is how the S&OP drives the budgeting and other financial planning that is performed.

The simplest way to convert the quantity-based plan into dollars is to multiply the quantity of units by the cost per unit. The result will be a spreadsheet that looks like the one we have already developed, except it will show dollars at cost instead of units. If we assign a cost of $42 per unit, our example will look like the one shown in Table 3.12. Looking at the S&OP in this dollar-based format can be very informative. Most executives are used to thinking in dollar terms, so this format will be comfortable for them. It may not provide the most useful information to all users, but it is a good starting point. You may be more familiar with viewing sales at sales price, but this can add complexity to this process. Sales dollars may not be equivalent over time due to sales promotions or other discounts. Production and inventory are not normally viewed in terms of sales price, so gross margins or another conversion would have to be added. This adds too much complexity and defeats the purpose of sales and operations planning. Keep it simple in the S&OP and for the monthly meeting. Individuals or departments can convert the plan as necessary, but keep the complexity out of your final document.

The second relevant aspect of the S&OP is to use the plan to drive budgeting and other financial planning. You have a plan that reflects your expected level of sales, your expected production, and your planned inventory levels. More importantly, these are tied together into one comprehensive plan. Sales are tied directly to revenue, and all budgeting starts with the expected revenue. From the sales plan you can determine the expected revenue. You know the costs or expenses associated with production, or you have estimates, so you can calculate the expected costs and expenses from the planned production. You also know the costs and expenses associated with holding inventory, so you can calculate the expected costs and expenses from the planned inventory levels. Of course, some of this analysis and financial planning may need to be developed if you are not currently doing it this way. But I think you will be able to see the benefit of budgeting and financial planning based on one comprehensive S&OP, rather than the method of having each department prepare their own budget in isolation then trying to fit them together into one comprehensive master budget.

3.6 Tying It All Together

As I've already stated, you will need a start-up phase during which you develop new procedures before S&OP becomes a regular part of your daily and monthly activities. Don't get too caught up in or worried about the start-up phase. There will be some extra work involved, but you should start to see some benefits almost immediately. If nothing else, you may see that different departments are working from different plans and wonder why. As sales and operations planning becomes an ongoing process you should see additional benefits that translate into better bottom line performance. As production is more closely aligned with sales you will gain better control over your inventory and you will be better able to react to sales or production that vary from the plan. The better you are able to react to these events, the more control you have over them, and the more control over the costs associated with them.

You can see from the previous examples that sales and operations planning leads to questions. Some of these questions are probably already being asked in your organization, but they are probably not being asked of the correct person. In the example, we show actual sales for January lower than planned sales and February actual sales higher than planned. At the S&OP meeting where January results and future plans are reviewed, we need to find out from the sales department why sales are down. This is often asked with an accusatory tone, which just causes friction and accomplishes nothing. Production needs to know if the planned sales are lost or if they have shifted to another time period. The finance department also needs to know, because it affects the budget and spending decisions.

We show production in January higher than planned. Why is this? Especially if sales are lower than planned, production that is above plan is a serious concern. Is it due to our production processes and procedures, or is it a result of some decision that was made at some point during the period? Again, we are not looking to assign blame but to discover causes that may be preventable in the future. We also show that we built up inventory to 39 at the end of January and 49 at the end of February. The variances in sales and production resulted in this inventory buildup, but it is important to note that the inventory is now well above our planned levels. These variances in sales, production, and inventory affect the entire company. The president or general manager, as well as the department heads, can review this information together and make better decisions than an individual department could make on its own.

In March, when we review February's results, we need to ask ourselves some questions, if we haven't already. Are the sales variances a trend or an isolated event? Have any sales been lost or was there a shift from one period to another? What caused the production overages? Are there supplier, policy, or process issues that need to be addressed? What about the inventory that has built up; is it a seasonal product that will cost more to store than to sell immediately at a discount? Is there a shelf life for the product? In the sales and operations planning meeting, we can answer these questions and plan for immediate corrective action. Many of the questions can be answered prior to the meeting and the actions taken or to be taken can be reported at the meeting. The point is, the company is working from one plan, there is communication between the functional areas, decisions are being made by the correct people, and they have the necessary information to make those decisions. This will translate into bottom line improvements when performed properly.

Now let's put together all the different elements (Table 3.13). Wow! Look at what we have now. We have sales, production, and inventory information on one page. We have planned, actual, and the cumulative difference for each, and it's all on one page. How often does the top management of any company get to see all this information, neatly summarized and presented together on one page? Not often enough, that's how often. This is valuable information. This table contains more information than many companies are used to seeing at one time. It is the type of high-level summary information that top management needs to see. Results are often reviewed by department, in isolation. That is not always sufficient information for many types of decisions. But, wait. Let's look at the same information in financial format.

Everything starts with the sales plan. Sales drive your business. All other decisions and plans are, in some way, derived from the planned or expected level of sales. Sales equate to revenue, and all decisions are based on the expected revenue, either short term or long term. Some people might say that you can start your plans with your expected expenses or cash outflows, then determine the level of sales needed to support these expenses. This may be true in some isolated instances or for some specific projects, but as an overall business planning strategy this is not the best method.

Table 3.14 shows our sample S&OP from Table 3.13 in financial format. Most executives will want to see the plan in both formats. Top management is generally more interested in the financial view of the S&OP, whereas the line managers are more likely to be interested in the volume levels. This is simply because each of these groups has a different focus. As I mentioned, getting started with the S&OP can be difficult and time consuming, but once the process is in place and is a normal part of your operations, it will get easier. It would be ideal to start with the sales plan, then derive the operations and financial plans from that. However, in reality, you may have to just start with whatever planning you are doing in these three areas, then bring them together, see how they compare

Table 3.13 Sales and Operations Plan, Family — Amps

	Jan	Feb	Mar	Apr	May	Jun	Jul	Aug	Sep	Oct	Nov	Dec
Planned sales	1,000	1,000	1,000	1,500	1,500	1,000	1,000	1,000	1,500	2,000	2,000	2,500
Actual sales	986	1,015										
Cumulative difference	(14)	1										
Planned production	1,000	1,000	1,500	1,500	1,500	1,500	1,500	1,500	1,500	1,500	1,500	1,500
Actual production	1,025	1,025										
Cumulative difference	25	50										
Planned inventory	0	0	500	500	500	1,000	1,500	2,000	2,000	1,500	1,000	0
Actual inventory	39	49										
Cumulative difference	39	49										

Table 3.14 Sales and Operations Plan, Financial, Family — Amps

	Jan	Feb	Mar	Apr	May	Jun	Jul	Aug	Sep	Oct	Nov	Dec
Planned sales	$42	$42	$42	$63	$63	$42	$42	$42	$63	$84	$84	$105
Actual sales	$41	$43										
Cumulative difference	($1)	$0										
Planned production	$43	$43	$64	$64	$64	$64	$64	$64	$64	$64	$64	$64
Actual production	$44	$44										
Cumulative difference	$1	$2										
Planned inventory	$0	$0	$22	$22	$22	$44	$66	$87	$87	$66	$44	$0
Actual inventory	$2	$2										
Cumulative difference	$2	$2										

Note: Dollar amounts in thousands.

with each other, and then make adjustments to them. At first, you may find that the three plans have very little in common with each other. This immediately brings up the question of how the business runs at all when all the different departments are operating with such disparate plans. Well, there's no need to get into that now. It should seem obvious, now that you are beginning to look at an integrated planning process, that the business could operate a whole lot better if everyone was operating from the same plan. That is what sales and operations planning is all about.

Once sales and operations planning has become a regular part of your process, the sales, operations, and financial planning will be done in concert, with the sales plan leading the way. At its simplest, the operations plan is a level of production that supports the sales. If you are able to produce all that you need in the same time period in which it sells, the production plan will exactly match sales. In most companies this is not the case, and this also ignores any safety stock. Because of production capacity, production lead times, forecast variances, and other factors the sales and production will generally not be equal on a month-to-month basis. In any given time period you will either build the exact quantity needed during that time period, build more than are needed in that time period to support future periods, or build less than are needed in that time period while using up existing inventory.

The production portion of the operations plan will reflect the quantity that you plan to produce in each time period. The inventory portion will reflect the total quantity of ending inventory that you are planning. There is a direct mathematical relationship between the sales, production, and inventory. For one particular time period, start with sales, then add production to arrive at the ending inventory (sales is a negative number, because it is flowing out). Ongoing, start with the prior period's ending inventory, subtract sales, and add production to arrive at the ending inventory. In some instances you may focus on your ending inventory and calculate the production needed to reach a desired inventory level. Be careful to consider all your production capabilities and constraints when working this way.

You can see in this example that we have built up some inventory because our sales were lower than expected in both January and February. Our production in January was lower than planned, but the sales were lower than expected by a greater amount. In March our inventory increases by 500 units. This was planned when we added production to March, June, July, and August because we cannot produce all that we plan to sell in October, November, and December. Our inventory increases again in June, July, and August due to the added production, but it drops in October because we are now starting to eat into some of that built-up inventory. In November and December the inventory drops more, and finally drops all the way to zero at the end of December.

Sales and operations planning allows you to plan your overall inventory levels and monitor them from month to month. Too often, inventory is allowed to increase or decrease as a result of decisions made in other areas. Inventory should be planned and managed as an integrated strategy with the sales and production strategies. Begin using sales and operations planning and you'll find that you have much greater control over your inventory.

3.6.1 Capacity Check

You may remember from Chapter 2 that each level of planning requires a capacity check. If we do not check capacity before moving to the next level of planning, we may develop a plan that is not feasible or achievable. A plan that cannot be achieved is worthless. It is worse than worthless; it is a liability. The people charged with executing the plan will be demoralized, knowing that the plan is not doable. They will ignore the plan and will act in ways that improve their situation, which may be at odds with the direction that upper management wishes to take. Also, all the time and resources used to develop the plan were wasted, which just adds cost to the company without any related benefits.

The capacity check that is done at the sales and operations planning level is known as resource requirements planning. At this level we are not concerned with details; we just need to know if we have the resources available to execute the plan. We are planning for aggregate volume of product families; we are not concerned with the detailed mix of products or detailed scheduling. We will do that at the master production scheduling and materials requirements planning levels.

Terminology is always a problem. The capacity check at the sales and operations planning level may be known as resource planning, resource requirements planning, rough-cut capacity planning, or some other term. I prefer to follow the APICS definitions, because one of their roles is to standardize terminology, which they publish in the 2005 *APICS Dictionary,* Eleventh Edition.

For resource requirements planning, we will work with aggregates and averages. For each product family, we can approximate the resources needed to produce a certain volume. Potential resources include labor, machine time, space, or other factors. From past experience, we have a pretty good idea of what it takes to produce a certain amount of product. We do not need to examine every resource. For some resources, you know you have enough without much, if any, analysis. For your company, space may be a resource you do not have to worry about unless you start to produce at well above your current or planned levels. For another company, space may be constrained and thus needs to be evaluated during resource requirements planning. In Hawaii, space is at a premium and can be a major concern for many manufacturers, whereas in New Mexico space is much more available and is not an issue for many firms. The point is, look at your critical resources, and don't worry about the others for now (just be careful that you look at the right ones).

Let's look at an example using machine time as one of the critical resources we need to evaluate. For this example, we will assume that the machine time refers to one machine in one workcenter. You may need to examine more than one machine or more than one workcenter, but we are going to keep it pretty simple here. Table 3.15 shows the average machine time associated with the production of the amps family of products. Note that we've calculated the average machine hours per unit from the previous year's data. Some companies capture and use more detailed information that they use in the sales and operations planning process. Again, remember that at this level we are looking at aggregate data. The average capacity may be sufficient for many firms. The costs of capturing and calculating more detailed information needs to be weighed against the benefit of using that information at this level of planning. It will differ from company to company.

From this information, we can calculate the expected machine hours needed for our planned production. Table 3.16 shows the machine hours we expect to need based on the planned production from the S&OP we've developed. The next question is, how much machine time do we have available? We just figured out how much we need; now we need to compare that to how much we have available (Table 3.17). As you can see, we have some concerns in several time periods. The April, June, September, and November columns show that we do not have enough machine time available to meet the planned number of production hours. Because the difference is small, management may decide that this is close enough to proceed, considering this is a rough idea anyway. However, if problems have arisen in the past with shipments not being made on time, management may question whether the plan as currently stated is feasible. You will also notice that we have a number of periods where the available machine time exceeds the planned machine time by a significant amount. The first question may well be whether some production can be shifted from the periods where capacity is in question to periods where capacity seems to be sufficient or in excess.

Complications may arise if another product family utilizes the same machines or workcenters as the amps family. If that is the case, we must perform the capacity checks for each product family, then total the requirements before comparing them to the capacity available. One dif-

Table 3.15 Machine Time, Family — Amps

	Jan	Feb	Mar	Apr	May	Jun	Jul	Aug	Sep	Oct	Nov	Dec
Production	856	875	923	987	944	980	943	899	910	876	1,001	989
Machine hours	213	233	183	235	240	220	223	193	181	200	241	229

Note: Total production, 11,183; total machine hours, 2,590; average machine hours per unit 0.23.

Table 3.16 Expected Machine Hours, Family — Amps

	Jan	Feb	Mar	Apr	May	Jun	Jul	Aug	Sep	Oct	Nov	Dec
Planned production	1,000	1,000	1,500	1,500	1,500	1,500	1,500	1,500	1,500	1,500	1,500	1,500
Planned machine hours	232	232	347	347	347	347	347	347	347	347	347	347

Table 3.17 Available Machine Hours, Family — Amps

	Jan	Feb	Mar	Apr	May	Jun	Jul	Aug	Sep	Oct	Nov	Dec
Planned production	1,000	1,000	1,500	1,500	1,500	1,500	1,500	1,500	1,500	1,500	1,500	1,500
Planned machine hours	232	232	347	347	347	347	347	347	347	347	347	347
Available machine hours	346	311	363	346	363	346	363	363	346	363	346	363
Difference	114	79	15	(2)	15	(2)	15	15	(2)	15	(2)	15

ficulty that may arise is that sales and operations planning is performed and reviewed by product family, and any capacity concerns may not become apparent until all capacity needs for the families have been added together. A solution may be to add a line that shows the total capacity required for each critical resource. See Table 3.18.

This presents us with quite a different picture of our situation. Whereas it first looked like everything was fine, and we could just make a few adjustments, now that we have added the machine time needed by all product families, the picture changes. Although this may be an extreme example, it makes the point that although sales and operations planning is performed by product family, you still need to look at the whole business. Although this may seem difficult and more complicated, the benefits of discovering capacity problems before they are likely to occur will far outweigh the costs of performing the reviews.

3.6.2 New Products

Just a quick note on new products: don't forget them when you are doing your sales and operations planning. New products require resources. You may need to schedule production time during the development and testing of new products, and you will definitely use production time when you start the production phase. New products may require a ramp-up phase that utilizes your production resources in increasing amounts. Be sure to account for this during the planning process. Along with this, as you develop new products, they may begin to replace some of your existing products. This means that the demand for the existing products will start to decrease. As demand decreases, production (supply) should also decrease so that you do not build up inventory of products that are losing market share.

Of course, the big question is, how do you plan for new products, especially as you look further into the future? It's never easy, but it is necessary. It is better to plan and be a little wrong than to not plan at all. Until you start producing the products, you won't know for sure the amount of resources needed to produce them. However, you should be able to come up with a pretty good idea. If the new products are similar to existing products, you can estimate the resources needed for the new products based on the resources needed for the existing ones. You can also use engineering estimates. If the new products are not similar to any existing items, you will have to rely on engineering estimates only. By engineering estimates, I mean all the information available from the engineers or designers of the new product, as well as the knowledge of the people working on the production line. You should tap into the wealth

Table 3.18 Total Machine Hours, Family — Amps

	Jan	Feb	Mar	Apr	May	Jun	Jul	Aug	Sep	Oct	Nov	Dec
Planned production	1,000	1,000	1,500	1,500	1,500	1,500	1,500	1,500	1,500	1,500	1,500	1,500
Planned machine hours	232	232	347	347	347	347	347	347	347	347	347	347
Available machine hours	346	311	363	346	363	346	363	363	346	363	346	363
Difference	114	79	15	(2)	15	(2)	15	15	(2)	15	(2)	15
Total planned hours — all families	310	326	434	412	415	419	393	400	422	418	402	399
Difference	36	(15)	(72)	(67)	(53)	(74)	(31)	(38)	(77)	(56)	(57)	(37)

of knowledge possessed by the people working, supervising, and managing the production line day-in and day-out.

3.7 What It Takes to Get Started

■ Commitment by top management
■ More education in sales and operations planning (see Appendix A for books on the subject and other resources)
■ Willingness and ability to communicate across functional boundaries

If you have these three things, you can get started. Commitment by executive management is the most important element. Top management has the ability to allocate the resources necessary to start, and become successful with, sales and operations planning. You will need more education in the sales and operations planning process, but don't let this slow you down. You can learn much of what you need to know along the way. Each month will be a learning process and you will get better at it the more you do it. Outside expertise or training can be useful, but don't put off starting until you find an expert or become an expert yourself. For one thing, you won't become an expert without doing it yourself.

A willingness and ability to participate in cross-functional communication is vital. At first, and for a short time, top management can force some of this on the functional managers, but it won't last and won't produce the best results. Everyone involved will eventually see the benefits of sales and operations planning, but they may not be sold on it up front. They must be educated in the benefits of sales and operations planning and in communicating with each other, rather than protecting individual fiefdoms. Again, outside expertise and training may be highly beneficial in teaching people how to communicate better, and the benefits of it.

Sales and operations planning can be used, and I would say should be used, with any of the other tools in your toolkit. No matter what other tools you are using, you will benefit from properly performing sales and operations planning. Sales and operations planning can be used by any size company. The complexity of the plan and process will vary with the size of the organization.

MANUFACTURING OPERATIONS

Chapter 4

Lean Manufacturing: Power Tools!

You have probably heard of Lean Manufacturing. Even so, you may be a bit confused as to what it encompasses, or you may be overwhelmed with all the terminology and techniques associated with Lean. If you have not heard much about Lean Manufacturing, you almost certainly have heard about Just-In-Time or the Toyota Production System, which are forerunners of Lean. In this chapter you will learn what Lean Manufacturing is all about, what it encompasses, and some of the terms and techniques used. As with the other tools in this book, the intent is not to teach you how to implement Lean in your company, but to provide you with enough information about Lean to evaluate it, especially in relation to the other tools presented. I will review just a few of the better known techniques and terms associated with Lean.

In a nutshell, Lean Manufacturing is a philosophy of how to operate your business. Lean Manufacturing looks and operates very differently from what is referred to as a "traditional" manufacturing environment. What most people are familiar with, and what most people envision when they think of a manufacturing plant, is the traditional environment. In the traditional plant, work, machinery, and people are broken down into various departments. Depending on the product produced, the departments may include cutting, mixing, forming, plating, packaging, and so on. In a Lean environment, most of these departments are eliminated and just about everything is rearranged. This isn't done just for the sake of change, but because the traditional way of doing things includes a lot of waste. This term, waste, is often misunderstood and misinterpreted. Waste will be discussed in more detail in the following section.

4.1 Lean on the Shop Floor

As the name implies, Lean Manufacturing is primarily a tool for production and the shop floor. As will be discussed below, many of the techniques associated with Lean can be utilized in other areas, but Lean started out as a tool to improve production. Lean Manufacturing is an evolution of Just-In-Time (JIT). The overriding principle of JIT and Lean is the elimination of waste. The term, elimination of waste, is tossed around quite a bit and is frequently misunderstood. This is a good place for us to start.

4.1.1 Elimination of Waste

If you get involved with Lean, you will hear a lot about waste and the elimination of waste. You need to understand just what is meant by waste and what is to be eliminated. The concept itself is very simple; waste is anything that is unnecessary, redundant, adds cost, adds time, or is anything else the customer is not willing to pay for. Some examples will help to clarify the concept.

In the traditional plant, products move from one department to another for the processing that is performed in each department. Let's use a simple example where we have a drilling department, a soldering department, and a final assembly department. In this example, the product (amp model XL 10) is produced in batches of 100 units. The processing will proceed as follows: a batch of 100 units enters the drilling department and waits to be processed. When the drilling equipment becomes available, or when this batch is scheduled to be run, the batch is processed. After processing, it waits to be moved to the soldering department. When the batch is moved to the soldering department it waits again for available soldering equipment, for an available worker, or until the batch is scheduled to be run. Again, once the processing is complete the units wait to be moved to the final assembly department. As you can guess by now, in the final assembly department the batch again waits to be processed. Then, when the processing is complete the batch waits again until it can be moved to the warehouse or the shipping department. See Figure 4.1 to see how this traditional layout might look.

When you work in this environment on a day-to-day basis, you tend to think that this is just the way it is and you don't question it. Even if you do step back and look at it, you may not consider some of the

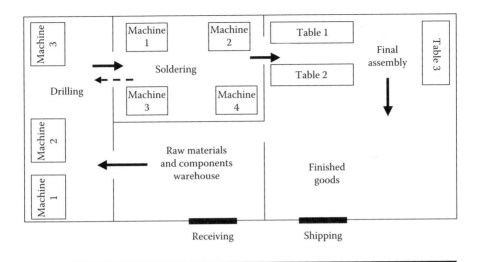

Figure 4.1 Traditional layout.

elements as waste. But just by using this simple example we can begin to discuss the waste inherent in the system. In this example, the two most obvious forms of waste are the movement between departments and the waiting before and after processing. Let's start with the movement between departments and identify the reasons that this is waste.

You may think that the movement of the products between departments is necessary, but you have a gut feeling that this is waste. Well, if the departments are in separate areas of the building, or in separate buildings, then the movement is necessary. The only way to eliminate the movement of the product would be to move the equipment to where the product is. Oh, now there's a thought! Keep that in mind. What makes the movement between departments a form of waste? Well, for one, it costs you money to move it. You have to buy equipment to move it. The equipment costs money to buy and to maintain and operate. Then you need people to move it, and these people cost money to hire and employ. You may say to yourself that you already own the equipment, it costs hardly anything to use, and you already have the employees, so you don't consider this waste and you can't do anything about it anyway. Well, you could have avoided the purchase of the equipment or you can avoid replacing it. You could sell the equipment. Even if it costs hardly anything to operate, it costs something, and are your profits big enough to ignore the small costs? How many small costs have to add together to be a big cost? You could have avoided hiring the employees, you could avoid replacing them, or you could put them to better use performing a task that increases your revenue rather than increasing your costs. Another question is, have you ever lost or damaged a product during its movement?

Sometimes, the processing steps don't move in only one direction. Sometimes the material or semifinished components move back and forth between departments for processing. Not only does this add waste in the form of additional time and distance, but it can also add confusion, damage, and lost materials. This, of course, adds cost without adding more value for the customer.

Why is the waiting before and after processing considered waste. First, while the items are sitting you have money tied up in inventory that could be invested in something that brings in revenue. Even though the waiting time seems very short, you may be surprised at how long things actually sit. The wait time can be days or even weeks and is often the largest element of your production lead time. Any time other than the time spent processing raw materials and components into a finished product that your customer will purchase is waste. Yes, you need to spend time on education and training, processing financial transactions, and performing other necessary activities, but in general if production is not working on products for sale, it's waste.

You might think it is impossible to eliminate all the waiting time before and after processing, but it has been done. You need to break out of your traditional thinking and start thinking Lean. Lean works toward flow rather than batch and queue. Batch and queue is the process described above in our example. Products are produced in batches and put in the queue (the waiting line) to await processing. Flow processing eliminates the batches and the queues, moving products through production one at a time in continuous movement. Process industries and dedicated assembly lines are examples of flow rather than batch and queue. Products flow continuously, either one at a time down the assembly line or contiguously like oil through a pipeline.

The point of this example is that Lean targets waste. When you begin a Lean implementation, you have to become relentless in identifying and eliminating waste. You cannot become complacent with waste and you cannot accept it. You need to question every activity and work to eliminate any waste in every process.

4.1.2 Cellular Manufacturing

This relentless elimination of waste uses several techniques. I told you to keep in mind the idea of moving the equipment rather than moving the product between various departments. This is the idea behind cellular manufacturing or work cells. In cellular manufacturing, all the equipment that is needed to process a product is brought together in one small area. In our example, the drilling machine, soldering machine, and final assembly would be removed from their current departments and placed right next to each other. The result is that the entire process of drilling, soldering, and final assembly can be completed in one area, in a shorter period of time, and maybe even by the same person. Manufacturing cells are often configured in a U shape, which takes up less space than if the equipment were in a line. The U shape also facilitates communication and cooperation among the workers assigned to the cell. Workers are usually cross-trained on multiple pieces of equipment or multiple processes within the cell. This can result in fewer workers needed per cell, workers able to help each other, and workers able to cover for absences. It also makes it easier to move from one machine or station to another, because they are closer together.

Manufacturing cells come in various shapes and sizes, but the idea is to improve operating results. Improvements in efficiency, reductions in lead time, and a reduced amount of floor space needed are common results. By bringing together all the equipment needed to produce a product, or family of products, the fundamental flow of the manufacturing process changes. These changes, along with other improvements, should allow smaller production batch sizes, improved quality, greater control of inventory, and greater control of the process. Smaller production batch sizes allow you to more closely match production with customer requirements. Reduced floor space needed for manufacturing allows you to increase capacity with a smaller capital investment, or allows you to utilize the freed up space for other purposes. Quality should improve because the responsibility for the finished product rests with the cell rather than with a department that is concerned with only one processing step. Greater ownership through the use of work cells also helps to improve quality. Greater control is gained over inventory and the production process because the process and area are compressed and more integrated. Figure 4.2 shows how our plant might look when we move our machinery into the work cell configuration.

Let's examine this cellular layout in a bit more detail. You'll notice a few things right away. Besides the equipment being rearranged into the U shape, you see that we have a lot more space than we had before. In the traditional layout you need space around each piece of equipment to hold the inventory before and after processing. You also need room to work and maneuver around and between the machines. And you need room to transport the materials and completed items between the various departments and through the plant. All of that wasted space is now exposed. Once it is exposed it can be exploited. It can be used for other purposes; ideally, for revenue-generating activities.

You will also notice that we have an extra piece of machinery in the corner. At first glance it might seem that we have decreased our capacity by taking one machine out of the mix. Actually, with these changes and

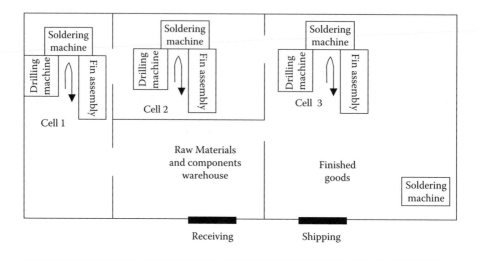

Figure 4.2   Cellular layout.

some other changes we will make, we have probably increased our capacity. For now, just let it sit there. As you see some of the other changes we make you can start to think about getting rid of that machine, hopefully getting some cash for it.

You can see the idea of flow when the machines are set up in these U-shaped cells. Because the machines are right next to each other, there is no need to accumulate a certain number of the items being worked on before moving them to the next operation. The work in process can flow one piece at a time from each operation to the next. Sometimes one person will perform all the work in the cell, doing all the processing for one item before moving on to the next item. The result is that the only partially completed item in the work cell will be the one that is currently being processed. The cell will not contain a pile of semiprocessed items. Wouldn't you rather have one complete item instead of several half-finished ones? This will tie in with the concept of producing to customer orders rather than producing items that have no current demand and will be placed in inventory.

4.1.3 Lot Size 1

We talked about moving items through the manufacturing process one piece at a time instead of in batches or lots. Let's look at some of the advantages of one-at-a-time versus larger batches. If you are using a machine or a production line to make one product, you cannot make another product at the same time on the same machine or line. The larger the batch or lot size, the more time must pass before you can make another product.

Let's say it takes five minutes to process each XL 10 on the drilling machine and our lot size is one hundred units. This means that it will take more than eight hours to process the entire batch through the drilling department (5 minutes × 100 ÷ 60 minutes/hour = 8.33 hours). We will have the drilling machine tied up with one product for more than eight hours. If we have three drilling machines in the drilling department, and they are all working on the XL 10, we will tie up the entire drilling department for almost three hours.

So what? So it takes more than three hours to process the batch and we're using all the machines in the department to process it. Is it a problem? Well, it depends, of course. After the drilling is complete, the batch is moved on to the soldering department. When the soldering is completed it moves on to final assembly. Say the soldering only takes three minutes for each piece and we have four machines, and final assembly takes three minutes each and we have three stations. It takes a little more than an hour to move through the soldering department (3 × 100 ÷ 60 minutes/hour = 1.25 hours) and almost two hours to move through final assembly. That's a total of more than five and a half hours. If we add setup times, move and queue times, and equipment downtime, with our one-shift plant, that's probably at least a whole day to process this batch through the plant — and that is if everything works as smoothly as we would like. I would like to point out two things here. It takes more than a full day to produce the product in this scenario, and during each operation we are tying up the equipment and thus cannot produce anything else while we are producing this batch.

Of course, you could run the batch through only one machine in each department, and that's probably what you would do. But, as you know, there are a lot more variables then we have talked about in this example. Besides equipment going down unexpectedly and equipment changeovers taking longer than planned, you have inventory shortages or defective parts and materials, you have employees who don't show up, you have late deliveries, you have work being stopped halfway through processing because of changing priorities, and all of the other little things that can happen during a regular day. This is just an example to illustrate the point. We could change the batch size, the number of machines, or the processing times. But keep in mind that there is a better way to operate.

Customers do not often order in nice predictable patterns. They are not going to order our batch of one hundred XL 10's one day, then a batch of another model the next day. They might order ten XL 10's one day, twenty the next day, and zero the following day. This varying pattern applies to all the different items we make. We protect ourselves by producing more than our immediate need and placing the rest in inventory. That's not a problem if you can afford to hold that extra inventory. You have to tie up cash in inventory. You have to pay for the storage space, insurance, and the equipment and people to store it, move it around, and count it occasionally. If any of it becomes obsolete, expires, or is damaged, you have to dispose of it. And sometimes it just plain goes missing. Wouldn't it be better to just make one, or a few, at a time as the customers want them?

In the traditional batch environment, the entire batch usually moves together. It's often harder than pulling teeth to break up a batch, unless you have a high-performance expediter (which is not necessarily a good thing). This means that if you're out of an item, in this example the XL 10, the customer might not be able to get one for more than a day — longer by the time all the paperwork is completed and the order is picked and packed and ready to go. If you could make one at a time it would take you 11 minutes, ignoring setups and movement and such. But in the batch environment with a lot size of 100, it takes about a day to complete the items. Your machines are tied up making one thing when they could be making another — the one the customer wants right now.

Is this the best you can do? Is this the best service you can offer your customers? Is this the most cost-effective way to operate? If you could shorten the time it takes to produce your products, decrease your investment in inventory, and serve your customers better, wouldn't you want to do it? Of course there is a little work involved in getting to the point where you can effectively produce one at a time, but it is certainly worthwhile to pursue it. Throughout the rest of this chapter we will discuss some of the ways to work toward producing in lot sizes of one.

4.1.4 Inventory Reduction

As mentioned above, businesses often use inventory as protection. If you are producing in batches, you size the batches so there is enough inventory in stock to cover your needs until the next time you make that item. You might size the batch so that you make two weeks' worth of inventory because you won't be making that item again for two weeks. At least that's the plan. You also make some extra units just in case, to cover those other things that pop up now and then.

Inventory is used as protection. It protects you not only from variances in customer demand, but also from your own operational deficiencies. Many inefficiencies and poor processes are hidden by inventory. Companies often do not realize they have less than desirable processes or performance because they hold enough inventory to cover up much of it. This extra inventory has another price. As mentioned above, you have cash tied up — you have to pay for storage, insurance, equipment, and people. You move it, count it, lose it, and maybe dispose of it. Because you don't have to write a check every time you move it or count it, and you're "paying for the space anyway," these costs are hidden. They are hiding right out in the open, but most people are not looking for them so they don't see them.

A good way to expose some of your internal problems or areas where improvements are warranted is to reduce your inventory. If you're producing the XL 10's in lots of one hundred because it takes you two hours to change over the equipment to produce another model, you're inventory is protecting you from inefficient setups. Reducing your inventory highlights the problem of your setups. If you reduce your inventory without reducing the time it takes to change over your equipment you may end up with different problems, such as not being able to make deliveries on time. You cannot just reduce inventory. Anticipate that the reduction will expose problem areas that you will have to fix right away. It's an incentive to make improvements. Don't make the mistake of reducing inventory but not making improvements. You wouldn't be the first to not make necessary changes, or not make them quickly enough, because a decision couldn't be made or the solution found. You wouldn't be the first, but you really don't want to be the next one.

Inventory should not be reduced arbitrarily either. Selected pockets of inventory should be designated. You probably suspect some areas need improvement, or know full well that they do. Start with these, and have some sort of controlled approach to the reduction. But you have to take some risk. Don't be too conservative. If you don't feel any pain you won't make the needed changes, or you won't make them soon enough.

The term *zero inventories* is often heard when talking about Lean. The idea is to reduce inventory to a level of zero, or near zero. That means you have no inventory. That's not totally realistic, because you will always have some inventory. If you are making something, you will at least have inventory during the time it takes to make it; but the concept is to reduce inventory to the bare minimum. I've heard stories of operations where all the raw materials needed for one day's production are delivered in the morning and all finished products are delivered to customers by the end of the day, resulting in no inventory left at the end of the day. I don't know if the stories are true, but they make the point. As you start to put all the pieces of the Lean puzzle together, you will see how you can move toward zero inventory. You may never get there, but if you don't aim for it as a target you probably won't even get close.

4.1.5 Pull System

Another difficult-to-grasp concept for many people is the pull system that is associated with Lean and JIT. The traditional manufacturing environment uses something called the push system. Let's start with the standard push system, then look at how it differs from a pull system.

When we talk about push versus pull, we're talking about the inventory and sales ordering processes. In the traditional, or push, system, the company produces products based on a combination of sales forecasts and actual customer orders. Inventory is produced with the expectation that customers will order and purchase the product at some point in the future. Sure, some customers order in advance of their needs, but many do not. It is up to the producer (you) to anticipate how many the customers will want and when they will want them, and schedule production accordingly. This results in the high inventory levels we've been talking about; the protection we need because of the difficulties and errors associated with forecasting. Thus, you often "push" the inventory on to your customers. You've made it, now you have to sell it. If the expected number of sales do not materialize, you offer incentives. You discount it, put it on sale, or maybe even beg someone to take it off your hands. You push it out the door.

Raw materials and components work pretty much the same way. You buy materials and make components to the same schedule, which is based on the forecast demand. You probably buy plenty of raw materials because of the time it takes to get them and because you get a discount for buying more. You buy more than you need immediately because you have to buy in certain lot sizes determined by your supplier, because you don't really know exactly how many you'll need, because you'll lose some or break some or they'll go bad, and a few extra just in case. The same goes for your manufactured components. You may end up pushing these materials on your production department. You tell them they need to use them because you've got too many. You beg them to take them off your hands. It's not a pretty sight.

The pull system is in direct contrast to the push system. With the pull system you do not make things in advance. Instead of making a bunch of something and hoping someone will buy them, you either make just a few or you don't make anything until you have firm commitments from your customers. Remember, in the traditional environment it may take you about a day to produce a batch of the XL 10's and you may be making two week's worth at a time. And you're hoping someone will buy them after you make them. But if you make them one at a time it only takes about 11 minutes to make each one.

How about instead of making two weeks' worth at once, you make only one day's worth, or just a few hours' worth, or even none ahead of time? If your customers want their items right away, you have to make a few in advance, but only enough for that day or a few hours. When the customers come in and take what you've already made, you immediately make more to replace the ones that were taken. With this system, you will be building more in line with actual customer demand patterns. Instead of making a batch and letting customers take from what you have on hand, you build your products in the same pattern as customers order them. Let's say your customers order a few a day; twelve one day, six the next, that sort of pattern. If you make a batch of one hundred every two weeks or so, your customer demand pattern does not match your production pattern. Getting your production more in sync with customer demand provides many advantages.

By building as customers order, you are pulling inventory through the system. Let's say your customers order an average of ten XL 10's per day. If you build three at a time you'll be making more than two hours' worth of inventory at a time (instead of two weeks' worth). When the actual orders come in, your inventory drops and that triggers the production of three more units. Now here's the key to the pull system: you don't allow any production to begin unless you receive a signal from the next operation telling you that you need to replenish the amount that was taken by customers. This same concept flows back through the component production and raw materials delivery processes. When finished goods are taken by a customer, or delivered to a customer, that sends a signal upstream to the previous operation that the finished goods need to be replenished. To assemble the finished goods you need to have the necessary manufactured components and raw materials available. As with the finished goods, the manufactured components are not produced in large batches. Just enough are made and kept available to assemble into finished goods when the signal is received. When those components are used, a signal is sent to replenish them. The same for raw materials, but instead of production it's deliveries from suppliers.

You keep only enough raw materials on hand to produce enough for your immediate needs. In this example, let's say that your supplier is willing to deliver your materials once a day. That means you need only one day's worth of materials delivered every day. But the supplier will not deliver until they receive a signal from you authorizing the delivery. This protects you from receiving materials you don't need yet if customers order fewer products one day. One of the tenets of Lean is that you work with your suppliers, partner with them, so that they will make more frequent deliveries of less quantity than usual.

As you can see, this is now a pull process. You pull products and materials from the previous step in the process only when you need them. You do not build up inventory in anticipation of future orders. In this example we built inventory of three units of the finished product, enough components to make those three units of finished product, and one day's worth of raw materials. But that's a lot better than what we were building up using the traditional methods. We talked about the lots of one hundred finished goods, but not much about the components or the raw materials. More than likely, the components were built in large lots too, and raw materials were purchased in batches enough for at least several weeks' worth of production.

In this example we still produced a few items in advance. What would happen if customers suddenly stopped ordering the XL 10, or if sales started to decline? In the traditional environment we would have a large quantity of products, components, and materials in inventory and we would likely not be able to react quickly to the change. Building to a forecast means we will most likely keep building until the forecast changes. That probably takes a while. First you have to discover the halt or slowdown of orders and find out whether it is temporary or permanent; then you have to communicate the change to the production department, which has to change the schedule. This won't happen fast. In the Lean environment we have a maximum of only three finished goods, the components to make three more, and one day's worth of materials. We won't make more finished goods or components until we get the signal to do so, and we won't authorize delivery of materials until we get a signal to do so. We're in a much better position under the Lean system.

In this example we still produced three finished goods at a time; we didn't get down to one. But in the Lean environment we would have produced those three one at a time in our work cell rather than producing and moving through different departments in large batches. Envision going through our example again, but producing only one at a time, or waiting until we get a customer order to produce any at all. We'll look at this a little more in Section 4.1.7.

You will find it difficult, to say the least, to produce in these small quantities unless you can reduce your equipment setup or changeover times. If you make only one or a few items at a time, you will make the same item more than once a day, with other items in between. You can't afford long, complicated, or difficult setups and changeovers under those circumstances. Section 4.1.8 covers setup reduction, with ideas on how to improve your setups.

4.1.6 Kanban

We talked about not producing items or authorizing delivery from vendors until a signal is received from the next step, or downstream operation, indicating that the items need to be replenished. *Kanban* is the term for that signal. Many people think a kanban is a container that is sent back and forth, signaling the need. The container is sent to the upstream workcenter, signaling for them to fill it and return it. That is just one example of a kanban. It is a convenient method for items that fit nicely into some sort of container and containers that move easily between operations. Containers are not always the best or most appropriate kanban. Kanbans are often cards. The kanban card is a common method of replenishment, especially for materials delivered from a supplier. It is a simple authorization for delivery of the item that the card represents. A kanban can also be a light, a fax, an electronic message, some lines on the floor, an empty shelf space, or anything else that is appropriate to your situation and environment. Because a container is familiar to many people, let's look at how that container might work as a kanban.

First, as you may have guessed, setting up the kanban system of replenishment requires a lot of work. If you use kanbans with your suppliers, you must work closely with them so that when they receive the signal (the kanban) they can immediately replenish the item indicated. This is quite different from a purchase order, even a blanket purchase order. With a purchase order, all sorts of approvals and steps are involved to get the ordered items out the door. With the regular purchase order system, the order has to go through order entry, it may need approval or credit checking, then it has to be picked, packed, dispatched, and shipped. With a kanban system, the kanban means send it. All the support needed to move the items to the customer has been developed and worked out during the system design. The only authorization or approval needed is receipt of the signal. The quantity is predetermined. If the kanban is a container, the quantity is simply the number that fit into the container. If it is a card or other signal, the quantity will be indicated or will be a preset amount. Either way, the quantity will have been determined in the design stage of the system.

The total number of kanbans in the system will also have to be worked out through analysis. The kanban system is simple in action, but the analysis to develop the quantity represented by each kanban and the total number of kanbans in the system can be a bit involved. This is not a free-form or wide-open system. The flow of materials and the quantities must be carefully worked out and controlled. Remember, we discussed the matching of production with customer orders. Customer will order products in various quantities and combinations. Therefore, you must produce in varying quantities and combinations. You also must remember that a certain amount of variation will occur from day to day or season to season. The kanban quantity and number of kanbans helps to smooth some of the variation while remaining flexible to produce to customer orders.

You must consider the capabilities of each production cell. Some may be able to produce many simple items quickly, whereas others may be designed to produce more complex items that take more time. Some components will be produced in different quantities than other components, even though you are striving to get down to that lot size of one. Some components may be used in more than one higher level component or finished good, so they will be needed in different quantities than those with less need. Your suppliers' capabilities play a role too. They may not have the same capabilities as you, or they may be limited on the number or timing of deliveries they can make to you.

4.1.7 Mixed-Model Scheduling

Mixed-model scheduling is not unique to the Lean environment, but it is fundamental to the Lean operating principles. Mixed-model scheduling is simply the production of a mixture of items rather than one item at a time. You're probably thinking that you always make a mixture of items, but the difference is in the quantity and timeframe for making them. Instead of making large batches of a product then changing over to produce a large batch of another product, you make small amounts of each item and repeat in some sort of pattern. The following example best illustrates this.

In the traditional environment, you may schedule the XL 10 in batches of one hundred, the RG 7 in batches of one hundred, the RG 9 in batches of two hundred fifty, and the TZ 123 in batches of fifty. As we stated, the batch of XL 10's will take about a day to complete. Assuming the same processing times for the other items, the other batches will take from almost three hours to about two days to complete. It will take you about three days to work through the schedule, if you schedule the items in sequence: XL 10, RG 7, RG 9, TZ 123. This is without adding changeover times, downtime, or efficiency and utilization into the calculation. Say you got a little behind for one reason or another. You just finished making the XL 10's and you sent them all out the door to fill backorders and new orders. You are now out of that item and it is not scheduled to be produced again for about three days. As soon as someone orders one you are immediately in a backorder situation, and the customer probably is not happy. Three days may not sound like much to some of you, but look at it from the customer's point of view. If the item is critical to them and they can't get it for three days, they are going to feel the impact quite severely. Also look at it from an organizational point of view. If you can be out of an item for three days and casually shrug your shoulders (as if it happens all the time), you need to seriously consider your effectiveness and your position in the market. If this is normal for you, you may find yourself out of the market. How effective are your competitors? If they can supply your customers without being out of stock, you have some issues with which to contend.

With a mixed-model schedule, you would produce a few XL 10's, a few RG 7's, a few RG 9's, and one or a few TZ 123's, then repeat that pattern. The actual quantity of each item and the pattern of producing them would be influenced by the daily demand pattern or the usual quantity and timing of customer orders every day. In our example, instead of batches of one hundred, one hundred, two hundred fifty, and fifty, let's try making five XL 10's, five RG 7's, ten RG 9's, and two TZ 123's. If we do this, we will work through the schedule in about two hours. We will repeat this pattern, continually making small quantities of each item.

Without taking into account changeovers, downtime, and so forth, this method will take us the same amount of time to produce the same quantity of each item as it would producing the larger batches. That is, it will still take us about three days to make one hundred of the XL 10's. But look at the difference. If you fall behind on an item somehow, if a customer cleans you out of an item sooner than expected, you will be out of stock for only a couple of hours instead of a few days. The impact on your customer service is tremendous. To change from the original batch size to a smaller batch size takes more analysis than just reducing the quantity. Besides all the other work it takes to reduce the batch sizes, you need to determine what the batch, or kanban, size of each item is going to be. You need to determine a batch size and the pattern of production. In our example we did a few simple things: we kept the same pattern of rotating through the four items in sequence, and we kept the batch sizes in the same relative quantities when compared to each other (we just reduced each item to about 10 percent of the original batch).

To develop the mixed-model schedule you really need to start from scratch. And although we're still calling them batches, in the Lean environment they aren't produced in batch mode anymore. We produce them one at a time through the production cell in a certain pattern. That pattern will include making several of the same items at a time, or making them one after the other, before switching to another item. We need to determine the pattern, or how many of each will be produced at a time before changing over the equipment to make another item, and the sequence. If we define a cycle as the time it takes to make at least one of each

Table 4.1 Schedule Comparison

Amps		Amps		Amps	
Model	Qty.	Model	Qty.	Model	Qty.
XL 10	100	XL 10	5	XL 10	3
RG 7	100	RG 7	5	RG 7	2
RG 9	250	RG 9	10	RG 9	5
TTZ 123	50	TTZ 123	2	XL 10	3
				RG 7	2
				RG 9	5
				TTZ 123	2

item, then during this cycle we may switch back and forth several times, making the same item two or more times. See Table 4.1 for an example.

It takes effort to determine your pattern or schedule. Your production pattern should match your demand pattern. This may be tougher in some environments or industries than others, but you can get them more closely aligned. Many organizations tend to look at demand patterns over longer time periods, such as years, quarters, months, or weeks. We need to get down to at least daily demand patterns. If you think that might be tough, just take a look at any restaurant. They have a pretty good idea of their daily patterns, as well as how it varies throughout the week. If they can do it, so can you. You may question the need for daily demand patterns and producing to this mixed-model schedule. The quantities and sequence you make will depend greatly on how your customers order and when they expect orders to be delivered or ready to be picked up. If you produce in large batches and hold a lot of inventory, you may not be as aware of your customer demand patterns as you should be. Your warehouse people probably have a pretty good idea, but the rest of the organization might not. That inventory is protecting you from the daily demand patterns. If you reduce the inventory, reducing that level of protection, you need to become more responsive to the pattern of demand over a shorter period of time. Instead of producing those large batches, moving the items into the warehouse, and having the warehouse personnel pull the items as customers order them, you will eliminate a couple of those steps. You will produce in smaller quantities, eliminate the move into inventory, eliminate the pulling from inventory, and just take the items off the line as the customers order them. Okay, you may keep a few in inventory, so that when you fill the customer orders your pull system moves into action.

Of course, you can't change the pattern of production — changing from batches of one hundred to batches of five, or fewer — without making other changes. One obvious point is the changeover or setup times. If your setup times are long, therefore costly, making a change like this will be a big mistake. But if you take these Lean techniques and put them together, you have an extremely effective method of improving your operations and satisfying your customers. Let's take a look now at how setups are viewed in the Lean environment.

4.1.8 SMED — Setup Time Reduction

SMED stands for single-minute exchange of die. Single minute means single digit, or the numbers one (1) through nine (9). Ten (10) is a double digit, or double minute, and that just won't do. Exchange of die refers to equipment or machinery setups or changeovers. So single-minute exchange of die means that an equipment or machinery changeover or setup should be accomplished in less than ten minutes. The changeover refers to changing from the production of one item to the production of another. Unless you have dedicated lines for producing only one item, in one configuration, you will have to perform a changeover or setup when switching from one item to another.

The downtime associated with changeovers is very costly and is often a source of debate and discussion. When scheduling production, changeovers and downtime are important factors to consider. Most companies try to minimize setups and downtime by scheduling similar items back to back, or scheduling those items with minimal changes after each other. At some point, though, you must schedule the big changeovers. You can minimize them, but you cannot eliminate them (at least not using traditional thinking). To move to a Lean environment and to reduce your changeovers to just a few minutes, you need to think differently, even creatively.

Even if you are not planning on jumping into Lean, quick changeovers or reduced changeover times are very valuable. Just think about how much easier scheduling will be or how much more flexible you will be if all your changeovers take only a few minutes. Think about the mixed-model scheduling discussed above. So where do you get started? The first thing is to analyze your current changeover procedures, then separate changeovers into two distinct components: external (elements of the changeover that are done while the equipment is still running) and internal (elements that are performed when the equipment is idle).

Chances are you do not have well-documented changeover procedures. You may or may not be surprised to find that people in your organization perform the same changeover differently. If that's the case, you should immediately question why. If one way is better than the other, why doesn't everybody do it that way? One of the reasons people perform the setup differently is that the procedure is not documented, so they do it the way they were taught or use the shortcuts they've discovered over time.

One common method of documenting the setup procedure is to film it. With the cost of digital video recorders these days, this is a cost-effective way to document, distribute, and archive the procedure. When the procedure is documented, it can be analyzed from a different point of view. You will probably notice some things right away that don't make sense or can be improved with little difficulty. Once the easy changes are made (or just discussed at this point), the challenge begins. If I had a dollar for every time I've heard "you can't do that," I'd be rich. You'll probably run into a lot of negative thinking like this at first, but don't let it slow you down. This is where the creative thinking comes in. Thomas Edison said something to the effect of, "I didn't fail. I just discovered five hundred ways not to do something." To make the big breakthroughs you have to throw a lot of ideas on the table, some of which might sound crazy at first.

One issue that will probably be brought up is the cost to make some of the changes or modifications to the process. This is a valid concern, but you have to weigh the cost of the changes against the cost of the downtime if you don't make the changes. If you have a setup that takes four hours to complete, requires a highly paid specialist to perform, and occurs several times a month, that costs you a lot. Then, of course, while the line is down you are not producing anything, which means you are incurring costs without associated revenue. You also incur costs when you're running, but you are also producing the products that will generate revenue (if they actually sell, that is).

Here is an example of the kind of crazy ideas I have. I worked in a commercial bakery and one changeover involved waiting for the ovens to cool down when switching from one type of product to another. Generally we ran the product that baked at the lowest temperature first, then increased the temperature in the ovens and ran a product that baked at a higher temperature. But we couldn't always do that, so sometimes we had to wait until the oven cooled. During the cooling time, the packaging line was changed over, cleanup was done, and breaks were taken. Sounds reasonable, but in a capacity-constrained plant, downtime is costly. So here's my idea. I don't know if it would ever work or is even possible, but I would have liked it to at least have been considered. Maybe it could be a starting point for other ideas that would work. Why not send some sort of heat-absorbing material, such as lightweight ceramic bricks, through the oven to cool it down quicker? I'm not an engineer, chemist, or physicist, but hey, it's an idea. You'll often find that the experts don't have the best ideas. Experts are often the victims of their own education and training. If they've learned to do something a certain way, it's not easy to change their thinking or be creative when looking at it. That's why it's good to bring people from outside to look at the situation and make suggestions. Let somebody from accounting or marketing watch the setup process and give their thoughts. Find someone from outside the company. Buy them lunch, show them the procedure, and ask them to comment on it. You might be surprised at what you hear. Keep an open mind and who knows where you'll go.

Now back to the internal versus external components of setups. External elements are those that are performed while the equipment is still running; in other words, while you are still producing a product. Internal elements are performed when the equipment is idle, when nothing is being produced. The run of the previous product has been completed and the production of the next product hasn't started. This is when you are spending money but aren't producing anything that will provide revenue to cover those expenditures. Perform most of the setup while the equipment is still running and the machine will spend less time being idle.

Often, the only reason a setup element is performed while the equipment is idle instead of while it is still running is because it has always been done that way. By recording your setups and analyzing them, you can often shorten the setup time simply by changing when you perform some of the activities. Of course, not all the elements can simply be changed from while the equipment is idle to when it is running, but it is a great place to start. After that you may need to make more substantial changes or get more creative. Some setup procedures may need to be performed when the equipment is idle because of the design of the equipment. It may not be possible or may be unsafe to perform the activity while the machine is running. You might have to make some changes to the machine or equipment itself. That may not be easy. But that doesn't mean you shouldn't consider it.

You want to get that setup time down to as short as possible, to single digits, preferably the low single digits. You need to look at every step of every procedure and look at how it is performed. Any time someone needs to use a tool, instead of just their hands, it adds time to the procedure. Sometimes it is just a few seconds, sometimes longer, but the time adds up. For example, instead of having to use a wrench or screwdriver, can you use a thumbscrew or a snap-on clamp? That will take less time and you don't have to worry about dropping any tools into the equipment (yeah, like that's never happened to you). You also won't need to worry about finding the tools or having them there when you need them. Any time you try to make an adjustment by sight or "feel," you add time, usually a lot of time because you have to readjust and readjust until you get it just right. Can you add a dial or etch some tick marks? That will decrease the time and need for adjustment.

You may have to make a major change or redesign of the machine or equipment, or you may need to buy a new machine. These are not light undertakings, but you need to evaluate the costs against the savings and improvements you will realize through reduced setup and changeover times. Time is money, and this is especially true when talking about setups. It costs you money whenever your equipment is sitting idle. If you have to produce in large batches because it costs you a lot of money to set up or change over your equipment, this is costing you even more money. And if you can't make a delivery on time because you are producing another product when you could be producing the one the customer wants, that is costing you a great deal of money. Think of the possibilities if you could set up or change over your equipment in less than ten minutes. Think of the flexibility and responsiveness you would achieve. Think of the reduction in inventory and all the activities involved with managing and controlling that inventory. Keep these thoughts in mind when you get to the point when you don't think you can make any more changes to your setup procedures. Get creative and break free from your traditional thinking.

4.1.9 Kaizen

Another technique associated with elimination of waste is kaizen. *Kaizen* is the term for continuous improvement. The idea behind kaizen is that every process can be improved, even if you have already improved it. The kaizen philosophy is continuous, small, incremental improvements. This means that you constantly examine your processes and your work rules, and make improvements every chance you get. If you improve a process, don't accept that it is the best it can possibly be and you're finished. Continuously question it and improve it. With regard to the elimination of waste, every time you improve a process you are removing more waste from the system.

Some people question why you should improve a process once it has already been improved. Others might question why the process wasn't fixed right the first time or why it wasn't perfected during the first analysis. Well that may not be possible for a couple of reasons. Just because you make a change or improvement to one process, doesn't mean you have done anything to change or improve the processes that feed into the one you have improved or the ones that are downstream from it. Maybe you have made a small improvement to the drilling process; you've mistake-proofed it, as in our example above. This is a great improvement, but it doesn't affect the soldering process (except that you won't be soldering any parts that were drilled incorrectly). Maybe at a later time, when you have made some improvement to the soldering process, you can go back to both processes and make an improvement that affects them both. Maybe you couldn't do that until the individual improvements were made.

Another thing is that you might not see, or be able to visualize, the perfect process or perfect form that an operation should take. After you make one improvement, maybe then the next opportunity for improvement becomes clear. Technological improvements may affect the pace of change and improvement. Maybe you cannot make further improvements using existing technology or techniques. New technological breakthroughs may then allow you to revisit your process and make a new improvement. So don't expect to be able to perfect a process the first time through. Plan to make improvements at every opportunity. Don't pass up the chance to make an improvement just because that's not what you are focusing or working on right now. Once you start walking down the path to a Lean enterprise, you will find that your employees discover new ways to do things and find opportunities for improvement along the way. Encourage them, and teach them how to go about making changes and improvements the right way. If the change is small and easily accomplished, let them do it without further discussion or approval. If it is a bigger change, or one that affects more than one area, develop a process to bring these ideas to light and make the improvements as quickly as you can. If you put up roadblocks to improvement, or delays in implementation, the ideas will dry up quickly. No one likes to beat their head against the wall trying to get someone to listen to them. That's not to say that every idea will be a good one and can and should be implemented. You need to explain that to people too. When employees propose ideas that you can't or won't implement, explain why. If you just dismiss their ideas, or worse, ignore them, you'll cause yourself irreparable harm. That's one of the quickest ways to get your employees to not only give up on you but also put up their own obstacles to change.

A different take on kaizen is the kaizen blitz. With a kaizen blitz, instead of small, incremental improvements to a process, you make a monumental improvement in a short period of time, usually just a few days. This doesn't mean you can't improve the process again, but later improvements will probably be smaller, because you have spent considerable resources reviewing and improving the process in a compressed timeframe.

Things that make a kaizen blitz different are the selection of the project to be undertaken, the assembly of a dedicated team to work on the project, and the focus of the team's efforts on that project. The assembly of the team and its focus on the project means that the team members must be relieved of their normal duties during the time they are assigned to the project. Depending on the project selected, this may be just a day or up to about five days. Typically, kaizen blitz events last two to three days.

A kaizen blitz consists of the following key steps:

1. Select the project
2. Select the project team
3. Define the project
4. Analyze the current process
5. Develop a plan for change
6. Implement the changes

Selecting the project requires thought and planning. Significant resources will be utilized for the project, so there must be a priority list of potential projects. The projects will require management support, because of the resources required and the fact that the project team may cross departments. Selecting the project team is a vital step. The team should consist of people who are closely tied to and familiar with the process being evaluated. Some team members might come from the areas that support the process or are affected by it.

The next step is to define the project. You may think that this should have already been done during the selection of the project, but project selection and project definition are not the same thing. During project selection, you pick the area or process to be improved or analyzed for improvement opportunities. During project definition you define the scope or boundaries of the project. Without a clearly defined scope, it's easy for projects to expand quickly and get out of control. You need to define where the project starts and where it ends. As you look at the process, you will find related areas that show opportunities for improvement, and you will be tempted to add those or include them in your project. Why not? You're already looking at it? But you need to stay focused and keep your projects under control. (See Chapter 8 for more information on project management and the effects of "scope creep.")

You need to understand the current process in order to analyze it. You need to know how it is being performed and why it is performed that way. One reason for selecting the team members you do is that they have a deep understanding of the how's and why's of the process. You need a good understanding of where you are to help you determine where you are going. This leads to the next step of developing the plan for change. This consists of where you are going and how you are going to get there. There will probably be a few steps in the journey. You need to know where you want to end up before you can determine the best route to get there. Then you must define each step along the way.

Finally, you need to implement the changes. You've defined the plan — you know where you're going and you know how you're going to get there. Now you have to do it. Do it, measure your progress, and measure your results. If you have discovered other opportunities for improvement along the way, document them and determine where in your priorities they fall. After you have implemented the changes and they are incorporated into your new way of doing things, check to make sure the new process is being performed as it is supposed to be and that the results are what you expected.

4.1.10 5S

5S is one of the first techniques you will use in a Lean implementation. If you use no other techniques associated with Lean Manufacturing, use the 5S to make your life easier. The 5S are five concepts for organizing your work areas. The developers of the Toyota Production System were Japanese. The 5S are five words, in Japanese, that identify workplace organization concepts. The five Japanese words begin with the letter S (the Anglicized words anyway; in Kanji there are no letters). When the Toyota Production System was brought to the United States as JIT, the five Japanese words were loosely translated into five English words beginning with the letter S.

In Japanese, the five words are

- *Seiri* — To identify the tools and other items needed at each workstation to produce the product and to remove any unnecessary tools or items from the work area. A lot of time is wasted when you must leave your workstation to look for a tool or other item needed to perform your work. You should clearly identify each tool that is needed to perform the work at each station. Those tools, and only those tools, should be kept at each workstation, and any others that are not needed should be removed from the area.
- *Seiton* — To neatly arrange and identify the tools that are needed. One method that is often used as an example is a pegboard on which you hang all the tools you need at the workstation. Each tool has a designated, identified space. The outline of each tool is drawn or painted on the pegboard, so you always know if a tool is missing or is not in the right place and you always know where to put each tool when you are not using it. The arrangement and identification of the tools will depend on the nature of the tools and the setup of the workcenter, but the idea is the same.

■ *Seiso* — To clean or to cleanup. Many a factory, warehouse, or office is just plain dirty or messy. Sometimes it's just annoying, sometimes it gets in the way of work getting done, and sometimes it's a health or safety hazard. No matter what, it's bad. Remember the adage, a place for everything and everything in its place. Keep the things you need, but straighten up and organize them. Put everything you need in a designated and clearly identified space, neatly organized. Get rid of everything you do not need. This is one of the most difficult things you will do. You will come up with all sorts of reasons (excuses) for keeping things: it's expensive, it's still worth something, you might need it some day, it doesn't take up much space, our bottom line will take a direct hit; the list could go on and on. Step back and take a deep breath. If you don't need the item right now, it's a waste to keep it. It probably gets in the way, takes up valuable space, and is already useless or worthless. Besides, it has probably already been paid for, so the money has already been spent. You might as well get something of value out of it. Maybe you can sell it or give it away, or get some scrap value out of it, but you need to get rid of it.

■ *Seiketsu* — To perform the previous three concepts — seiri, seiton, and seiso — at regular and frequent intervals to maintain order and cleanliness and avoid complacency. Once you clean and organize everything, it won't take long for it to get cluttered and dirty again. You should perform these tasks on a regular basis to reorganize and reclean the area. It's easy to put something aside when you don't need it right away. You think you'll get back to it soon or you'll need it again soon. The chances are great, however, that you will not need it again any time soon. There's a reason you don't need it now, and you probably won't need it again at all, let alone anytime soon.

■ *Shitsuke* — To turn seiri, seiton, seiso, and seiketsu into habits. With seiketsu, you clean and organize at regular intervals to keep from becoming complacent about it. Eventually you need to get into the habit of performing these tasks constantly and consistently. Instead of putting aside something unneeded, then performing your regular cleanup, why not just get in the habit of getting rid of things right away when you don't need them? Instead of waiting for the end of the day or end of the week to do a thorough cleanup of the area, get in the habit of cleaning up after every operation or as part of the regular work process. If your operation changes such that you need a new tool and don't need the one you've been using, don't wait to change your tool arrangement. Do it immediately. Make the necessary changes to the system you use to identify your tools and their designated locations.

In English, the corresponding terms are sort, set in order, shine, standardize, and sustain. I suggest you use the original Japanese words. These are new concepts and a new way of doing things, so why not add some new words to your vocabulary? If you think that learning or using words from a foreign language is too difficult or won't be accepted, I say, burrito, baguette, and sauerkraut, or Toyota, Peugeot, and Volkswagen.

4.1.11 Value Stream Mapping

Prior to the actual implementation of a Lean system, you need to review where you are and where you want to go. Value stream mapping is a technique to help you with this review. You are probably familiar with flowcharting and may have even used one to diagram a process. A flowchart simply shows the steps in a process, with decision points, information flow, physical flow, and other information. Value stream mapping goes a step further and identifies the points in the process that add value to the product (physical product or service). Value is defined as aspects of the product that the customer is willing to pay for. In a production process, assembling parts into a finished product adds value because the customer is willing to pay for a functional, assembled product rather than a bunch of parts that they would need to assemble themselves. Moving parts and partially processed assemblies between departments is not a value-added process, because the customer does not want to pay for the additional costs associated with this movement. Value stream mapping helps you identify these non-value-added steps in the process, which allows you to then remove the waste from the system. Value stream mapping also includes information flow that is necessary to produce the product or provide the service.

One way to map your process is to put together a relatively simple flowchart as a starting point. You can then get into more detail for the various steps in the process as needed. Once you create your basic flowchart, you can set up a table listing some of the attributes associated with each step. Attributes you may want to consider include the size of the batches at each step of the process, the amount of queue time and wait time, the processing rate, the scrap or yield rate, and the cumulative processing time. This will give you a pretty good idea of where to look for some great opportunities for improvement.

When developing your value stream map, you should identify each processing step as belonging to one of three categories. These three categories are

1. Value creation
2. Non-value-added, but necessary
3. Non-value-added, not necessary

The value creation steps are the actual processing steps, when materials are processed or assembled into the product that is wanted by the customer. The non-value-added, but necessary steps are those activities that do not add value from the customer's point of view but are necessary for the company or cannot be eliminated at this time. It may be possible to eliminate them at a later date, but they are necessary at this time. These activities include accounting functions, accounts payable processing, order entry activities, and other activities. The third category includes those activities that do not add value from the customer's point of view and are not necessary for the business. They can be eliminated immediately to improve the process. These activities include the queue and wait times, warehousing of excess inventory, movement of materials or semifinished products between departments or between processing steps, and all the other wasteful activities discussed throughout this chapter.

Although you probably already have a good understanding of your processes, it is always helpful to put it down on paper, and a graphical format helps to visualize the entire process easily. As you make improvements to your process, you can show the changes on your map. This makes everyone involved feel good about the improvements and gives them an extra incentive to keep making progress.

4.1.12 Poka-Yoke

Poka-yoke is mistake-proofing. What this means is that you want to design your processes so that they can be performed only one way; the right way. You want to design your machines and equipment so that the materials you run through them can be loaded only one way; the right way. For instance, in our example of the drilling operation, we don't want to make any mistakes when drilling. One mistake might be to drill a hole in the wrong location. To prevent that from happening, we might design our drilling setup so that the material to be drilled can be inserted into the machine only one way.

To drill a board for our XL 10 amp, we may have to drill holes in specific locations relative to the left and right sides of the completed board. Even if you clearly identify the left and right sides, if the boards are loaded into the drilling machine in the wrong direction, the holes might not be drilled in the correct spots (Figure 4.3).

Even the most diligent employee will make a mistake occasionally. A new employee, or someone having a bad day, might make lots of mistakes.

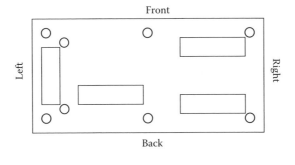

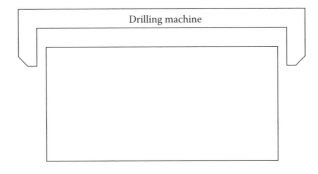

Figure 4.3   Board layout.

These mistakes cost money. The material might have to be scrapped, or perhaps it can be reworked or used somewhere else, but some waste has been created. Even more important, we have also wasted valuable production time producing a defective part. Whatever our planned batch size was — one hundred or one — we have not produced what was planned. This means we must produce this item again sooner than we had planned. One board out of a hundred might not sound like much, but what if you needed every one of that one hundred? And what if the defective part wasn't detected right away? What if it went through to the next process or even somehow ended up in the finished product? You are creating more waste at every step and wasting valuable time at every step. You may be able to recover a portion of a defective or scrapped item, but you can never recover the lost time.

So we need a way to ensure that we cannot possibly load the materials incorrectly. We want to ensure that we load the boards in the drilling machine the same way every time. If we do that, it will not be possible to drill the holes in the wrong spot. One possible solution is shown in Figure 4.4.

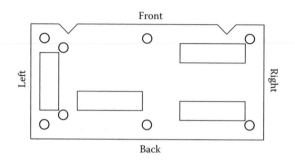

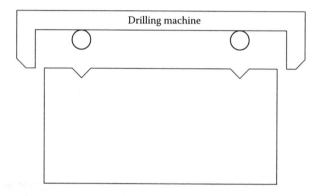

Figure 4.4   Board layout solution.

In this example, simply adding a couple of notches on the front side of the board and a couple of guides on the drilling machine allows the boards to be loaded into the machine only one way; the right way. Well, somebody could still load them in backward, so maybe you need to add something so that if the notches and guides are not lined up, the machine won't run. The idea is to design the process so that it cannot be performed incorrectly. It is mistake-proof. That's poka-yoke.

4.1.13 Quality

Lean Manufacturing and quality go hand in hand. Quality at the source is a founding tenet of Lean. You often hear Total Quality or Six Sigma associated with Lean, but regardless of the terminology, the principles are the same. The idea is simple. When you are producing products or providing services, you want to make a quality product the first time through. Any defects, scrap, or reprocessing is a waste and should be eliminated.

One activity that is all too common in many organizations is the processing of a defective product. This happens when one step in the process either creates a defect or fails to discover one, and this defective part or product continues to the next processing step. In our example of the drilling process of the XL 10, let's say we haven't implemented any of our changes or improvements yet. What might happen is that one board is drilled incorrectly. The operator may fail to detect this defect if there are a hundred boards being processed at a time. This defective board is then passed on to the next step: soldering. At the soldering step, the operator might notice the defect and scrap that board or might not notice it and just send it through the soldering process. This could continue through final assembly, and depending on whether any testing is performed, it might be delivered to a customer. Let's examine this scenario in more detail.

The first step is pretty obvious. By some error, we've made a defective part: a board with the holes drilled in the wrong place. The board is defective. It probably needs to be scrapped. Maybe it can be salvaged for something else, but probably not. So far we've wasted processing time creating a defective part, we're throwing away the money we spent on the material, and we have to spend more money to get rid of the defective part. Depending on the quantity of the items that are defective (often more than one part gets processed incorrectly) and the nature of the product, disposal costs can be quite substantial. This is only the first step in the process. At each step where the defective part goes through further processing the costs increase. If this defective part goes through the soldering process, we waste that processing time too, as well as the wasted materials used in that process. The farther along in processing, the more difficult and expensive disposal likely becomes. If it makes it through the entire production process and to the customer, costs just explode. What is the cost of a lost customer if they receive a defective product then switch to you competitor? You probably are paying to store defective items in inventory (remember, we haven't made any improvements yet), ship defective items to your customer, then have to process the return, pay to ship the item back, and probably put it back into inventory until somebody decides to do something with it. Add up all these costs and multiply them by the total number of defective items and you will likely come up with a very big number.

That is waste, pure and simple, and there is no reason for it. A few simple fixes can prevent the defects in the first place, or detect them as soon as they happen so they don't accumulate throughout the system. Quality at the source means that we make each part and each item correctly the first time. Each operator is responsible for their operation and every part they process. They are responsible for their equipment, their machines, and their tools. They are responsible and accountable. But they also have the authority they need to prevent the defects from occurring or from spreading or increasing.

Chapters 5 and 7 discuss quality tools and Six Sigma, so those details will not be discussed here. But, we will discuss how to prevent defects and how to prevent any that occur from spreading. The first thing, of course, is training. Not only do you have to train your employees how to perform their tasks correctly, you have to train them how to monitor their results and how to detect defects. You want them not only to detect defects but also to detect when the process is getting out of control. You must train them to know what to do when they discover defects or an out-of-control process. Once they detect a defect, they should not let that part go any farther; that's pretty easy. But when they discover more than one defective part or they find the process has gone out of control, they need to stop processing more parts until they find and correct what is causing the defects. In our drilling process, maybe the drills or drill bits are worn out, causing the holes to be drilled out of spec. If the current process calls for the drill bits to be replaced on a regular schedule, and the operator does not have the authority to change them at any other time, we will continue to produce defective boards until it is time to change the bits.

Hopefully, those sorts of situations are rare. It's probably more likely that individual operators cannot change the bits themselves; they have to find a machinist or obtain authorization to do it. We will not make more defective parts, but we waste time getting authorization or finding someone who can change the bits. What you really want to do is give the operators the training and authorization to detect any defects immediately, stop processing when necessary to prevent making more defective parts, and take the necessary corrective action. Ideally, you have designed the process so that it is mistake-proof. Even so, you might still have the problem of the drill bit wearing out before it is due to be changed. You need some way to either check or test each part as processing is completed or sample the parts after they are processed. If defects are discovered, or you are getting out of tolerance, the operators should be authorized and trained to take corrective action. If they discover the problem we have described, they should halt the production of more parts, change the drill bit or bits, scrap any defects, and continue processing. This eliminates or minimizes defective parts and minimizes the downtime needed to correct the problem. This saves both time and money. It also gives employees a higher level of satisfaction. They have control and authority over their jobs and take a great deal of pride in what they accomplish.

Some companies are hesitant to give employees this level of control and authority. They fear employees will abuse their authority or will not make the right decisions and cost the company a lot of money and disruption. Well, you're not going to make these changes without the necessary training and education. That might take a while, and they might have to build up the trust, but that should not stop you from getting started. Besides, you already trust your employees a great deal just by having them perform their regular jobs. If you don't trust your machine operators to make the right decisions, how can you trust them to handle the materials, equipment, and machines they operate every day? If you don't trust your warehouse workers, how can you let them drive forklifts handling thousands of dollars' worth of products and equipment every day? You already have some level of implicit trust of your employees, and you can't watch them every minute of every day. Give them the education and training they need, put in all the controls you need (you need controls no matter how much you trust your employees), and let them do their jobs. You will get much more out of it than you put in.

4.2 Terms and Techniques

Some of these terms and techniques are discussed in more detail throughout this chapter, but they are provided here as a handy, quick reference guide. These are some of the terms you will encounter as you get more into Lean and begin implementation:

- *Cellular manufacturing or manufacturing cells* — Processing areas where all the equipment and machinery necessary to build a product, or family of products, are brought together. Manufacturing cells are usually configured in a U shape.
- *5S* — Five concepts for cleaning and organizing your workspace. The concepts are boiled down to one-word statements that all begin with the letter S:
    - *Seiri (sort)* — Identify the tools and other items needed at each workstation and remove everything else.
    - *Seiton (set in order)* — Neatly arrange and identify the tools.
    - *Seiso (shine)* — Clean or cleanup. Get rid of dirt and messiness.
    - *Seiketsu (standardize)* — Perform the previous three concepts regularly and frequently to maintain the order and cleanliness and avoid complacency.
    - *Shitsuke (sustain)* — Turn seiri, seiton, seiso, and seiketsu into habits.
- *Five why's* — The technique of asking "why?" five times when analyzing a problem. The intent is to get to the root cause of the problem; thorough and deep analysis is required.

■ *Just-In-Time (JIT)* — An evolution of the Toyota Production System. A philosophy of eliminating waste in a system, it is characterized by a number of techniques and methods of performance. One concept is the delivery of parts and materials only when they are needed, or just in time, and not holding inventory of the materials and parts.

■ *Kaizen* — A Japanese word that means continuous improvement. The idea of kaizen is to make small, incremental improvements to a process over time. Kaizen assumes that a process can always be improved, even if it has already been improved.

■ *Kaizen blitz* — A process improvement event in a compressed timeframe.

■ *Kanban* — A signal that authorizes production or delivery from a supplier. The signal can take various forms. A kanban can be a card, a container, a space on the floor, a light, or any other signal that authorizes production or delivery of an item. The kanban also identifies the quantity to be produced or delivered.

■ *Milk run* — A method of scheduling and routing the pickup of materials or delivery of products on a regular route. The term comes from the delivery of milk from the milkman (no longer commonplace). The idea is to have a regularly scheduled and fixed delivery or pickup route. Instead of making irregular deliveries of large batches, you make regular (daily) deliveries of small quantities to support just-in-time production.

■ *Muda* — Waste. This is what you want to eliminate, in all its forms and in every area of the organization. The seven muda identified by Taiichi Ohno are

  – *Overproduction* — Producing products before the customer wants them.

  – *Waiting* — Refers to items waiting to be processed.

  – *Transport* — Moving items from one place to another.

  – *Overprocessing* — If products or tools are not designed properly, excess processing has occurred.

  – *Movement* — Refers to workers having to move from their workstation to look for tools, parts, products, and so forth.

  – *Inventory* — Any items that are not needed for immediate use in production or immediate delivery to customers.

  – *Defective parts* — Any item that is produced that is not in perfect condition for use by the customer. This includes materials or components.

■ *Poka-yoke* — Mistake-proofing.

■ *Sensei* — Teacher; it is a term of respect.

- *Single-piece flow* — Production of one item at a time rather than a batch of predetermined quantity. It allows greater flexibility to respond to changing customer requirements.
- *SMED* — An acronym for single-minute exchange of die, a concept where changeovers are completed in less than ten minutes. Single-minute means single-digit numbers, 1 to 9.
- *Takt time* — The rate, or pace, at which items are produced. Ideally, this rate equals the rate at which customers use the items.
- *Total Productive Maintenance (TPM)* — An equipment and machine maintenance system to ensure that the equipment and machines are always operational when needed to produce products.
- *Toyota Production System (TPS)* — The method of production developed by Toyota, which is the basis for JIT and Lean Manufacturing.
- *Value added* — Value is always viewed from the customers' perspective. Adding value means you are performing a processing step that customers either cannot perform or are willing to pay you to perform so they don't have to do it themselves. Each step in the production process is either value added or not, and you should strive to eliminate the non-value-added steps.
- *Value stream mapping* — A technique that identifies the value-added parts of a process. It is similar to flowcharting, except value stream mapping highlights those steps in a process that add value to the product and it incorporates information flow.
- *Visual control* — The organization of the workplace, including inventory, in such a way that anyone can tell the status just by looking at the space. Eliminates the need to look up information in files or a database, or having to ask people or search for information or the status.

4.3 Planning in a Lean Environment

Because of the pull system, the use of kanbans, the flexibility of quick changeovers, and small lot production, there is often confusion over the role of planning in a Lean environment, especially the roles of master production scheduling and materials requirements planning. Let's clear up the confusion right here. You still have to plan. The plans might look a little different, but you still have to do it. The execution is vastly different, but there is still planning and there is still execution. You also still have to measure your performance and use that information to make improvements (the feedback loop). The real question is, how do you do the planning in a Lean plant?

If you remember the planning and the levels of planning from Chapter 2, you'll see that all the same elements needed to plan in the traditional environment are present in the Lean environment. If you haven't read Chapter 2 yet, or just need a reminder, the levels of planning are

- *Strategic planning* — Long-term strategic direction
- *Sales and operations planning* — Volume, at the item family level
- *Master production scheduling* — Mix, at the individual item level
- *Materials requirements planning* — Raw materials and manufactured components

We're not concerned with strategic planning here, but let's look at sales and operations planning, master production scheduling, and materials requirements planning in the Lean environment. We'll start with sales and operations planning. Sales and operations planning is concerned with aggregate volume by product line or item families. At this level we're not even interested in looking at which individual items we're going to make, let alone the detailed scheduling of when we'll make them. We're concerned with balancing supply with demand at aggregate volume. In the sales and operations planning process no real differences exist between the traditional environment and a Lean environment. You should be able to respond to your customers better in the Lean environment, so your sales volume might increase. You still need to check your plans for available capacity, but you can still do that in a Lean environment. With changes in your processes, such as cellular production, smaller batches, and quick changeovers, your production capacity should increase. So, there's no real difference in planning for sales volume and no real difference in planning production volume. Differences may appear in some of the details that roll up into the sales and operations plan (S&OP) because of the implementation of Lean techniques, but overall there is no difference in sales and operations planning in traditional and Lean environments.

How about at the next level of planning, master production scheduling? What do we do at that level? We convert, or break down, the volume from the S&OP into the individual item mix that makes up that volume. The time periods, or time buckets, are usually shorter than those used in the S&OP, but we're still going to plan production of individual items in specific time periods. In the Lean environment, because of our increased flexibility and the use of the pull system, the schedule will not be fixed as far out as in the traditional environment. But we will still have a fixed schedule for a certain period of time. Remember, mixed-model scheduling is a bit different from the traditional batch system, but we still work with scheduled production. All this really means is that our time fences are going to move in, closer to the present, in the Lean environment. We still must develop our schedule, or planned production, but the fixed schedule will be a very short time period, the liquid period will be much longer, and the slushy period will be somewhere in between (probably pretty short too). So no real differences here either. So far, no real differences in planning between the traditional and Lean environments.

Let's not forget our capacity check at the master production scheduling level. We must check the feasibility of our plan. We can still do that. We know how much we can produce in the Lean environment, just as we do in the traditional environment, so we can check the schedule against available capacity. Again, because of the changes we have made, capacity should change, but it should increase and we can still measure it. Not a problem.

Now we're down to the materials requirements planning level; surely there are differences here. Well, let's look at what we do in materials requirements planning and see if we have to do anything different in a Lean environment. Materials requirements planning takes the finished items from the master production schedule (MPS), looks at inventory levels, lead times, and other information, and calculates the quantity of raw materials and manufactured components that are needed to produce those finished goods. It also tells us when we need to purchase the materials or make the components. We still need to know what materials and components are needed to produce the finished goods, and we still need to know the quantities. But what about the timing? The timing is determined by the expected usage of materials, current inventory levels, expected deliveries, and lead time. All these elements are still present in Lean, they're just different now. The expected usage depends on the production of the finished goods, which we have from the MPS, but with different time fences and shortened lead times. We still have inventory, we just have a lot less of it; maybe none some of the time. You might think that you don't have expected deliveries because of the use of kanbans and just-in-time delivery, but you do. If you expect delivery to be made exactly when you need it and only in the quantity that you need, then that is your expected delivery and you can use that in your planning. In the Lean environment we expect lead times to be shorter, both for production and delivery, but you still have a lead time. So it looks like all the elements of planning are still present in the Lean environment, so the planning should be pretty much the same.

Yes, the planning is very similar. The real difference is what we do with the information we get from the plans. From the planning we just discussed, we have expected sales, or demand, and we have the necessary materials and components to produce the goods that will be needed. We will use this information to determine the schedule and the rate of production in our production cells, the size and number of kanbans, and expectations for our suppliers. The mixed-model schedule has to be derived from something. We do not simply pick the numbers and the pattern out of thin air. It is based on our expected needs, which we have come up with through the planning process. The same applies to the quantity and size of the kanbans; we determine them by our needs as a result of our plans. It requires a bit of work to determine the mix and pattern of the production cells and the size and number of kanbans, but it comes directly from the expected need, which comes from the planning process.

In the traditional production environment you should still develop strong and close relationships with your suppliers. That's not unique to a Lean system. The difference in the Lean environment is that you want your suppliers to be able to support your changed and improved operations. You want them to deliver smaller quantities more frequently. You might even want them to deliver the items directly to your production line, or cells. And you don't want them to deliver the items until you give them the signal, the kanban. But your suppliers cannot support you if they do not have the information they need. Just like you, they need to know the expected quantity and timing of their demand. You can provide this from the results of the planning process we have been discussing. The frozen period of your schedule is very short, but it is still there. And you have an expected or planned level of usage based on your expectations and plans, so you can share this with your suppliers. With this information they can develop their plans, their production rates and mix of items. Their plans will have different time fences than before, but they will have a good estimate of volume and timing from the information you have given them. You must commit to a certain level of purchases, because they need to do their own planning and need to commit to their suppliers. But you can also agree on the changes to deliveries that are allowable or acceptable, changes in both quantity and timing, because of your need for flexibility and their requirements.

Another change that might be appropriate is the authority for purchases, or the release of kanbans. This is where some changes to the purchasing process might take place. Instead of the need for formal purchase orders to be issued, the kanban will be sent. This is usually done by the line workers themselves rather than the purchasing department. The role of the purchasing department changes from order processors and purchase order issuers to relationship managers. The purchasing professionals establish the relations with the suppliers and work with them to develop the kanban system and the allowable and acceptable changes to the planned schedules. This eliminates a lot of the waste inherent in the traditional purchasing system, where the need for materials is identified by the planning or production department, purchases are checked and authorized by a supervisor or manager, and a request is sent to the purchasing department. Then there is more processing, checking, and authorizing before the item is actually ordered or the purchase order issued. When you set up all the parameters in advance, then allow the production line to authorize the delivery, you remove a great deal of waste from the system. This is good for everyone involved and will help you reach that next level of performance.

4.4 Maintenance

One of the seven muda, or types of waste, is waiting. Whenever items are ready to be processed, but they are waiting for whatever reason, that is waste. Equipment breakdowns or scheduled maintenance can often be the causes. Maintaining your equipment is good. Shutting down your equipment to perform maintenance while items wait to be processed is bad. Equipment that breaks down when items are waiting to be processed is even worse.

As you move toward Lean operations in your organization, it becomes increasingly important to eliminate equipment downtime. In any production environment it is important to maintain your machinery and equipment, but as you reduce the waste in your systems and start to produce in smaller batches that are more aligned with your customer demand, proper equipment maintenance becomes vital. If you use inventory to protect yourself, in the traditional environment you can afford a little downtime to perform scheduled maintenance on your equipment. But as you reduce your level of protection, you must keep your equipment up and running so that you can use it when you need it.

For a long time, equipment and machine maintenance consisted primarily of fixing it after it broke down. Machines were often run until they could not run anymore; then they were patched together, repaired, or replaced. This approach doesn't make operational or economic sense, so eventually a better system was developed. Nakajima et al.* refer to the old approach as breakdown maintenance and consider it the first stage in the development of maintenance programs. The next stage of development was preventive maintenance. Preventive maintenance con-

---

* *TPM Development Program: Implementing Total Productive Maintenance,* Nakajima, S., Ed., Productivity Press, 1989.

sists of periodically inspecting machinery and equipment so that any conditions that would cause breakdowns or the production of defective parts are detected and, therefore, corrected before the breakdown or defects can occur. Preventive maintenance entails routine or daily maintenance activities. Relevant activities might include developing maintenance standards, developing maintenance plans, maintaining maintenance records, developing maintenance budgets, and actually carrying out the maintenance activities. Predictive maintenance is another stage and falls somewhere in this continuum of maintenance development somewhere in here too.

The next development in maintenance programs was productive maintenance. Productive maintenance includes the activities of preventive maintenance plus the concepts of maintenance prevention and maintainability improvement. Maintenance prevention is pursued during the design of machinery and equipment. Equipment and machinery should be designed to operate maintenance free; in other words, so that it requires minimal, if any, maintenance throughout its life span. Maintainability improvement is similar to maintenance prevention, but it is performed on existing equipment. Existing machinery and equipment is modified to prevent or minimize maintenance and make any required maintenance as easy as possible.

Total Productive Maintenance (TPM) includes all the activities discussed so far, with the addition of what are called small group activities. This means that maintenance is performed by the equipment and machine operators themselves. Many organizations require specialized maintenance personnel to maintain equipment. Of course, in some instances this is warranted, but as you can see from the concepts of maintenance prevention and maintainability improvement, it is avoidable in many cases. You may argue that your union contracts require maintenance to be performed by maintenance specialists, machinists, or some other specially designated position. That's an excuse and an obstacle that can be overcome. You might not be able to make the change immediately, but you can make it (and you can make it so that both you and the union are satisfied).

As with many changes, transitioning toward TPM requires training. Obviously, you cannot have the machine operators start doing their own maintenance without the proper training; so start training them. This is where your current maintenance people come into play. They become higher level trainers and advisors rather than "simply" maintenance personnel. The best, and easiest, place to start the operators is with the basics. Maintaining the basic conditions of the machinery and equipment is something they can start doing right away. They might already be doing some of this, in which case you are off to a great start.

Maintaining basic conditions includes cleaning, lubricating, and ensuring that all parts are properly in place. Cleaning is not just a quick brushing or dusting of the equipment, but a thorough and detailed cleaning combined with an inspection. You probably (hopefully) clean your equipment every day. Depending on the process being performed, you may clean it after every production run or even during production. Your cleaning needs to move up to another level. It should involve opening the equipment, moving parts to get under them or behind them, and anything else necessary to achieve the level of cleanliness and inspection that is required under TPM. You might be surprised at the quantity of potential problems or defects you find during the cleaning and inspection process. You shouldn't be, though, if you haven't gotten this detailed in your cleaning and inspection before. As you do this regularly you will discover fewer and fewer problems, because you will find them and correct them, and you will find defects and problems before they can cause breakdowns. That's the reason you're doing all this.

Besides cleaning, you need to lubricate and oil your machinery. Moving parts depend on lubrication to function properly. Oil and grease reservoirs must be kept filled and free of dust and dirt. All lubrication points must be checked and filled as needed, and all lines must be kept free of clogs, cracks, leaks, and general deterioration. Lack of lubrication or improper lubrication can cause sudden seizures of equipment or can shorten the life of the equipment. Either way, it's a bad thing.

Check your equipment to make sure all parts are properly in place. Among the many potential problems that can occur if a part is missing is injury; it is a safety issue. Besides making sure all the parts are there and where they are supposed to be, you must tighten and secure all the bolts, screws, knobs, handles, and belts. A loose bolt or belt can cause breakdowns or otherwise damage the equipment. It can also result in the production of defective parts, and again, it can be a safety issue. Train the operators to know what to look for when cleaning, lubricating, and inspecting the equipment, then have them get started.

Performance measures are important in all areas of the organization, including equipment maintenance. Measuring the performance, or effectiveness, of the TPM program is important for two reasons. One, it allows us to determine the areas where we need to focus attention, and two, it tells us how we are doing. Several measures exist for evaluating machine and TPM program performance. These measures indicate how well the maintenance program is performing by analyzing the amount of downtime, the speed or rate of the equipment, and the quality of the products that are produced. If the equipment is down because of breakdowns or other maintenance issues, isn't running at the correct speed, or is not able to produce quality products, we need to know so we can make corrections. Some of the measures we can use are shown below.

An overall measure of performance is overall equipment effectiveness:

$$\text{Overall equipment effectiveness} = \text{availability} \times \text{performance rate} \times \text{quality rate}$$

where

$$\text{Availability} = \frac{\text{Amount of Time actually Used}}{\text{Amount of Time Available}} \times 100\%$$

Availability is the same as utilization in capacity planning (see Chapter 3).

$$\text{Performance rate} = \text{net operating rate} \times \text{operating speed rate}$$

$$\text{Net operating rate} = \frac{\text{Output} \times \text{Actual Cycle Time}}{\text{Amount of Time Acually Used}}$$

Output is the total number of items produced.

$$\text{Operating speed rate} = \frac{\text{Designed Cycle Time of the Machine}}{\text{Actual Cycle Time}}$$

Cycle time is the amount of time to produce one item, or one unit.

$$\text{Quality rate} = \frac{\text{Number of Good Units Produced}}{\text{Total Number of Units Produced}}$$

The total number of units produced includes all the good units plus all the units that are defective. Defective units include any unusable units that are produced during the setup process. As with any measure, trends over time are more important than a snapshot.

4.5 The Lean Enterprise — Not Just for Manufacturing

Lean techniques can, and have been, used successfully beyond the shop floor, in the office and other areas. Production is a process, and so are all your other functions. Lean Manufacturing techniques have been used in distribution centers, banks, hospitals, nonprofit service providers, and many other organizations to make improvements and enhance performance. Once you learn how to use the techniques, you can apply them to almost any situation. Adopt the concepts and adapt them to fit your organization.

One important area where Lean techniques have proven to be beneficial is research and development (R&D). R&D is a vital function, but it is often not very efficient. You might think that the very nature of research and development makes it inherently inefficient, but it doesn't have to be that way. Sure, you have to come up with new ideas, test them, build prototypes, and all sorts of other activities. And very few of these new products or designs will ever make it to market. But that doesn't mean you can't remove a lot of waste from the system. The first step to implementing Lean in the R&D area is the same as on the shop floor: analyze and map the current process, then look for opportunities for improvement.

Look at all the steps in the current process of designing or developing a new product. You will probably find many similarities to the traditional manufacturing operation. Ideas or designs move from department to department or from function to function within the R&D department. These ideas or designs probably move in some sort of batch mode, meaning that they don't move until there is a certain quantity of them to move to the next function or department. Traditional business practices tell us not to move things one at a time. But why can't you make the R&D functions more like the production cell in a factory? Well, you can. Take a look at the movement and see what functions you can group together. Things often "need" approval at every step of the process, before the design can move to the next step. That sounds like waste and a candidate for elimination.

I'm sure there are very legitimate sounding reasons for having approvals all along the way. For instance, you don't want to waste the time of the next person or function working on something that doesn't meet standards, doesn't fit the concept you're looking for, or simply won't work. The first question I would ask is, why and how do ideas or designs get as far as they do before they are checked to see if they meet your requirements? The person working at the first stage should know or understand the standards, the general parameters for the new idea or design, and should be capable of developing a product (or service) that works, at least up to that point. If that person needs input from someone else during this stage, the person or people who provide the input should be there to provide the input immediately. It is a huge waste for someone to spend time working at this initial stage then waiting until they are done to receive feedback, input, or approval. Get the people together so there is no need for approvals at the end of this step, so that they can work together to get it right the first time and in a shorter time.

Continue looking at who, or how many functions, you can put together to eliminate the handoffs that are common in R&D. Some companies have design engineers, industrial engineers, and manufacturing engineers, all handing off designs back and forth to each other, making changes or corrections and complaining about what the others have done. Why not have them work together so their expertise can be combined, rather than having them work against one another? And why not do some cross-training so that each specialty has a little better understanding of the requirements of the others? Instead of having the design engineers come up with wonderful designs that are handed off to the manufacturing engineers, who then make changes so that the item can actually be manufactured, you can have the design engineer and manufacturing engineer work together in a design cell. This will eliminate the handoff and all the time that is wasted developing a design that will be changed significantly in the next step. If the two functions have an understanding of the other's requirements, there will be fewer changes and passing back and forth. If they sit next to each other, there won't be any passing back and forth; if there is any question, they will simply ask while they are working on it.

This idea of the design cell can be taken further. What other steps can be incorporated into the cell? Are there drawings that need to be produced, artwork that needs to be developed, paperwork that needs to be completed, or other tasks, administrative or technical, that can be incorporated into the cell? Remember the concept of the manufacturing cell. The cell eliminates the separation of processing functions into specialized departments where items are processed in batches. This greatly reduces, or eliminates, the movement of partially completed items between those departments, reducing or eliminating the waste of that movement. The items are processed one at a time or in very small batches, increasing flexibility. Quality is improved because the cells take ownership of the products instead of just handing over a completed batch to the next department. All of these improvements can be made in the R&D area just as they can in the production area. And the results will be equally impressive, so don't wait to get started.

The same concepts can be brought into the office, the warehouse, and sales or customer service. All of these areas are composed of various processes, just like the production department. Analyze the current processes and map the flow of materials and information. Look for opportunities to combine functions or jobs, eliminate steps or functions that add cost but don't add value, cross-train employees, and place people and equipment close together to eliminate wasted movement.

There are tremendous opportunities for improvement if you utilize many of the concepts we have talked about here. The 5S can be implemented in any area. The basic idea is simply to clean up the area, get rid of any junk that isn't needed or isn't being used, organize everything you do use and need, and keep this up. Anytime someone has to search for a tool or piece of equipment, it's a waste, whether it's an office worker searching for a stapler or a warehouse worker searching for a pallet jack.

Make sure everyone has the tools they need to do their job. It's cheaper to buy staplers for everyone in the office than it is to have people wasting time looking for one when they need it.

Maintenance in the office, in the warehouse, and throughout the organization is just as important as maintenance of the production equipment. You don't want your copy machine to go down when you're in the middle of making copies for an important meeting with your boss. You don't want your lift truck to break down while you're loading a truck and the next scheduled truck is on its way in. Develop and implement a TPM program for all the equipment throughout the organization.

Remember, Lean is a philosophy, not any particular technique. Adopt the philosophy and utilize the techniques as appropriate in every area of your organization. You will achieve spectacular results, improve operations, and become much more efficient, effective, and productive.

4.6 The Lean Supply Chain

Your company does not stand alone. You have suppliers and customers. And your suppliers have suppliers, and your customers have customers, and so on and so on. You cannot become a truly Lean organization without the help of your suppliers. Your suppliers play a pivotal role in the success of your organization and the success of your Lean implementation. You also need to work with and educate your customers, to truly become Lean. They will need to change some of their habits and their ways of ordering and accepting deliveries to become a part of a Lean supply chain.

Reductions in inventory are both a result of Lean and a technique used to become Lean. One technique of Lean is to reduce, or eliminate, stores of materials inventory. To be able to do this, you need your suppliers to deliver your materials in smaller quantities on a more frequent basis. If your suppliers are not able to support these deliveries, you will be unable to achieve maximum results with your Lean implementation. To become fully Lean, you need suppliers who can supply materials based on your needs and schedule. They may be able to supply you in this way, but they may do it by simply holding more of your materials, rather than you holding them. This does nothing to reduce your materials costs and will probably raise your suppliers' costs. Your carrying costs will go down, but theirs will go up, so the net effect is zero, at best. This will either increase the price the suppliers charge you or reduce your suppliers' profit. In either case, your relationship with your suppliers will suffer, which will eventually decrease effectiveness of the supply chain.

From another point of view, maybe your customers are becoming Lean. They will pressure you to improve your delivery performance to match their needs. If you do not have a Lean operation, or are not working toward it, your profits will suffer or your relationship with your customers will suffer. This will have long-term consequences for your company.

A lean supply chain will reduce costs throughout the chain. These reduced costs can be passed along to all the customers in the chain. This should also result in an associated increase in profits. You may ask, how does passing along cost savings increase my profits? You can pass along less than the total amount of savings, keeping some of the savings for yourself. Or you can pass along all of the savings, expecting an increase in total sales. The idea is to work on achieving a Lean supply chain, because that is much more competitive than any single entity operating Lean.

To get to a Lean supply chain, you will have to share a lot more information with your suppliers and customers than you are probably used to. You will probably have to share information further up and down the chain than you are used to. This will get more comfortable in time because you will also be developing closer relationships. The level of trust and confidence in your supply chain partners will increase as you move ahead. It won't happen overnight, but you have to start somewhere, and you might have to be the one to take the initiative and take some risk. No risk, no reward.

4.7 A Little Lean

As Rebecca Morgan states in her article, "A Little Lean" (*APICS — The Performance Advantage*, July/August, 2002), you don't have to undertake a complete Lean implementation to reap tremendous benefits. Many of the tools and techniques associated with Lean, and discussed throughout this chapter, can be used independently. Start with the 5S to clean and organize your workspace and enjoy the results. Even if you go no further, you will earn the rewards of your efforts. Lean is a powerful tool and an important philosophy. A full Lean implementation will completely transform your organization. But if you're not ready for a full implementation, or you don't have the resources or support, you can implement many of the techniques in your own individual space or the processes that you control. Take advantage of the tools available to you and start your journey to a Lean enterprise where you can. When other people see what you have done and what you have accomplished, it probably won't be long until they're beating down your door to get you to help them accomplish the same things.

Chapter 5

Six Sigma for the Little Guy: Uh-Oh — Math

I'm guessing you may have absolutely no idea what a sigma is, let alone what six of them are or what you do with them. You've probably heard of the Six Sigma quality program, however, although you may not know much about that either. Not to worry! In this chapter you will learn just enough about Six Sigma to be dangerous, and you'll be able to impress your friends at parties.

Six Sigma is similar to Lean Manufacturing in that it is a well-defined methodology that uses a variety of techniques to achieve the desired improvements to an organization. Actually, Six Sigma and Lean Manufacturing complement each other so well that many people think you can't have one without the other. Lean Six Sigma, or Lean Sigma, is a growing evolution of the Lean and Six Sigma movements. Six Sigma is based heavily on, and is an evolution of, quality programs such as Total Quality Management.

There are two things you absolutely must know to understand Six Sigma. The first is that only the customer can really define what is a quality product. You must take the customer's point of view when you design your products and processes. The second thing you must understand is variation. Every process has some degree of variation. (See Section 7.2 on statistical process control for more discussion on variation and the two types of variation.) If you measure and cut ten boards, the boards will vary in length. This variation may be very small, and for these boards it may be insignificant, but there will be some variation. If you can buy into these two ideas (customer focus and variation), it's all downhill from here. But there *is* the math; statistics mostly, sometimes referred to by poor college students as "Sadistics." You may be able to delegate most of the number crunching, but you need to have at least a rudimentary understanding of the concepts behind the math. And you need people who understand the details and can use them.

Don't let the first few sections of the chapter get to you; they're a little math-heavy, but hopefully understandable. It's important to have an understanding of the basic math and statistics that form the foundation for Six Sigma. Many times the statistical foundations are glossed over because some people either have difficulty grasping the math behind it or don't have the time or inclination to learn more about it. That's okay; it's not a requirement to learn all the details. I'm just a believer in having at least a rudimentary understanding of the underlying principles of any system before trying to adopt it. After we get the math out of the way, we will talk about how the rest of the system builds upon that foundation.

5.1 Sigma

We might as well jump right in and get it out of the way. Just what is that sigma thing anyway? First, sigma is a Greek letter. Greek letters are used in mathematics as symbols to represent things. The one you are probably most familiar with is pi.

In statistics, you look at the results of the measurement of an attribute of interest. For example, if you take a group of people and measure each person's height, you have a list of all their heights. These are the results of the measurements of the attribute, height, in which you are interested. The measurements you have taken come from either a population or a sample. A population is an entire group, such as all the people in the United States or all of the output from a factory. A sample is a subset of the population. You may have heard the term *random sample,* meaning a random selection of a certain quantity from the population. Samples are used because it is usually too difficult to work with, measure, or perform calculations on the entire population. A sample is more manageable because there are fewer items to work with or measure. You would have a difficult time measuring the height of every person in the United States or every item produced by a factory. An adequate sample size is a quantity that allows you to say that the results of the sample are representative of the entire population. You have to have a large enough sample for the results to be "statistically significant," or representative of the population, but you don't want a sample that is so big it is difficult to work with or unnecessary.

Within each population or sample, the results of your measurements will vary. The amount of variation will depend, in part, on how precisely you measure the results. Back to our example, let's say the people in your sample vary from 5'2" to 6'3". You determine that the average (sample mean) height is 5'8". The height of each individual person in the sample will vary around the average. Sigma is a measure of that variation from the average. If you measure all the variations (that is, how much each person's height varies from the average), you can also calculate the average variation. Now you have the average height of the group and the average variation from that average height. One measurement of this average variation is the *standard deviation*. That's another term you've probably heard of but of which you might have only a vague understanding. Table 5.1 provides a visual presentation of the sample heights.

You see that the average variation is different from the standard deviation. They are close, but they are different measures so the results are slightly different. The standard deviation is a more precise measure, so it is more useful. Because most of these calculations are done on computers, it is easy to calculate the standard deviation. That's wonderful, you say, but how does the average height of a group of people help me with the quality of my products? Good question. Let's see if we can answer that.

The letter, or symbol, sigma is used in statistics to represent one standard deviation, meaning the standard deviation multiplied by one. Two standard deviations is the standard deviation multiplied by two. And so on and so on, until you get to six standard deviations, or six sigma.

You know that the standard deviation, or sigma, is a measure of the variation in a group of measurements. Specifically, Six Sigma refers to six standard deviations from the mean within a population or sample. When you produce a product, or a part or component, there will be slight differences, or variations, between that one and others you make. Some variations are small and have no effect on the item. But some will be significant enough that the item is considered defective and must be reworked or scrapped. Going back to our production of the XL 10 model of amps (from previous chapters), in the drilling process the holes that are drilled will be different sizes and in different locations on the board. Hopefully, most of the time the variation from board to board will be so small it is hardly measurable. Sometimes, however, the variation will be big enough to cause the board to be defective. If the drill bit gets dull, it won't cut properly; if for some reason the bit breaks and no hole is drilled, that's a big variation; or if the machine isn't properly set up, the wrong size bit might be used. These are all considered defects. The

Table 5.1 Sample Heights

Person No.	Height	Deviation from Avg.
1	5'7"	1"
2	5'11"	3"
3	5'9"	1"
4	5'2'''	6"
5	6'1"	5"
6	5'5"	3"
7	5'6"	2"
8	6'3"	7"
9	5'5"	3"
10	5'10"	2"
Avg.	5'8"	3.3"
Standard deviation		4"

variations from the average or correct size and placement of the holes is so great that they are unacceptable.

Whether you process a physical product you are making or a service you are providing, you can measure the results. If you're drilling holes in a board, you can measure the size of the holes and their placement. If you're processing accounts payable, you can measure the accuracy of the posting and processing. Now, there are two aspects of these results that we are concerned with. We have the variation from the average, which we've just discussed, but we have another equally important aspect.

Every process is designed to perform to some standard or to meet some specification. For our drilling process, we have specifications for the diameter of the holes and their placement on the board. When we measure the diameter of the holes, we can determine the average diameter of the holes and the variation as measured by the standard deviation. We can also compare the diameter of the holes to the specifications. We cannot determine whether a board is defective unless we know the specifications. If we don't know or don't have the specifications, anything could be acceptable, even no hole. If the specs call for a diameter of 3.22 ± .04 mm, any hole out of that range is a defect.

Another point that needs to be mentioned is the precision or sensitivity of the measurements. You will notice that in our example of height all the measurements are reported to the nearest full inch. If we used a ruler or yardstick that has marks only for inches, that's as close as we can report. If we had a measuring instrument that was more precise — marked in one-eighth-inch increments, let's say — we could get a more precise measure. It is important that the instruments you use for your measurements are capable of measuring to the degree of sensitivity or precision that you need. Some processes require very precise measurements. Sizes in the micron range for some technological processes are common. Specialized and precise equipment is required to take measurements at this level. Other processes allow for a much wider tolerance. An inch or two variation in the width of a highway isn't a big deal (as far as I know), and the tools to make measurements at this level are widely available.

That brings up another related topic: calibration. The instruments you use to take your measurements must be properly calibrated. When you go to the market, you want their scales to be properly calibrated. When you buy a pound of coffee, you want two things: you want a pound of coffee and you want to pay for only a pound of coffee. If the scale is not calibrated, you may find out later that you received only 15 oz even though you paid for 16 oz. With the price of coffee these days, you want to get what you paid for. Of course, if their scale is calibrated in your favor, you'd like that, but they wouldn't. Calibration is a whole field of study in itself, but what you need to know is that you must calibrate your equipment and do it properly. You can't calibrate a scale that measures to half an ounce if your calibration tools (weights, in this case) are capable of only full ounce measures.

We've talked about variation around an average measure and variation from a standard or specification. Both of these variations can be measured through the calculation of the standard deviation. That's all well and good, and very interesting, but we need to put these two measures together to get the really big bang out of them.

5.2 The Normal Distribution

Before we go any further, let's take a little side trip here. We need to talk about the normal distribution, sometimes referred to as the *standard normal curve* or other similar term. This will help you better understand all this standard deviation and sigma stuff. When we measure some attribute of a population or sample we end up with a set of results. We discussed that above. When we look at all the measurements, the average, and the variation, we see some sort of pattern.

One pattern you might expect to see is that most of the results are clustered around the average and few results are far from the average. We looked at only ten people in the example of people's heights, so we can't see much of a pattern there. So let's look at another example. Let's say we have the data for all the holes drilled in our boards for a month. If we drill 2,000 boards in a month, and there are eight holes drilled per board, we have 16,000 holes, and 16,000 data points to look at. Let's say

Table 5.2 Drilling Results

Specification	3.22 mm
Average	3.25 mm

Hole Size Range	No. of Holes
0.00–2.96 mm	10
2.97–2.99 mm	15
3.00–3.02 mm	50
3.03–3.05 mm	125
3.06–3.08 mm	200
3.09–3.11 mm	450
3.12–3.14 mm	700
3.15–3.17 mm	1,000
3.18–3.20 mm	1,500
3.21–3.23 mm	2,400
3.24–3.26 mm	3,100
3.27–3.29 mm	2,400
3.30–3.32 mm	1,500
3.33–3.35 mm	1,000
3.36–3.38 mm	700
3.39–3.41 mm	450
3.42–3.44 mm	200
3.45–3.47 mm	125
3.48–3.50 mm	50
3.51–3.53 mm	15
3.54–3.56 mm	10
Total	**16,000**
Standard deviation	**0.08 mm**

we measured every one of those holes and recorded the results, which are summarized in Table 5.2.

You can see from the table that most of the values are clustered around the average and that the further you get from the average, the fewer results you see. This is a common pattern for many things you might measure in your organization, because you are trying to produce items that meet the requirements or are within specifications. If you measured the number of items your warehouse workers picked per hour you might find a similar pattern when you recorded the results and summarized them. That's because if you have a standard or expectation for the number of items picked per hour, you're probably going to come pretty close to that number most of the time.

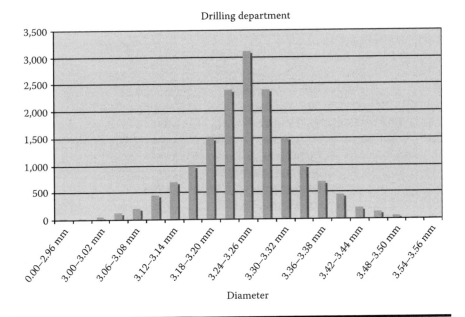

Figure 5.1 Histogram.

Now let's look at our results a little differently. Let's look at them graphed as a histogram. A histogram is simply a bar graph to show data, as seen in Figure 5.1. For a slightly different view, we can add to the histogram the line we would see if we plotted the values in our summary table (see Figure 5.2). If we remove the histogram and leave just the curve of the data points, we have the graph as shown in Figure 5.3.

Well, we've got a bunch of pretty pictures and graphs, but what do we do with them? I'm trying to build you up to visualize how we can use the average and standard deviation to improve our processes. Simple graphs such as these demonstrate how we can use data to measure our processes and measure our defects to improve the processes. Let's look at a few more graphs, building upon what we've done so far. To review, we measured the output of the drilling operation, then summarized the results (Table 5.1) and created a histogram, or bar chart, of those results (Figure 5.1). Then we showed that histogram with the curved line you would see if you graphed the results with a line graph instead of a bar chart (Figure 5.2). Then we took away the bar chart, leaving just the line graph representation. Try it yourself. Type the summary information into a spreadsheet and create a graph. It will look just like the graph you see in Figure 5.3.

Let's add to that graph some of the other information from our summarized data. The average of 3.25 mm is the highest point on the

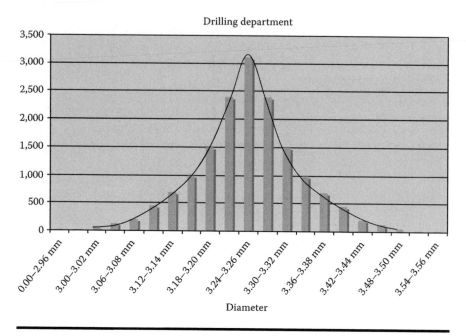

Figure 5.2   Histogram with graph line.

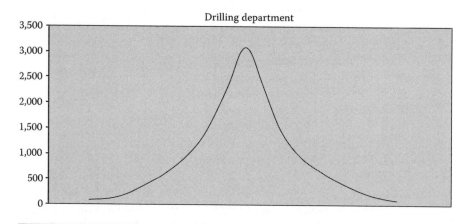

Figure 5.3   Graph.

graph, so we will add that. We have calculated that the standard deviation is .08 mm, meaning that the average variation from the average is .08 mm. So on average, the diameter of the holes ranges between 3.17 and 3.33 mm. We will look at this in a little more detail in the next section, but Figure 5.4 shows how this looks on our graph.

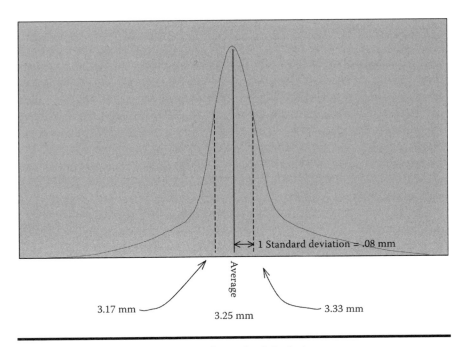

Figure 5.4 Graph with average and standard deviation.

This pattern of data distribution is known as a normal distribution.[*] Other patterns of data distribution create graphs that look quite different, but we are interested in this normal distribution because of its application to the business and manufacturing environments. This pattern has been found to exist in many situations, so we will assume this is the pattern our data would show if we took the trouble to collect it and graph it. We make this assumption with the data we are going to study with our processes under Six Sigma.

This type of data distribution has been thoroughly studied, and it has been found that 68.3 percent of all the values in the population fall within one standard deviation (plus or minus) of the average. That is, 68.3 percent of the total number of measurements taken (16,000 holes) will fall within the range of the average (3.25 mm diam.) plus or minus one standard deviation (±.08 mm). The normal distribution is symmetrical around the average, so when you look at the values within one standard deviation, you look both above and below the average. For 16,000 measurements, 10,923 (68.3 percent) would be within the range of 3.17 to 3.33 mm. Table 5.3 shows the percentage of values that would fall within the ranges of the standard deviations, up to six. It also includes the quantity of values

---

[*] If you are statistically astute, you will notice that the numbers used in this example are not exactly normally distributed, but they work well enough to be used here.

Table 5.3   Standard Deviation Percentages

Standard Deviation	% of Values	Theoretical Quantity	Actual Quantity
1	68.3	10,923	10,900
2	95.4	15,272	15,200
3	99.7	15,957	15,900
4	99.994	15,999	16,000
5	99.9999	15,999.99	16,000
6	99.9999998	15,999.99997	16,000

that would fall within the ranges if our results were normally distributed (column 2) and the actual quantity of values in each range from our drilling example (column 3).

Our results are pretty close to the quantities you would expect to find in each range. This confirms that we are safe using the assumption that our data is normally distributed. In practice, you probably won't test this with your data; just go with the assumption unless you strongly suspect otherwise. As you can see, the greatest concentration of values, or results, is close to the average. As you get further away from the average there are fewer and fewer values. It drops off very quickly. Look back at the histogram (Figure 5.1) or graph (Figure 5.3). You can see that the greatest concentration of values is close to the average (the center or high point of the graph), and the further you get from the average the fewer values you find. Remember, you are trying to drill a certain size hole, so you are going to come pretty close to that most of the time. You have processes in place to help you drill the same size hole every time; for example, using the same size drill bit every time and adjusting the machine the same way every time. Chances are slim (hopefully) that you will drill a hole that is twice the size you want, or that you won't drill a hole at all. That's what the graph shows. Going back to the height example, most people are close to the average height. There are a lot of people between 5'2" and 6'3" tall. There are fewer people shorter than 5'2" and taller than 6'3", and there are even fewer people shorter than 4 feet or taller than 7 feet. (Despite what you see in the NBA, there aren't a lot of people around who are 7 feet tall.) The normal distribution puts some numbers to this pattern, and we can use them to help us manage and improve our processes.

The normal distribution is a probability distribution, which means that the graph represents the probability or likelihood of an event happening. In the drilling example, we are actually measuring the probability that the diameter will be a certain size or fall within a certain range. The probability that the diameter of one of our drilled holes falls within one standard deviation, plus or minus, of the mean (average) is .683, or 68.3 percent.

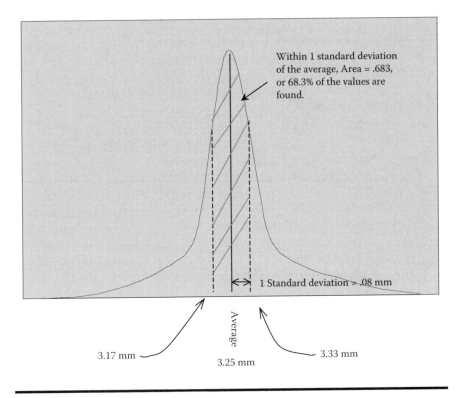

Figure 5.5   Curve denoting area within one standard deviation.

In statistics we talk about the area under the curve. In the graph of the normal distribution (Figure 5.3), the area under the curve is 1, or 100 percent. In other words, the probability of a value falling anywhere under the curve is 1, or 100 percent. The probability of a person being between zero inches and a billion feet (or infinity feet) tall is 1, or 100 percent. The total area under the curve is 1, so any portion of the area under the curve is less than one. The area between −1 and +1 standard deviations of the average is .683 (−1 refers to the area below the average, +1 refers to the area above the average). The area between −2 and +2 standard deviations is .954. You can see that this is the same as the values in the table of standard deviation percentages (Table 5.3). See Figure 5.5 for a graphic representation of the area we're talking about.

Again, this is all wonderful and interesting, but it is also important. We will use these fun facts to help us understand and use a Six Sigma program and improve our processes. Keep these facts in mind:

- We assume, unless otherwise indicated, that our data, or the results of our measurements, are normally distributed
- The normal distribution is symmetrical around the average (the mean)
- The standard deviation is the average variation from the mean
- The percentage of values found within one standard deviation of the average, plus or minus, is 68.3 percent, or the area under the curve within one standard deviation, plus or minus, of the average is .683.
- The total area under the curve is 1
- The percentage of values within x standard deviations of the average, plus or minus, are shown in Table 5.3
- Standard deviation = sigma

5.3 More about Those Sigmas

If six sigma is six standard deviations from an average value, how can this be related to an entire performance-enhancing program for your company? The Six Sigma methodology uses this statistics-based foundation to build the system. We discussed the normal distribution and the standard deviation in relation to our processes. That's all well and good; but we're missing something if we measure process performance in isolation.

Another key component of our process is the customer's requirements. Let's talk about measuring things in that context and go back to our example of drilling holes. Our specifications define the requirements of the diameter of the holes. For other processes, products, or services, the customer might define the specifications and tolerance directly, but in this example the customer's requirements for the finished product flow from the specifications for the diameter of the holes in the board. The tolerance is the allowed level of variation in the finished product. Finished product means the output of whatever process step you are measuring. For our drilling example, the specifications call for a diameter of 3.22 mm, plus or minus .04 mm. Any hole with a diameter that falls outside this range is considered a defect.

If the specification is 3.22 mm, but our average is 3.25 mm, does this mean we have a problem? Possibly, but we have to examine this a little further. Our specification includes an allowable variance of .04 mm. This means that the allowable diameter is 3.22 ± .04 mm, or a diameter from 3.18 to 3.26 mm. Our average of 3.25 mm falls within this range, but our average variation, or standard deviation, means that some of the holes will fall outside this range. Our demonstrated results show that our diameters will range from 3.17 mm to 3.33 mm. The data falls out of the acceptable range on both the low end and the high end. Some of the holes are too small and some are too big.

We can use the characteristics of the normal distribution to determine how many of the items we process are defective, and in turn determine our "sigma level." Remember that the total area under the curve is 1 and that the area under the curve within one standard deviation of the average is .683. We will use these facts to figure out how we're doing with our drilling operation.

First, we split our results into two parts: those below the average and those above the average. Below the average, one standard deviation from the average of 3.25 mm is 3.17 mm. But our tolerance allows for a diameter of no less than 3.18 mm. The same thing occurs above the average. One standard deviation above the average is 3.33 mm, but the acceptable tolerance is only 3.26 mm. Using our knowledge of the area under the curve, and tables that have been developed by all those people who have studied the normal distribution so thoroughly, we can determine the percentages of items that are out of tolerance.

Refer to Figure 5.6 during the following discussion. Below the average, 34.15 percent of the values fall within one standard deviation of the average; in other words, half of 68.3 percent. The lower limit of the tolerance in our example (3.18 mm) is less than one standard deviation from the average, so less than 34.15 percent of the values will fall between the average and that lower limit. It turns out that 31.06 percent of the values fall within the lower limit of the tolerance (3.18 mm) and the average (3.25 mm). On the side above the average, again, 34.15 percent of the values fall within one standard deviation of the average, but the difference between the average and the tolerance is smaller than the area below the average. On the above-average side, 5.17 percent of the values fall between the average (3.25 mm) and the tolerance (3.26 mm). Totaling the percentages for the areas that are within tolerance below and above the average (31.06 + 5.17 percent), we see that 36.23 percent of the holes we drilled are within the tolerance; 36.23 percent of our total items processed are acceptable. Fewer than 37 out of every 100 items we process through the drilling operation are acceptable. That means that 63.77 percent of the items we process through drilling are defective.

Almost 64 percent of our items are defective. That's not very good. And that's only one measurement on one process. We have plenty of other opportunities for defects. In the drilling operation we have assessed only the diameter of the hole; we haven't looked at the placement of the holes or any other potential problems. Before we get into measuring the defect rate, let's continue with what we have so far.

We have two particular problems with our drilling operation. First, our average diameter is not equal to the specified diameter. Although it is not

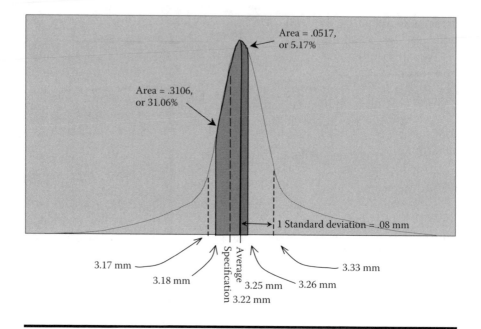

Area = .0517, or 5.17%

Area = .3106, or 31.06%

1 Standard deviation = .08 mm

3.17 mm

3.18 mm

Average Specification

3.25 mm   3.26 mm

3.22 mm

3.33 mm

Figure 5.6   Curve denoting data out of tolerance.

necessary for them to be exactly equal, in our example the average is closer to the upper limit of the tolerance than it is to the specified diameter. We want our output to be within the specifications. Because every process will have some level of variability, it is usually better if the average is equal to, or very close to, the specification.

Second, our standard deviation is so large that the range of our output is outside the range of the acceptable tolerance. Even if the average was equal to the specification, our variation is so great that we would be out of tolerance a significant amount of the time. We want to reduce the variation in the process so that we are within the tolerance range almost all the time.

Now we're getting into the heart of Six Sigma. We want to reduce the variation in our processes, but we can't just look at them in isolation; we have to look at the results in terms of our customer's specifications. We could have the most repeatable process, with no variation, but if we aren't meeting the requirements or specifications, we're just wasting time and money. Six Sigma is a framework for identifying requirements, always looking from the customer's point of view, analyzing performance through a data- or fact-driven methodology, and taking corrective action to improve processes, reduce variability, increase repeatability, and meet or exceed customer expectations. The rest of this chapter discusses some of the tools and methods used in Six Sigma.

5.4 Six Sigma Performance

We've discussed the basis of Six Sigma: the normal distribution and standard deviation. But we still need to build on that to define a Six Sigma level of performance. I mentioned briefly the defect rate, and in our drilling example we had a high level of defects. But we only looked at one aspect in one process step in the production of the product. That's too narrow a focus if we want to look at our entire operation and determine our level of performance.

In Six Sigma we talk about defects per million opportunities (DPMO). Part of this definition requires that we measure the defect opportunities, or how many opportunities there are for a defect to be present or to occur. Let's look at two of the steps in our production process: drilling and soldering. Here is a partial list of defect opportunities:

- Drilling
    - Hole diameter out of tolerance
    - Improper placement of hole
- Soldering
    - Improper placement of solder
    - Too much solder
    - Too little solder
    - No contact between soldered components

Even this partial list presents many opportunities for defects to occur during the processing of this product. You need to analyze your process to determine the number of defect opportunities. Be careful not to go overboard when coming up with your list, though. The list needs to be realistic. You should not list potential defect opportunities that aren't really likely to occur, and you should not list duplicates or the same defect that could occur at two different points in the process. If you've listed lightning striking the plant and traveling through the electrical system to the equipment to render the circuit board unusable, you've probably gone too far in your defect opportunity listing.

Counting defect opportunities isn't always as simple as it might first appear. In drilling, should you list a defect opportunity for each hole, or just "hole diameter out of tolerance" as one opportunity no matter how many holes there are per board? I would count this as only one opportunity. Here's why: you don't want to make this tracking and analysis too complex or cumbersome to perform, and it's better to start out simpler and move to more complex as you get better at both performing the analysis and making improvements. Besides, when you start to develop your Six Sigma program, perform the analysis, and begin to make improvements, you will find that your level of performance under Six Sigma measurement standards is not at the level you think it is. After having said all that, we're going to say that in this example, we have eight defect opportunities per board. Because we've been using all 16,000 holes in our examples, we want to keep them in this example too.

In our example and calculations from the previous section, we determined that 63.77 percent of the items processed were defective. This means that 1,275 out of 2,000 boards were defective (2,000 × .6377 = 1,275.4).* Expressed as yield, the yield rate is 36.23 percent (1 − defect rate = yield). We determined that 1,275 boards were defective. We didn't determine the total number of defects. A board that is defective can have more than one defect, as noted in the defect opportunities. And reducing the total number of defects doesn't necessarily reduce the defect rate. Just because you've solved the problem of holes being drilled in the wrong location on the board, doesn't mean you've solved the problem of holes that are the wrong size.

The next step is to determine the total number of defects and compare that with the defect opportunities. To stay consistent with our example, let's say that drilling the proper diameter hole is our final challenge, that we've made progress in the other areas. So we'll say that the total number of defects was 10,203 (.6377 × 16,000 = 10,203). Our defect opportunity is eight opportunities per board, or 16,000 defect opportunities (8 opportunities per board × 2,000 boards = 16,000 defect opportunities). So 10,203 defects per 16,000 defect opportunities is .6377 defects per opportunity (10,203/16,000 = .6377). Converting this to defects per million opportunities, we find that we have 637,688 DPMO (.6377 × 1,000,000 = 637,688).

If we convert this to a sigma level of performance, we find that this is a little less than a 1.125 sigma level. The conversion from DPMO to sigma level can be done by using a conversion table (See Appendix B). From the table, 634,650 DPMO is equal to a 1.125 sigma performance level. Because we have a slightly higher DPMO, we have a slightly lower sigma level of performance. We have a ways to go to get to a six sigma level of performance. A six sigma level of performance is the equivalent of only 3.4 DPMO. Yes, you read that right, 3.4 defects per million opportunities.

You may be wondering why you must perform all these calculations and conversions to determine the DPMO and the sigma level of performance. One reason for measuring performance as defects per million

---

* This is assuming that the defects are spread evenly throughout all items processed. The defect rate calculated by the holes that were out of tolerance is actually the defect rate for the holes drilled. For this example, we assume that these out-of-tolerance holes are distributed evenly among the 2,000 boards.

opportunities is that it is a comparative measure; that is, we can compare different processes and different products or services using a standard measure. A simple process or product has fewer opportunities for defects than a more complex process or product. Unless you standardize the measurements, it would be difficult or impossible to compare the processes. It is important to standardize the measures so that you can compare their performance against each other. For one thing, you might want to focus your attention on the process that is at the lower level of performance.

I must make a note regarding the yield rate. In our example we did not discuss any rework. Throwing something away costs us money, but processing something twice costs us money too. We don't want to waste any materials and we want to process items only once. Besides, with our variance, we are unlikely to get it right during reprocessing a second (or third) time. Another useful measure of yield is first-pass yield. First-pass yield is the yield rate with reworked items removed from the calculation. With drilling, if a hole is too big there is not much you can do about it, but if it's too small you can redrill it to get it to the right size (hopefully). If your process allows for rework, your first-pass yield rate will most likely be lower than your final yield rate.

The goal of Six Sigma is to improve your performance by reducing the level of variation in your processes to a level that results in only 3.4 DPMO, which equates to a six sigma level of performance. Is this even possible? And how do you get there? I'll discuss the possibility and how to achieve it in the remainder of this chapter.

For now, though, if you look back at our graph (Figure 5.6), you can picture what a six sigma level of performance would look like. If you were operating at a six sigma level of performance, the variation, as measured by the standard deviation, would be so small that six standard deviations would fit within the tolerance defined in the specifications (Figure 5.7).

You will still have some variation in your processes; you can never get rid of all variation. But by using the Six Sigma methodology and the tools associated with Six Sigma, you can reduce the amount of variation. You can reduce it to such a small amount that even the items that are six standard deviations away from the average (which aren't very many to begin with) are within the defined tolerance or the standards you have set.

5.5 DMAIC/PDCA

The study of statistics is, in part, the study of variation. The discipline of Six Sigma is the relentless pursuit to reduce variation. Several steps must

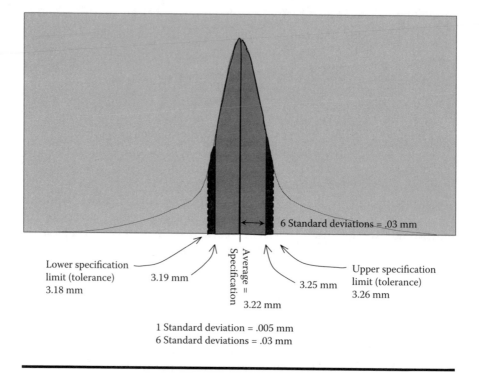

Lower specification
limit (tolerance)
3.18 mm

3.19 mm

Average = Specification

3.22 mm

3.25 mm

Upper specification
limit (tolerance)
3.26 mm

6 Standard deviations = .03 mm

1 Standard deviation = .005 mm
6 Standard deviations = .03 mm

Figure 5.7   Six sigma-level graph.

be taken to make this pursuit worthwhile and valuable. The starting point for Six Sigma is the DMAIC cycle. DMAIC stands for:

■ **D**efine
■ **M**easure
■ **A**nalyze
■ **I**mprove
■ **C**ontrol

DMAIC is based on the plan-do-check-act, or PDCA, cycle, which might be more familiar to you. We need a starting point, and we need a feedback loop so that when we analyze and improve our processes we don't become complacent. You might be done with it for awhile, but as business conditions change so must your processes.

Let's look more closely at the DMAIC cycle and see what's involved in each step. We start with define. In the define stage, we're setting the stage for everything that follows. At the beginning we don't even know what we're doing yet. We must define the problem. Before that, we have to identify the problem. This is one of the most basic steps, but one that many people fail to execute. What really is the problem? What is missing, not working right, or not delivering the results we want? This is where we narrow the focus and become more precise in the problem definition, rather than trying to work with generalizations or a scope that's too broad. For our example, let's identify the problem as too many holes drilled out of tolerance.

Okay, good start. Next we state the requirements, or what the end result should be. In our example it's pretty clear; our specifications and tolerance are already defined. We need to drill holes that are 3.22 mm in diameter, ±.04 mm. It's not always that easy to define the requirements, but even if it is you should probably verify them, to be sure that is really the requirement and not just what you've been working with or what's been acceptable up until now. If you're looking at a process for the delivery of a service, rather than the making of a physical product, it might not be as easy to define the requirement or the end result.

We've identified the problem and what the end result should be. These must be very clear and must be documented. Documentation formalizes the improvement project and gives it clarity. Then we set the goal or target. These also need to be very clear and need to be documented. We will build the improvement on this goal. The goal comes directly from the requirements or end result, stating it in a way that says you're going to get there. We might say that the drilling process will drill holes that are within tolerance 99.99997 percent of the time. Or we might say that the drilling process will be improved so that the average diameter is equal to the specified diameter and the standard deviation will be no more than .005 mm.

Whoa, we've got something to work with now. We have identified the problem, clearly stated the requirements or the end result, and set a goal or target to work toward. If you get through this initial stage, you're already ahead of many others. But you've still got a lot of work to do.

In the measure phase you do just what you think; you measure things. You measure the results of the process on which you are working. If you do not have adequate measures in place already, you must establish the measurements. You might have to determine the proper measurements needed to evaluate the process, and they might be different than any measures you have in place. Don't rely on just the measures that you're already using. Establish and conduct measurements that evaluate the process in the context of the problem you've identified. In our example, measuring the rate of drilling (how many boards are drilled per hour) is not relevant to the problem we've identified. Then again, the rate may affect the quality, so that might need to be explored. The point is, though, that the measures should be relevant and appropriate to the identified problem. The measurements should help you solve the problem and improve the process. Don't just measure for measurement's sake.

These measurements will also verify that the problem you have identified is actually a problem. It could be that what you thought was a problem was really performing as it is supposed to, and if that's what you find, that's okay. It just means there is a different problem, possibly a problem with perception or misunderstanding of what the process should be doing. The measurements will also let you focus even more, to pinpoint those areas that are the cause of the problem. And that leads us to the next phase, analyze.

In the analyze stage you make a concerted effort to discover the root cause or causes for the problems that have been identified. It is vital that the root causes are discovered. If not, the problem cannot be fully corrected. If you don't dig deep enough you might find that you're treating a symptom rather than the actual cause of the problem. One method often used to find the root cause is the five why's. Ask the question, "why?" five times when delving into the cause of a problem. For example, to discover the cause of the drilling problem, we might use the following line of questions:

- Holes are out of tolerance — Why are they out of tolerance?
- Drill bits wear out quickly — Why do they wear out quickly?
- Bits are made of wrong metal type — Why are they built of the wrong metal type?
- Supplier ships wrong bits — Why does the supplier ship the wrong bits?
- Correct specifications weren't communicated — Why not? System not in place to ensure supplier communication or compliance

We've discovered one reason for the results found in the drilling operation, but there are probably more. Go through this process to find the most important root causes affecting the process. Again, don't go overboard. You're looking for a few of the most important causes of the problem that's affecting the process you're examining. After you make improvements you might want to revisit this process and find other causes for continuing problems, but you can't find every problem at once. Focus and prioritize your efforts; don't try to do everything at once.

After you have analyzed the root causes, you should review them along with the measurements you are performing to make sure they are in line with each other. If the measurements and the root causes don't align or don't make sense together, there is a disconnect that needs to be resolved before moving forward.

Now you are into the improve stage. This is where the rubber meets the road. You've done all the work identifying the problem, measuring the results of the process to build a foundation upon which to work, analyzing and verifying the problem, and developing the goals for the process. Now it's time to put all those plans into action and make the improvements. That sounds great, but it's not always as easy as it sounds. You need to adhere to a process improvement methodology that works. You can't just force through your vision of the way things should work. If you've tried that before you're aware of the difficulties and likely results.

The various methodologies on process improvement have one thing in common: the idea of institutionalizing improved performance. It is nearly impossible to make an improvement that becomes established as the new norm if you force it on the people who will make it happen. It's been done, but it's rare and it's not the most effective method. You need to choose and use an improvement methodology that works for you and your organization. You've probably made improvements before, so you likely have a system that works well for you. If you're fairly new to process improvement, you need to find a well-defined methodology that you can use, that fits your corporate culture, and that works for you.

When making improvements, you need to go back to the root causes you discovered in the previous steps. You want to attack those root causes. That is where the real improvement will come from and the drastic changes that are needed will be made. The root causes of the problems must be eliminated. In our example, we determined that there is not a system in place to ensure that requirements are communicated with suppliers and that the suppliers are complying with those requirements. We need to address that problem. We need to address it directly and aggressively. Simply announcing a policy to communicate requirements to suppliers and checking on their compliance is not a solution. That's not even a Band-Aid. There's probably already some sort of policy like that in place; maybe even procedures written down somewhere. Obviously, though, it's not working, because you have identified this as a root cause.

You need to attack this issue and enact a system and procedures to eliminate it. Why aren't the requirements being communicated? Who's responsible for doing it? Why aren't your suppliers working to ensure they understand the requirements? What documentation is required to prove that all necessary requirements have been shared with suppliers? What performance measures are in place to measure the process and the people who perform the process? How strong are your relations with your suppliers? How often do you communicate with them? What are your problem resolution procedures? What are your suppliers' performance measures? How much do you know about their capabilities? How do they measure their own performance, and what are their standards?

You need to ask questions like these and address all the issues that they bring up. If measurements and standards aren't in place, you need to put them in place. If you don't have strong and close relations with your suppliers, you need to work on establishing them. If you don't know your suppliers' capabilities, standards, and performance measurement systems, you need to find out. You need to dig deep, ask questions, and develop solutions.

Once you've done all that and developed your proposed solutions to eliminate the root causes, you need to test them. Until you actually put the improvements in place and work them a while, you cannot be sure they are really improving anything. So put the improvements in place, let the new system run for a while, and test the results. If something needs to be tweaked or changed, go ahead and make the changes and continue testing. When you are satisfied that the new process or procedures work as planned, it's time to institutionalize the changes. If you haven't already done so, you need to standardize the procedures. This reduces variation and ensures that everyone performs the work the same way. The improved process becomes the new norm. All along, you need to measure results and monitor performance. You don't want fluctuations or unwanted changes to creep in over time, and measurements will help ensure that doesn't happen.

The final stage, control, is an extension of the improvement stage. The process and procedures have been institutionalized and standardized and have become the normal way of doing things. Everyone is trained and any new employees are trained in the new way. Measurements are in place and control procedures have been established and put in place to catch and prevent fluctuations or variation. Any new problems that arise over time are addressed and formal procedures to uncover and correct small problems before they become big problems are in place. Of course, conditions change. The business climate changes, new products are developed, consumers' tastes change, suppliers change (or their capabilities change), and you must be prepared to change too. The DMAIC system is often portrayed as a circle, or a closed-loop system, because eventually any changes or improvements you have made will have to be revisited and reviewed (Figure 5.8). You don't want to make changes just for change's sake, but you must be prepared to make changes when they're needed. Through active monitoring and dynamic measurement and evaluation systems, you will be prepared to make changes when conditions warrant them.

5.6 Customer Focus

As I mentioned above, the discipline of Six Sigma is the relentless pursuit to reduce variation, and this pursuit must be worthwhile and valuable.

Figure 5.8   The DMAIC cycle.

To be valuable, customer requirements must drive the system. Otherwise, you are looking at processes in isolation, not as an integrated whole. Customers are not interested in the performance of one particular process or one part of the process. They are interested in the finished product, which requires the integration and interrelation of all components of the system. Because the customer defines a quality product (or service), we must start there to know what needs improvement and where to focus our efforts.

You need to focus on the customers. This does not mean you pay lip service to customer service or that you establish a customer service department. This means you need to determine exactly what your customers' requirements are. You can't define your customers' requirements and you can't define a quality product. Only your customers can. Then you need to do everything you can to deliver the product to the customers per their specifications. Their specifications include the technical specifications, performance of the product, price, delivery, method and timing of invoices, after sales service, and every aspect of the product itself and their contact with you.

Too often, companies think they know what customers or the market want. This is especially true for mass-produced or commodity items, but is also true for some custom-designed and -produced items. Customers may state their needs, but the supplier may tell them that "we do it this way," or they will misinterpret the customers' requests. This is not a customer-focused way of doing business. True, you cannot be all things to all people, but most companies can do a far better job of determining who their customers are and what their needs are.

If you make a highly customized or engineered product, it is relatively easy to determine your customers' requirements, what they need, and what they want. They are working with you in the development or configuration of the product. You do, however, need to take care to fully understand the customers' needs and wants. They might not clearly state everything, or they just might expect you to know what they want. You need to make sure you ask pointed questions and restate what they have told you and have them verify your understanding of what they have said. You also need to incorporate some sort of measurement system to capture customer requirements and ensure that they are incorporated into the product and the delivery and service of the product.

If you make a higher volume or mass-produced product, it takes more effort to uncover what the customers want and what they expect from the product and from you. In this case you need to do market research. Some possibilities include customer surveys, focus groups, user groups, and face-to-face interviews. You need to really understand what functions and features the customers want, how they expect the product to perform, the results they expect from using the product, and how they expect to get it, pay for it, service it, and return it. And you might have more than one type or level of customer. Your immediate customer might not be the final consumer; there may be several layers of distribution between you and the final consumer. You need to know and understand the needs of every link in the supply chain, because any defect or difficulty experienced at any point in the chain can have an adverse effect on another link and the final consumer. You need to design not only the product, but also the packaging and the delivery methods, to ensure satisfaction throughout the supply chain. If your customers want pallet loads, but their customers want cases, and their customers want individual units, you need to design the packaging and the stacking so that it is easy for each succeeding customer to perform their breakdown and repackaging for delivery to their customers.

Beyond understanding your customers' needs and providing them what they want, you must be obsessed with providing outstanding service. This is a pet peeve of mine. Increasingly, U.S. producers are finding it difficult to compete on price. But anyone can compete on quality and service. You have to become almost fanatical about the service you provide. Every day I encounter businesses that seem to go out of their way to provide me with poor service. I often stop doing business with these companies. Take a moment to think about the service you provide, to your internal customers and external customers. Do you return phone calls promptly? Do you remedy complaints immediately? Do you listen to your customers' concerns, or dismiss them without much thought? Do you tell your customers that you can't help them but you can direct them to the person who can? Or do you find out what help they need, find the person who can help them, then explain to that person yourself what the customer needs? And do you follow up to find out if the customer was satisfied?

These are just some things to think about regarding customer service. You can find many examples of organizations that provide outstanding service. You will probably find that these companies are also very profitable. You need to ask yourself if you can afford to provide service that is less than outstanding. If you have trouble competing on price, and you are having trouble competing on service, you are in trouble, period.

The point, and the relation to Six Sigma, is that you are responsible for the entire customer experience, not just a functioning product. Customer focus is central to Six Sigma. Without focusing on the customers and understanding their needs and wants, you will never be able to achieve a Six Sigma level of performance. Quality is defined by the customer. The customer ultimately determines the amount of variation that is acceptable. So you need to fully and completely understand exactly what your customers expect and want to improve performance relative to their requirements. Only then can you expect to achieve the level of performance that will bring you toward Six Sigma levels.

Remember, the goal of Six Sigma is to reduce variation in your processes so that the customers' requirements are met on a consistent basis. You will still have variation, but the amount of variation will be reduced to a point that it fits within your customers' specifications. You must keep this in mind while using the techniques and methodologies associated with Six Sigma. It's easy to get caught up in all the details and day-to-day activities in analyzing, measuring, defining, and improving your processes. It is of no use, and an extreme waste of time and money, to go through all these steps and design a fantastic process that in the end does not give your customers what they want.

5.7 Just the Facts

How does Six Sigma achieve results? Two key methods are (1) the use of facts and data rather than anecdotal evidence and (2) process improvement. And, of course, the emphasis on the customer is always present. Six Sigma uses a variety of techniques to collect, analyze, and use data to identify areas for improvement and to focus improvement efforts. A number of process improvement techniques are employed.

Statistical and quality techniques are the primary basis for gathering, analyzing, and using the data that drives the improvement processes. First, you need to determine what to measure, then you need to determine how to measure it. It sounds simple, but it may not be so simple when you try to do it. This is an important step and needs to be done right. After determining what to measure and how to measure it, you need to perform the measurements and collect the data. Then you need to analyze the data and use it on an ongoing basis. I won't go into detail here, but look at Chapter 7, which covers some of the various quality tools. Six Sigma uses advanced statistical tools, some of which are briefly discussed below, and there are plenty of references available to learn more (see Appendix A). Again, this book is designed to provide an overview of the various tools in your toolkit, not to provide the details of implementing these tools.

As stated in the previous section, the basis for Six Sigma process improvements is the DMAIC model. In every phase of the DMAIC cycle, facts, based on data that is captured through the measurements taken, are used to drive decision making. Using intuition, gut feelings, and guesstimates to make decisions will not lead to Six Sigma levels of performance. You may be able to make some initial improvements or make a discovery using these methods, but they are not sustainable and will not bring you up to the levels you desire.

As you delve deeper into Six Sigma, you will constantly hear about fact-based decision making. Understanding customer requirements depends on knowing the facts about what your customers need and want. You cannot guess at what they want; you need the facts. Finding the root cause of problems depends on knowing the facts. You need data to support your hypothesis. Analyzing a process and the performance of the process depends on knowing the facts. You cannot determine the variability of results, the average, the standard deviation, the capability, the specification limits, or anything else about the process without hard, reliable data.

"Just the facts, ma'am." We're on a Six Sigma mission.

5.8 Green Belts, Black Belts, and Beyond

Six Sigma requires a substantial investment in resources and a substantial level of knowledge about the techniques used. Project management skills are also required, because most improvement activities are undertaken as well-defined projects. Employees will be involved with Six Sigma activities at various levels. Some people will be involved in the intimate details of mapping processes, developing measurements, analyzing results, crunching numbers, and other tasks. Other people will be involved with more than one project or activity at the same time. They will have a higher level of understanding of the application of the techniques; they will coach, guide, and direct others and act as project managers. At higher levels, a person may be responsible for a number of projects or activities. They will act as a trainer, coach, and expert resource.

In Six Sigma, these different levels of activity, experience, and knowledge are identified using a system borrowed from the martial arts. Different colored belts are used in many forms of martial arts to signify different levels of expertise and experience. Six Sigma borrows the colored belt system to identify different levels of training, responsibility, and expertise.

The primary levels and associated belts are green belt, black belt, and master black belt. Different organizations may define the requirements for each level differently, and some organizations have added more "belts." To attempt to standardize this system, testing has been developed to certify the various levels. Numerous organizations can provide the training, testing, and certification. The amount and levels of training provided, the amount of experience required, and the testing itself are not yet consistent or standardized among all the various organizations that provide training. Do your homework before committing to any particular program or organization. Having said that, here is a look at the training involved and the type of performance expected at the different belt levels.

Green belts are generally team leaders or project leaders. At this level, the person is charged with the primary responsibility to complete the project with the desired results. The green belts won't do all the work themselves; they'll have a project team consisting of members with various skills and expertise needed to complete the project. The green belt is the team leader who brings the team together, organizes the work, assigns tasks, and performs the tasks that are required to bring the project to successful completion. The project leader makes sure the project scope and requirements are clearly defined. Green belts ensure that the work is documented, provide status reports and updates, and identify any gaps in skills in the project team. Green belt training includes team dynamics, team building, project management, and related skills, but is not heavy on technical skills or statistical techniques. To move beyond the green belt, many organizations require that a person successfully complete a certain number of projects as the team leader.

Black belts act as coaches and support to the green belts. They provide needed technical expertise as well as higher level management skills. As mentioned, to attain the black belt designation, many organizations require that a person successfully complete a certain number of projects as a green belt. Black belt training is heavy on the technical and statistical tools that are used to analyze and measure processes and performance. Other management skills, such as change management, project management, and team development skills might be emphasized also. Different organizations define and utilize the role of the black belt differently, and you need to find the right structure for your organization.

The master black belt acts as a coach, change agent, technical expert, project identifier, and other higher level roles. Depending on how you structure your Six Sigma roles, the master black belt might be used only as a high-level technical expert, or he or she may be used more as a coach, facilitator, and intermediary. The master black belt might play the role of identifying potential projects and coordinating projects, developing budgets, assigning resources, and working with the various sponsors, champions, and process owners, which brings us to the next level.

The Six Sigma program needs a sponsor or champion who will advocate and communicate the benefits of Six Sigma and the improvement projects that must be undertaken. Individual projects often need high-level sponsors to facilitate cross-functional projects. Resources must be allocated, projects must be prioritized, budgets must be developed and approved, and other administrative tasks must be performed in order for Six Sigma to become an integral part of the organization. These higher level sponsors and champions serve a vital function in an organization that is serious about Six Sigma. Figure 5.9 shows a possible structure of roles in a Six Sigma organization.

5.9 Project Selection

Six Sigma is very much a project-based system. Although many of the concepts behind Six Sigma permeate the organization, most of the work is conducted as projects. Performance improvement throughout the organization is achieved by improving individual processes, including cross-functional processes. From the beginning of the Six Sigma journey there will be plenty of projects from which to choose. As you work on one project, you'll find several more that you want to work on. As you improve one process, you'll find a number of other processes that need improving. There needs to be some method to determine which projects should be undertaken, or some way to prioritize projects.

You will, of course, probably work on more than one project at a time. Even in smaller organizations, two or more projects are likely to be in progress concurrently. Of course, they might not all be Six Sigma projects, but if you get into the Six Sigma mindset you will have identified many opportunities for improvement and you will want to work on a number of them at the same time. Unfortunately, depending on your company's size, you probably can't work on all of them at the same time. You may be able to work on only one at a time, at least if you want them to be done right. You will have a list of potential projects that you want to complete, and you will have to select among them and prioritize them. So let's look at how to do that.

Take a step back and remember what Six Sigma is all about and what you are trying to accomplish with the Six Sigma program. You are trying to improve your processes so that variability is reduced and customer requirements are met. Using these two criteria to select and prioritize projects is a good starting point. Prioritize your list of projects by the

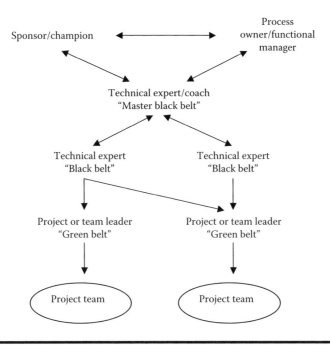

Figure 5.9 Six Sigma hierarchy.

impact they will have in these areas. Another thing to consider when selecting projects is your current capabilities, or your ability to successfully complete the project and make the needed improvements. You may not have the abilities or resources available to complete certain identified projects at this time. That's okay. Leave the project on your list, but identify the resources needed to complete it successfully. When you acquire the resources, in house or outsourced, you can undertake the project.

You may decide that you want to place a greater emphasis on one or the other criteria: customer satisfaction or reducing variability. Reducing variability might not directly affect customer satisfaction, but it will have an indirect effect. Consider our drilling operation. Scrapping all the boards with holes that don't fall within tolerance does not directly affect our customers' satisfaction, except that they will get a finished product that meets their specifications. The indirect connection is that our customers will get a good product but will pay a higher price than they would have if we hadn't had to scrap any boards because they didn't meet specs.

In a case like this, if defective products are somehow getting to the customers, you might be able to identify two projects. One project is to improve customer satisfaction by ensuring that defective products don't reach the customer. Another is to improve the drilling process so that the operation results in fewer defective parts. You need some sort of criteria to evaluate and prioritize these two projects, assuming you don't have the resources to work on them at the same time.

You may consider a number of variables when deciding between the two projects. What is your strategy and what is the current condition of your organization? If you have a significant market share or your competitors aren't performing any better than you in the customer satisfaction department, maybe working on the internal process and reducing scrap and costs is your priority. If scrap and costs are in line with your competitors, but you're looking for more market share or trying to develop more customer loyalty, then it might be more appropriate to work on the customer satisfaction project.

Somehow you will develop criteria and a methodology to prioritize and select projects. The main concern is to undertake projects you are capable of completing successfully, in a reasonable amount of time, and that fit in with your strategy and the direction you wish to take.

5.10 More Six Sigma Tools

Six Sigma utilizes many of the same tools used with other, more traditional quality systems: statistical process control, control charts, cause-and-effect diagrams (also known as Ishikawa diagrams or fishbone diagrams), checksheets, and others. Many of these tools will be covered in Chapter 7, so they won't be repeated here. A few of the more advanced Six Sigma tools will be introduced here, just so you will have heard of them if you decide to pursue Six Sigma in your organization.

If you remember from the discussion of sample size at the beginning of the chapter, one question that must be addressed is whether the sample size is large enough to be representative of the entire population. A related issue is whether the results that are found are "statistically significant." Sometimes it's pretty easy to get a good idea of what your data is telling you. A trend of sales volume month to month over the last two years would show up pretty clearly in a graph of the sales data. Other times, however, it's not very obvious — obscure even — what the data is telling you, or if it's telling you anything at all. It is often necessary to perform tests of the data or results, known as tests of statistical significance.

Common tests of statistical significance are confidence intervals, coefficient of correlation, chi-square, t-test, and analysis of variance (ANOVA). Remember, we use samples because either we cannot get the data for the entire population or it is prohibitive to use the entire population. However, whenever we select a sample, there is a chance that the results of a test from the sample might be different than the results we would get if the test were performed on the population. If we measured all the holes drilled in a year and measured the average and standard deviation, the result might be different than if we randomly selected only a few hundred holes and performed the same measurements. The confidence level, and confidence interval, is a measure of how likely the results of the sample reflect the results of the population. For example, if we calculate a confidence level of 98 percent, there is a 98 percent chance that the results from the sample and the population would be the same.

Correlation refers to the relationship between two items (two variables). If there is a correlation, it means that one item, or variable, has an effect on the other or is affected by the other. Housing prices and mortgage rates are correlated. When interest rates rise, housing prices tend to fall. A relationship can be tested, and the strength of the relationship can be determined. The coefficient of correlation measures the strength of the relationship between the two variables. Regression is related to correlation and is a measure of the extent or amount that one variable depends on the other. It is usually used to calculate or forecast future results, or the value of one variable given the value of the other variable. Some terms you've probably heard associated with regression analysis are independent and dependent variables. You calculate the dependent variable from the independent variable, meaning that you predict the dependent variable when given the value of the independent variable. One depends on the other.

The chi-squared test (pronounced with a "k" sound, not "ch") determines whether actual results fit the expected results. As you collect and analyze data over time, you will develop a pretty good idea of what to expect when you take new samples. We've made an assumption about our data being normally distributed (if graphed it would show the normal curve described previously in the chapter). When we test samples we expect the results to be normally distributed. The chi-squared test helps to determine whether that assumption is correct.

Even though a sample might be representative of the entire population, the average (mean) and the standard deviation of the sample might not be exactly the same. The t-test calculates the standard deviation of the sample when the standard deviation of the entire population isn't known already.

When you perform analysis of samples, you obtain the mean, standard deviation, and other statistics. The analysis of variance is a way to compare the averages (means) from several samples. It is used to determine whether the differences in the sample means are significant.

These are all pretty advanced techniques that probably sound meaningless at this point. I mention them here just to introduce the terms and topics so that when you start getting more into Six Sigma you will have heard of them before.

Design of experiments is a method to test solutions in your improvement activities. Many of the statistical techniques just described are used in conjunction with the design of experiments. When you want to test or experiment with possible solutions, you need a well-defined system to perform those tests and experiments. Because you often must experiment with and change several variables, you can't be haphazard or unclear about what you're doing. You need to define the variables to be tested and how you are going to test them, develop a matrix of the variables and the changes or combinations in them that will be tested, and clearly document and analyze the results. In our drilling example, you might want to test the type of metal the drill bits are made of, the speed of the drill, and the number of holes drilled before replacement. You might want to test three types of metal, five drill speeds, and ten quantities of holes drilled. With these variables you have 150 possible combinations of changes ($3 \times 5 \times 10 = 150$). You don't want to randomly make changes to one, two, or all three variables and see what happens. You want to change one variable while leaving the others alone and analyze the results as you methodically change all the possible combinations that you have defined up front. If you don't perform the experiments properly, you could end up with a worse mess than you started with.

Another important and very useful tool is failure modes and effects analysis (FMEA). FMEA is used to identify potential problems and develop measures to reduce or eliminate them. When you develop a product, service, or process, or improve an existing one, there may be problems with the finished result. It is good practice to have a system to identify these potential problems so they can be reduced or eliminated before the product or service is finalized. As with design of experiments, you want a well-defined system or methodology for identifying and addressing these potential problems. FMEA is such a system. It generally follows the following steps:

1. Identify the product, service, or process to be analyzed.
2. Identify the possible or potential problems with the product, service, or process.
3. Assign a numeric score for:
   - Severity of the problem
   - Probability the problem will occur
   - Probability the problem will be detected
4. Determine the overall score for each possible problem. Multiply the three scores together to come up with the overall score.
5. Develop solutions to reduce or eliminate the problem.

Potential problems are ranked by overall score, and those items with the highest score are worked on first.

5.11 Where to Start

Much of the discussion in this chapter was a bit technical, and you might feel a bit overwhelmed with this Six Sigma stuff. But don't set it aside just because it sounds complex or difficult. There are a lot of good concepts and useful tools behind the Six Sigma system. Focus on the concepts and the ideals that are the foundation of the system. Utilize some of the tools and incorporate them into your daily activities.

Implementing Six Sigma, or moving toward becoming a Six Sigma organization, is a worthwhile goal. You will reap tremendous benefits if you adopt the concepts and use the tools. But if you don't have the resources or aren't ready to completely jump in, you can and should still adopt the concepts and use some of the tools.

Remember two of the main concepts behind Six Sigma — customer satisfaction and reducing variability. If your organization is relentless in the pursuit of customer satisfaction, whatever other tools you use to achieve a higher level of satisfaction will pay off. If you focus on improving your processes by reducing the variability of the output, it will pay off. Focus on your customer and deliver to them what they want. Develop processes that consistently produce products or services that meet specifications. Six Sigma is just a well-defined system to get you where you want to be with respect to customer satisfaction and product quality. It's another tool in your toolkit. Use it when appropriate.

Chapter 6

Theory of Constraints: Think!

One thing that confuses people about the theory of constraints (TOC) is that there are two different aspects to it. The more familiar aspect is the manufacturing improvement tools that include the drum-buffer-rope methodology of managing constraints. The other aspect, which is becoming more widely known and used, is the TOC thinking processes. The thinking processes are powerful tools, but they do require some time and effort to understand and use.

The theory of constraints, also known as constraints management, was developed by Dr. Eli Goldratt. His theories were introduced to the masses in his best-selling book, *The Goal*. *The Goal* introduced and explained the drum-buffer-rope methodology and the five focusing steps. The thinking processes were also introduced but were not explained in any great detail. Some companies were able to use the concepts introduced in *The Goal* to make dramatic improvements in their operations. Others were not able to use the concepts or were confused by them. For those companies that were not able to use *The Goal* to make improvements, the problem is not with the concepts of TOC or with the people who read the book. The book was written in the form of a novel and is an introduction to the concepts, not an implementation guide.

This book is also not an implementation guide. This book is designed to give you an overview of the various tools presented so that you can then make a more informed decision as to what tool or tools to use.

There are plenty of materials and organizations from which you can learn more if you move forward with implementation. (See Appendix A for some references.)

One thing that is neglected when TOC is discussed is that many of the other tools presented in this book need to be, or should be, used during the process of the five focusing steps. As will be discussed below, the five focusing steps are used to identify and "eliminate," or break, a constraint or constraints. During the elimination phase, you may need to use a variety of tools to make the improvements.

6.1 A Note on *The Goal*

The title of the book, *The Goal*, has specific significance. TOC is a management philosophy and was developed in the context of a manufacturing organization. Its origins are based in production scheduling and the attempt to optimize the scheduling of a production plant. The question is asked, what is the goal of the organization? The answer is, to make money, now and in the future. This is important, because ultimately, the fundamental purpose, or goal, of most companies is to make money. This is true for most nonprofit organizations too; the only difference for a nonprofit is where the money goes and what it is used for. All the activities of the organization must somehow help the organization achieve the goal. The concept of throughput and throughput accounting is based on the goal of making money and will be discussed in more detail below.

6.2 Drum-Buffer-Rope

Although the drum-buffer-rope methodology for managing constraints is a technique that will be used after the constraint is identified during the five focusing steps, I will discuss it first, because it may be more familiar than the five focusing steps to many people. As I've stated before, this book is targeted to small manufacturers, so I assume most readers are manufacturing something. Drum-buffer-rope will be presented in a manufacturing context, but it is applicable to any process. You need to keep this in mind as you begin to identify and break constraints. Your constraints may not be in your manufacturing process.

So what exactly do we mean by constraint? A constraint is simply something that prevents a system from performing at a higher level. In a manufacturing context, a constraint is something that prevents the company from producing as much as it needs. Notice that I didn't say to produce as much as possible, but as much as is needed? You may not want to produce as much as possible to achieve your goals. This relates to the concept of throughput, which will be discussed below. Another term that is used is constraining resource. The constraining resource is the machine, workcenter, tool, person, or even policy that prohibits more output.

In the manufacturing process, a series of steps are performed that take the various raw materials and components and processes and assembles them into the final product. Each step in the process has different capabilities, or different capacities. Much of the time, companies look at each step in the process separately, rather then the entire process as a whole. Many improvement initiatives are aimed at increasing the efficiency or productivity of one or more steps in the process. In fact, many measurements of an organization's performance, and a manager's performance, are based on the efficiency or productivity of the individual steps in the process. The theory of constraints points out that this way of thinking is wrong, wrong, wrong.

Figure 6.1 shows the series of activities from our example in Chapter 4, with the capacity of each workcenter. The drilling workcenter is the constraint, because it is the resource that limits the output of the entire system. In case this isn't clear, let's walk through it. Of course it's easier to see in this simplified example where the activities are lined up nice and neat. In a traditional production environment the activities are not always so neatly lined up, and that's where some of the confusion comes in.

The theory of constraints says that you must look at the entire system as a whole, and that the optimization of any one step in the process will not necessarily help you achieve the goal. This is often hard for people to accept, but if you step back and think about it, it makes sense. Using the example from the chapter on Lean Manufacturing (Chapter 4), we have a simple, three-step process of drilling, soldering, and assembly of the XL 10. In this example, the capabilities of each step are as follows: the drilling process can complete 12 pieces an hour (five minutes per piece = 12 pieces per hour), the soldering process can complete 20 pieces an hour (three minutes per piece = 20 pieces per hour), and the assembly process can also complete 20 pieces an hour.

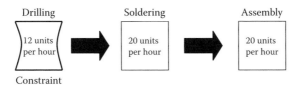

Figure 6.1   Constraint.

The maximum output of this three-step process is 12 pieces per hour, which is the rate of the first step in the process, drilling. If you could get a great deal on a new piece of equipment to double the capability of the soldering process, why would you even think about it? Increasing the capacity of the soldering department does absolutely nothing to increase the output of the system. To increase the output of the entire system, you must increase the capability of the drilling process. This is the part of the system with the smallest capacity.

If you can't yet see why the maximum output of the system is only 12 pieces an hour when the soldering and assembly processes are capable of 20 pieces an hour, I'll walk through it with you. First, let's assume that the products flow through from one step to another as single pieces. In other words, as soon as the processing of each piece is completed, it moves to the next step, instead of waiting until an entire batch of some number of items is completed and moved as a group. Now let's start sending pieces through the system. Let's send 20 pieces through the system.

How long does is take the 20 pieces to get through the first step, drilling? Drilling can produce 12 pieces per hour. So to produce 20 pieces, it will take about one hour and forty minutes (20 pieces ÷ 12 pieces per hour = 1.67 hours, or 1 hour and 40 minutes). Because the pieces are flowing through the system one at a time, as they complete the drilling process they immediately move to soldering. The pieces are leaving the drilling process at a rate of 12 per hour. But the next process, soldering, can handle 20 pieces per hour. Soldering can produce 20 pieces per hour, but they're arriving at only 12 per hour. That means there will be some idle time at the soldering machine. The assembly process can also handle 20 pieces per hour, and the soldering process *could* send 20 pieces per hour to assembly, but the pieces are coming out of soldering at only 12 per hour (because that's how many were sent into soldering).

All 20 pieces are eventually completed, but at the rate of only 12 per hour. You may still be thinking that because the last step has the capability of processing 20 pieces per hour, the output of the system is 20 per hour. But look at it for a second. The pieces are leaving the drilling process at a rate of 12 per hour, so they are entering the soldering process at the same rate, 12 per hour. The assembly process is capable of completing 20 pieces per hour, but it is receiving items at a rate of only 12 per hour. If only 12 pieces per hour are entering the process, only 12 pieces per hour are leaving the process. The process could complete 20 pieces per hour if 20 pieces per hour where entering this step, but they aren't. For the assembly process to complete 20 pieces per hour it must receive 20 pieces per hour from the previous step.

As you can see, putting any resources into increasing the capacity of the soldering or assembly process is wasted. You need to focus your efforts on the drilling process — the step with the lowest capacity. Figure 6.2 shows the system with an increased capacity in the assembly process. As you can see, the constraint is still in the same place, so we have wasted the effort to increase capacity in assembly.

Let's look at this another way, in case you still think you can get 20 units per hour out of the system. Let's build up some inventory and see what happens. Let's say we've built up inventory and put it in the soldering and assembly areas so that those processes can run at their capacity (Figure 6.3).

What happens now that we have this inventory? (We won't worry about how it got there.) Let's take it a step at a time. Assembly can process 40 units per hour and there are 80 units of inventory ready to be processed. So assembly will process 40 units per hour and 40 units per hour will come off the line. If we look at just the assembly process, we could run at this maximum rate for two hours.

Next let's look at soldering. Soldering can process 20 units per hour and there are 80 units ready to be processed. That means that soldering can produce at its maximum rate for four hours. Soldering is running at its maximum rate, which means that 20 units per hour are leaving soldering and entering assembly. In two hours there will be 40 units waiting to be processed through assembly. The original 80 units take two hours to be processed through assembly, so just when those 80 units are being completed there will be another 40 units to be processed. That means, then, that assembly can run at its maximum rate for three hours.

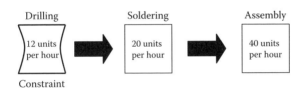

Figure 6.2   Constraint 2.

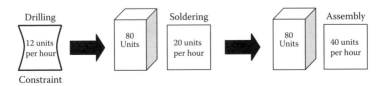

Figure 6.3   Inventory.

With the inventory we had built up, we can run assembly at its maximum rate for three hours and soldering at its maximum rate for four hours. After three hours, though, assembly can't run at its maximum rate anymore. After three hours, all the inventory has been processed and we're left with whatever is coming out of soldering. That's 20 units per hour. So after three hours, soldering is still running at its maximum rate, but assembly can run at only 20 units per hour even though it has the potential to process 40 units per hour. And what about after four hours? Soldering has finished processing the stack of inventory, so it is now limited to running at the rate that is coming out of drilling, which is 12 units per hour. So after four hours we're back to running at 12 units per hour, which is the limit of the constraining resource.

For a while we had fooled ourselves into thinking we could get more output from the system. We magically put in some inventory that allowed two of the machines to work at a higher rate. But how could that inventory have gotten there? To build up inventory the machines would have to shut down, or slow down, for a while. If the machines shut down, there is no output. With no output for some period of time, then increased output for a few hours, the average will come out to 12 units or less per hour. If the constraining resource runs constantly and the others don't shut down too long, you'll get the 12 units per hour out of the system. If the constraining resource shuts down or slows down, the output of the entire system goes down.

Let's change things around and put the constraining resource at the end instead of the beginning (Figure 6.4). For example, we'll change the capacity of the drilling and soldering processes so they are the same: 40 units per hour. This means it will take 1.5 minutes to process each unit in drilling and soldering and five minutes in assembly (originally it was five minutes in drilling and three minutes in soldering and assembly).

Now if we start sending products through the system, we'll be able to process 40 units per hour through both drilling and soldering, but then they'll arrive at assembly, where the capacity is lower. What's going to happen? Inventory will start piling up in front of assembly. In a typical

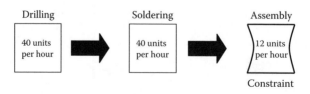

Figure 6.4 Constraint moved.

plant, the idea is to keep every machine, workcenter, or department working as much as possible. Downtime is bad! You've paid lots of money for the machines and you're paying the workers, so you've got to keep the equipment running. Also, many measurements and incentives are based on efficiency and machine utilization times. If you're the manager of the drilling department and part of your evaluation is based on machine utilization and efficiency, wouldn't you want to keep running at the maximum rate? Sure you would. But what would happen downstream, and to the system as a whole? Let's take a look.

If we send materials through the system at a rate that keeps the first two workcenters running at full speed, we'll start piling up inventory in front of assembly, as we mentioned. We'll also be processing different items through the system, so we'll have different items piling up in inventory. That leaves us with a prioritization dilemma. How do we know which of the piled-up items to work on first? My guess is that priorities will change; you'll start processing one item, then stop and switch to another item when someone starts yelling that they need that one, not this one. But let's hold off on that problem for now.

This is all wonderful, you say, but what about this drum-buffer-rope stuff? Let's start talking about that now. You might think that the first thing to do is increase the capacity of the constraining resource. This would, theoretically, increase the output of the system, but you need to do some checking before you do this. A couple of questions should immediately come to mind. One, are you actually achieving the potential output of 12 pieces per hour? Just because the system is capable of producing at this rate doesn't mean it actually is. Equipment downtime, either planned or unplanned, due to breakdown, maintenance, worker shortage, changeovers, or even lack of work, can cause the output to not be as planned or expected. You should examine these causes and see if you can do anything about them to increase your output. Then you need to ask yourself if you really need to increase your output anyway. Are you selling everything that you make, or are you making items just to add them to inventory? Of course, you may have valid reasons to increase your inventory, but you should examine them closely.

As I said, the constraining resource limits the output of the system. The constraining resource is the drum. The drum sets the pace. Think of *Ben Hur* and the guy on the ship beating out the pace of the strokes on the big drum. Using the drum-buffer-rope method, the drum sets the pace for the entire system. The drum is the constraint in the system because it is the least capable step in the system. As you can see from our example in Figure 6.4 , the assembly department sets the pace for the entire system. We're going to use that drum. We're going to use it to control ourselves, to keep us from overloading the system or building up unwanted inventory (notice it is unwanted inventory?).

Because the drum sets the pace for the system, we want all the steps in the system to be based on that pace. The drum is going to determine the release of materials into the system. If you release materials into the system at the rate that the drilling and soldering departments can use them, you'll end up with a large pile of inventory in front of assembly. The assembly department won't be able to process them fast enough. As you get to a more complex system than our example, releasing materials at the pace of the drum (the constraining resource) becomes even more important.

We understand the drum, now what about the buffer? The buffer is an inventory buffer. It is the amount of inventory you keep in front of the drum. If the drum, or the constraining resource, doesn't have anything to work on, for whatever reason, the capacity of the entire system is lowered. The purpose of the buffer is to help ensure that the drum always has something to work on, that it is never idle. In our example, the buffer will be in front of the assembly step. We don't want assembly to run out of work, so we keep some inventory in front of it so it always has something to work on. You need to plan and monitor this buffer, not just let it happen. You don't want to pile up too much inventory, because that causes its own problems, and you don't want to let the buffer get down to zero. We can keep our buffer stocked at the level we want by processing more or fewer items at the preceding step or steps. If we want to increase the buffer, we'll increase the rate or amount we process through the system until the amount of inventory we're shooting for builds up. If we want to decrease the buffer, we'll slow down the rate or the amount that we process through the system.

Finally, we have the rope. The rope ties the drum, the pace-setting operation, to the release of materials into the system. You don't want to release work into the system at a rate that is greater than the pace of the drum (unless, of course, you need to add some inventory to your buffer). The rope is a signal that controls the flow of work into the system. When scheduling the release of materials into the system you must monitor your constraining resource (the drum) as well as your buffer(s). This may be difficult to accept, but there will be times when you're not releasing materials, or more work, into the system. You don't want one or more of your machines or departments to work. Keeping everyone and everything busy is so ingrained in many manufacturers (and other organizations) that this is often very difficult to overcome. This is especially true if managers are evaluated, and rewarded, on the efficiency or productivity of individual workcenters or departments. But remember, we are con-

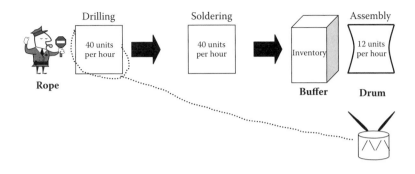

Figure 6.5   Drum-buffer-rope.

cerned with the system as a whole, not with individual workcenters or departments. Let's see what our system looks like now (Figure 6.5).

We're concerned with the system as a whole. You have to remember that. The output of the entire system is only as good as the output of the constraining resource. Increasing output, performance, or efficiency in any area other than the constraint is a waste of time and money. Idle machines and idle people are necessary at times. That doesn't mean people can just sit around twiddling their thumbs. There are many things that people can do when they're not processing products through their area. They can clean and maintain their areas and machines and equipment. They can be educated and trained, cross-trained, or sent to help out in other areas. I'm sure you can come up with endless ideas for using these people effectively. How about this? They can work on increasing the performance and capacity of the constraining resource. Wouldn't that be a great use for them?

The example we've been using has only three processing steps. It's a pretty simple process. Of course, most plants aren't that simple. If you're operating in a traditional environment, you probably have a number of different departments with several machines in each department. You have many different products, several different families of items, and various subassemblies and manufactured components. You have a complex production schedule, conflicting and changing priorities, and maybe a team of expediters.

In an environment like that it's not always easy to identify the constraint. You probably have a pretty good idea, though. If you're not sure, the first place to look is the area where inventory is backed up.

No matter how complex your plant is, the concepts we've discussed here work just the same. You might have to have more than one buffer, but there will be one constraint (at least there will be only one primary constraint), and that constraint will set the pace for the entire system. The

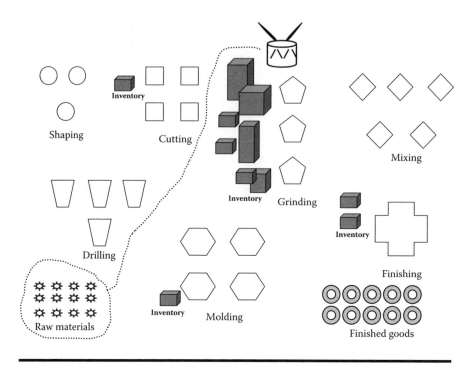

Figure 6.6 Complex system.

constraint, or drum, will control the flow of materials into the system by use of a rope, or some sort of signal. See Figure 6.6 for a more complex system, but still using drum-buffer-rope.

The release of materials into the system is controlled by the constraint (grinding). Not all items are processed through the constraining resource, so materials for those items can be released as needed. However, you have to be careful. A nonconstraining resource can feed into the constraining resource, and you don't want to overload a nonconstraint so that it can't feed the constraint. We'll discuss that a little more later.

6.2.1 Buffers and Buffer Management

We've called the buffers inventory buffers, because we keep inventory in front of the capacity-constrained resource to prevent downtime of the constraint due to a lack of work. It might be better, though, to refer to the buffers as time buffers. The same issues that arise in capacity management arise in buffer management. You're dealing with a mix of products and you need to have some sort of standard measure to help you manage and measure capacity or buffers. Time is often a useful standard.

We're using the processing of the XL 10 to demonstrate. The XL 10, in our revised example, takes three minutes per unit in drilling, three minutes per unit in soldering, and five minutes per unit in assembly. Another item, say the RG 7, might take four minutes per unit in drilling, five minutes per unit in soldering, and eight minutes per unit in assembly. If we talk in terms of units, a buffer of 100 units really means different size buffers for the two items; 100 units of the XL 10 translates to 8.3 hours of work in assembly, and 100 units of the RG 7 translates to 13.3 hours of work in assembly. If the buffer is to protect the constraint from running out of work and shutting down, we're concerned with the amount of work in the buffer, not the number of units. That's why the concept of a time buffer is very useful.

An important question is, how big should the buffers be? To answer that, let's consider again why we have the buffer. The buffer is protection; protection of the constraint. We don't want the constraint to run out of work, because the constraint determines the output of the entire system. And how does the buffer get created in the first place? The resources feeding the constraint feed the buffer. The constraint should be processing items at a constant rate (ideally, anyway), because we're focusing on it and making sure it is running constantly (except for necessary downtime). This means that variations in the output of the feeding operations affect the size of the buffer.

If the feeding operations have problems that cause disruptions, the buffer won't get fed and the buffer will shrink. If you want to increase the buffer, all you have to do is increase the output from the feeding operations. That's not a problem because they have greater capacity than the constraint. The size of the buffer needs to be based on the variations in the output of the feeding operations, the problems that cause the buffer to not get fed and to shrink.

The size of the buffer needs to be at least the length of time (remember, it's a time buffer) to recover from a certain amount of disruptions in the feeding operations. As is discussed in Chapters 5 and 7 on Six Sigma and quality, variation tends to follow an established pattern. That means that the length of time of the disruptions and the frequency of the disruptions will tend to follow an established pattern. You can use these patterns to determine the size of the buffers.

If the variation in output is very small, small enough so that you can recover from the disruptions before the buffer is affected, you might not even need a buffer. As the variation in the length of time of the disruption or the frequency of the disruptions increases, the size of the buffer will have to increase. Also, as with any variation, there will be some outliers or rare events. If you had some major event like a complete breakdown of a piece of equipment and it took you two weeks to replace the equipment, that's probably (hopefully) a rare event. You can't protect yourself from every possibility, so you have to choose some level of protection that you're comfortable with. Take all this into account when determining your buffer. Of course, the easiest place to start is with an estimated or arbitrary buffer size.

There's nothing wrong with making an educated guess to start with, but at least put a little effort into it. The starting point isn't so vital, but what you do after that is. Once you determine your buffer and build it, you have to monitor it and manage it. You need to monitor the actual size of the buffer with this planned size that you've come up with. The actual size of the buffer will vary because the output of the feeding operations will vary. The output of the feeding operations will vary for two reasons. Normal variations will occur from uncontrolled disruptions, and planned variations will come from your scheduling and from your efforts to increase or decrease the size of the buffer to keep it in line with your plans.

Buffer management refers to the monitoring of the buffers and the actions you take to control them. You want to monitor the buffers both as a measurement of your performance and as a control mechanism. If the buffer doesn't vary, it means that you're not eating into it, which means that it's not really doing anything. If you're not eating into it, you're not using it, and it's not protecting you. It's just sitting there; probably keeping someone's peace of mind, but not really doing anything. Actually, that's not true; it's doing something. It's just not what you want it to do. So monitor your buffers, manage them, and change them when it's appropriate to change them.

We've discussed one of the most well-known aspects of TOC (the drum-buffer-rope method), but there are some important steps you might need to take before you even get to this point. Let's look at another aspect of TOC, one that helps us get to the stage where we're using the drum-buffer-rope method: the five focusing steps.

6.3 The Five Focusing Steps

The usual starting point for action is a serious problem or crisis. Some organizations are proactive and have systems in place to review processes and make improvements before problems start, but most of the time it's some sort of big problem that gets us started. In most cases, it's more of a reaction than action. Something bad happens, someone gets yelled at, and people try to do something. That something is often just a quick fix, which doesn't really solve anything.

Ideally, you'll review and analyze your systems and processes regularly to make changes and improvements before problems manifest themselves. Even if you don't do that, and you're faced with a problem that you need to solve, the five focusing steps are a good place to start.

The five focusing steps are used to determine where and how to spend the time and energy to make improvements. You need to determine what to change, what to change to, and how to make the change, within the context of achieving the goal of your organization. The five focusing steps are:

1. Identify the system's constraint(s).
2. Decide how to exploit the system's constraint(s).
3. Subordinate everything else to the above decisions.
4. Elevate the system's constraint(s).
5. If, in the previous step, a constraint has been broken, go back to Step 1, but do not allow inertia to cause a system constraint.

6.3.1 Step 1 — Identify the System's Constraint(s)

This seems pretty straightforward, but it is not always as simple as it appears. Manufacturing processes don't tend to be simple and problems are not always clear. Usually the problem starts with complaints from the customer; for instance, the order was not shipped on time, the order was incomplete, the customer received defective items, the promise dates don't meet the customer's needs, the lead times are too long, and other similar issues.

Instead of actually trying to solve the underlying problem, expedite mode often takes over. Schedules, if they exist in the first place, become meaningless. Orders on the floor are reprioritized based on who screams the loudest about what they want done. Partially completed batches are stopped and set aside so a new "hot" order can get through that workcenter right away. Vendors are called. They're "bribed" and pleaded with to get the materials that were ordered on the truck today and to get materials that haven't been ordered yet in your warehouse tomorrow. You know the drill.

These are signs of a system gone crazy, and you have probably seen it in action. There must be a better way. Instead of running around expediting and fighting fires, you need to make some changes to your systems and processes. If not, you'll be running around like this forever. Things might slow down for a while, but pretty soon another customer will get fed up and start making noise. Then you're right back to the fire-fighting mode. You need to make some changes. First you need to determine what to change. You can't guess; you need to know what specifically to change. But then, before you start to do anything, you need to know what to change to. If you don't know where you're going you could end up anywhere. Finally, you need to determine how to make the change. That's often the hardest part. You know what you need to do, but how do you go about doing it? We'll get into that a little later.

A good place to start looking for what to change is to look for the processing step that has inventory piled up in front of it. That's a good indication of a constraint, but do some further checking to be sure. Constraints generally take one of three forms: policy constraints, resource constraints, and material constraints. Policy constraints are probably the most common. They should be the easiest and cheapest to overcome, but that's not always the case. Policy constraints include things such as batch sizes, order release policies, and so on. If you process items in certain lot sizes, do you know why the batches are that size? Probably not. "Just because," or "we've always done it that way," are likely answers. Why are priorities set the way they are? Why are items processed in a particular order? The answers aren't always clear, and these policy constraints could be affecting the output of your entire system. You need to find out what is causing the constraint.

Resource constraints are not as common as you might think. Problems are usually associated with how the work is sent through the system, not a particular resource within the system. Resources include machinery, tools, people, and other things needed to produce your products. Resource constraints are easy to overcome, at least in theory. The constraint within the constraint is the decision to acquire more resources and to clearly identify and justify the need for the additional resources.

Material constraints are not common, but they do occur. Be sure to verify that a material constraint is really a material constraint and not a policy constraint. Is the material not available or not available in enough quantity, or have you just not anticipated, planned, or ordered the material in time? That's the difference between a material and a policy constraint: real unavailability or lack of planning.

Constraint	Example
Policy	Batch size, order release policy, prioritization policy
Resource	Machine, tools, people, money
Material	Material shortages, unavailability

6.3.2 Step 2 — Decide How to Exploit the System's Constraint(s)

Now you need to figure out what you're going to do to overcome the constraint. This is sort of the process redesign phase. You must determine what the improvement will look like. If you need to develop new procedures or develop new policies, this is where you do it. If you need to acquire new resources, or modify existing resources, this is where you figure all that out. During this phase you need to keep the goal in mind and understand the concept of throughput.

The solution to overcoming the constraint is determined, in part, by the type of constraint. Whatever the constraint, the improvement or the process redesign is pretty much the same. Because it is likely that a policy constraint is the cause of the problems, the solution is more of a process change or the implementation of a new process. The first thing to do is analyze the existing process. Mapping it is probably a good idea. It's hard to change something if you aren't even clear about its current state. Many people think they know what the current process is, but until you map it out and put it down on paper, it's really unknown.

After you document the current state, you can look for ways to improve the process. This is one area where many of the other tools in the toolkit can be put to use. Maybe a constraint looks like a resource constraint because you can't process enough material through it to fill customers' orders or meet customers' lead times. But maybe this is really a policy constraint, arising from the policy of operating in a traditional production environment. Instead of continuing to operate in this mode and trying to force a solution by adding another shift or another piece of equipment, maybe a move to cellular manufacturing and the use of Lean techniques would be appropriate.

Maybe you're having trouble prioritizing orders and scheduling the plant because your information systems are not adequate for your needs. The constraint might be a lack of information or poorly managed information that can be overcome with an improved information system — an enterprise resource planning implementation. Six Sigma methodologies might be used to identify system constraints and develop the improved processes. If inventory is missing or can't be controlled, resulting in a constraint, a cycle counting system might overcome this constraint.

6.3.3 Step 3 — Subordinate Everything Else to the Above Decisions

What does it mean to subordinate everything to the above decisions? Because the constraint determines the performance of the entire system, all efforts need to be focused on the constraint. You don't need to worry about improving the other parts of the system, because improving them will have no effect on the performance of the whole system. You do, however, need to make sure that the rest of the system is synchronized with the constraint, so that the constraint does not become idle.

Subordinating means that the rest of the system supports the constraint. Nonconstraining resources feed the constraint. You need to manage the nonconstraints so that they properly feed the constraint with enough, and the proper, work. You don't want to feed too much — that's exactly what we're trying to avoid — but you can't let the constraint run out of work either. Materials releases, schedules, and work order prioritization of nonconstraints must be synchronized, or subordinated, to the constraint. All efforts are focused on maximizing the effectiveness, efficiency, and productivity of the constraint. That's subordination.

6.3.4 Step 4 — Elevate the System's Constraint(s)

To elevate the system's constraint means to make it so that it is no longer the constraint. After you have done all that you can do to maximize the throughput of the system by focusing all your efforts on improving the constraint, only then should you make any investment to increase the capacity of the constraint. If, in the example we've been using, the constraint was assembly and we've done everything possible to maximize throughput, we may need to add another assembly machine or station to increase the total throughput of the system.

Say you've implemented a Lean Manufacturing system, instituted work cells and adopted a pull inventory system to overcome the constraint and you still need to increase throughput. If this is the case, then you should consider adding machines, cells, people, or shifts to increase capacity. You shouldn't do this before you've done everything else you can do to overcome the constraint.

6.3.5 Step 5 — Go Back to Step 1?

If, in the previous step, a constraint has been broken, go back to Step 1, but do not allow inertia to cause a system constraint. Finally, after you have made all these improvements, "broken" your constraint, and increased throughput, you need to go back to Step 1 and start over. The warning to not let inertia become a constraint means that you shouldn't just keep on going, doing what you've been doing. You need to make sure you have correctly identified the constraint and that you identify any new constraint that may have unexpectedly appeared as you worked through the process.

After you've gone through the first four steps, identified the constraint, made the improvements, and overcome the constraint, a new constraint will appear. It has to. If you've made wonderful improvements and increased your throughput and your capacity to the point where you have extra capacity, you still have a constraint. Remember, the goal is to make money now and in the future. You want to keep making money, to grow and increase value. Sales volume that doesn't match the available capacity is now a constraint. You need to overcome the constraint of not enough sales to utilize the available capacity.

6.3.6 Change

One important issue inherent in all of this discussion is that things have to change. Change isn't easy for an organization. Change management is a discipline that is severely lacking in many organizations. Making the changes and managing the changes have to be handled effectively for any improvement to take place. So how do you change?

The common perception is that people are resistant to change. That's not true. People love to change. They change all the time. The problems arise when you try to force people to change. They don't like it and they'll do just about anything to resist it. The question then becomes, how do you get people to want to change and to make the change that you want them to make?

One idea is to get people to "buy in" to the changes that you want to make. This method has some merit, but it is still very passive. "Okay, we agree that we need to change. What now?" That doesn't tend to bring about the changes that are needed. Begging, pleading, and bribery are other methods that are practiced, but as I'm sure you know, they are not very effective. So what then? How do you get people to change?

Well, why do people change and why do they want to change? People change when there's a benefit to them. (What's in it for me?) The benefit can be tangible: money, easier work, shorter hours. Or the benefit can be intangible: increased status, greater satisfaction, a feeling of control. People will probably change a process if they will make the same amount of money working fewer hours or if the work is easier to perform. Some people will make a change if they can receive a new, more important-sounding title. If they feel a greater sense of accomplishment, a feeling that their efforts count for something, they'll want to make a change. If the change is their idea, or they think it's their idea, they'll be excited to change. If they are the ones in control — it's their idea and they say how it's going to be done — they'll be fighting to make the change. They'll be upset and disappointed if things don't change.

That's the trick, to get people to take ownership and control over the change; to have them make it their idea to change; to get them to believe that only change will do, that the current state is just not acceptable. Dr. Goldratt advocates the Socratic method and the use of the thinking processes to bring about needed changes. We'll discuss these methods in

Section 6.5, but first let's look at an aspect of TOC that we touched on above but didn't get into in depth.

6.4 Throughput and Throughput Accounting

Sometimes it is not easy to determine if you're making money or not. Financial accounting rules and cost accounting rules do not make the determination easy or clear, at least not for the average person. Being profitable, on paper, doesn't mean you're making money. Having a positive cash flow is probably a better indicator, especially for the small manufacturer.

The theory of constraints suggests that there is a better way to determine if you are making money (that is, achieving the goal). The concepts of throughput and throughput accounting are presented as an alternative to traditional cost accounting methods. Although many people agree that throughput accounting is superior for determining whether you're reaching the goal, it has not yet become widespread. Until throughput accounting is accepted by the accounting standards organizations, regulatory agencies, and is taught as part of the curriculum in university accounting programs, it will have a difficult time gaining wide acceptance. This, of course, does not mean that you cannot or should not use it. Any organization is free to use whatever measurements they want to help them determine if they're making money. The challenge is to reconcile the results from throughput accounting with the required cost and financial accounting reporting.

So what is throughput? If you're trained in or familiar with traditional accounting, the concept of throughput takes some rethinking. If you are not familiar with accounting, you probably should learn a bit, but I don't wish that pain on you. Throughput is the rate at which money is generated by an organization.* This sounds simple enough, and it really is. The difficulty lies in reconciling this with the complexities and rules associated with traditional accounting practices and changing people's thinking. Look at the definition again: the rate at which money is generated. If you do not have sales, you are not generating money and you do not have throughput. Throughput is not the total revenue generated from sales; throughput is the money generated. This means that throughput is the money received from the sales minus the money spent making and selling the items that were sold. The difference between throughput and net profit is that, in traditional accounting, net profit is based on product costs that include allocations of overhead and labor that are not treated the same way in throughput accounting.

---

* Throughput is not output. Remember, to have throughput you have to have sales. If you're just producing items and putting them into inventory, you have output, but you don't have throughput (Figure 6.7).

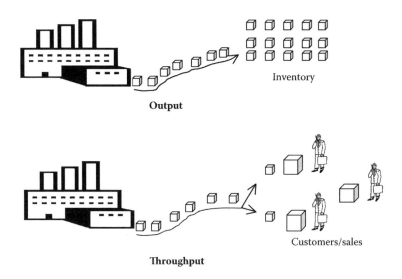

Figure 6.7 Output versus throughput.

In TOC, two other measurements are used along with throughput: inventory and operating expense. In TOC, the concept of inventory is quite different than in traditional thinking. In TOC, inventory is all the money spent, or invested, to purchase things needed to produce the items to be sold. It includes all assets, such as equipment, machinery, and buildings, as well as the materials and parts, but does not include labor or overhead. The operating expense is defined as the money spent turning inventory into throughput. Operating expense includes all the labor and overhead, sales commissions and other related costs, and all other supplies and related items.

In TOC, net profit is calculated as throughput minus operating expense (NP = T – OE). The return on investment is calculated as net profit divided by inventory (ROI = NP ÷ I, or ROI = (T – OE) ÷ I). These measurements are a bit different from the traditional method of calculating them, but they are very useful as management tools to evaluate the performance of your company. And that is their purpose, to give companies better tools for evaluating performance. Financial accounting and cost accounting have their place, but the information provided by them might not be adequate to help you achieve the goal.

The measurements of TOC evaluate the system as a whole. (Throughput means generating money; it doesn't evaluate any single area.) Traditional methods tend to be used to measure the performance of individual areas of the company, rather than the system as a whole. As was discussed in the section on drum-buffer-rope, the performance of the system is what's important. Measuring individual parts of the system, as a prelude to making improvements to them, is pointless, unless you are working on the constraint.

6.5 The Thinking Processes

The five focusing steps are used to focus your efforts on the correct problem. Drum-buffer-rope is a method for scheduling the plant and managing production and inventory. The thinking processes are used to identify core issues, develop the improved process, and overcome obstacles along the way. You need to know what to change, what to change to, and how to effect the change. The thinking processes are methodologies that were developed to use logic to effectively and thoroughly work through these steps. The thinking processes aim to present logical thoughts and arguments on paper so they can be reviewed, discussed, and revised as necessary. Logic diagrams, which look a little like flowcharts, are used in the thinking processes.

6.5.1 Evaporating Clouds

Although the Socratic method* is very helpful for identifying root causes, something more is usually needed to find a solution to the identified problem. The root cause is often a conflict between two opposing forces. The process of evaporating clouds, also known as the conflict resolution diagram, works to find a solution to the conflict. Advocates of TOC believe that compromises or tradeoffs are not necessary or desirable to solve conflicts. Rather, the belief is that win–win solutions can usually be found.

It is necessary to clearly state the problem, and getting it down on paper makes it easier to visualize and keep things clear. The evaporating clouds method is a way to state the problem and put it in a visual format so the objective, required conditions, prerequisites, and the conflict are easily identified and documented. Defining the problem clearly makes it easier (presumably) to figure out the solution. Figure 6.8 shows the general form of the conflict resolution diagram and an example.

But what is meant by evaporating clouds? On the surface, the term evaporating clouds seems to refer to overcoming or eliminating the conflict — making it go away. That's true in a way; we want to make that cloud of conflict evaporate, but not quite in the way you think.

---

* The Socratic method is a method of teaching by asking questions, rather than by lecturing. The learner develops the answers, instead of having them given. As related to root cause analysis, the cause is found by asking a series of questions that, when answered, leads to the root cause of the problem being evaluated.

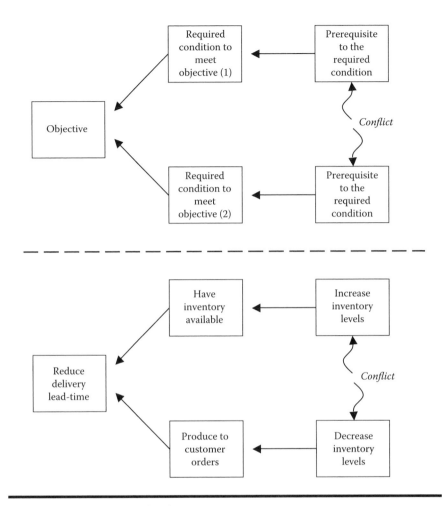

Figure 6.8   Evaporating clouds.

Normally, this situation (Figure 6.8) would immediately inspire a compromise solution; in this example, some sort of middle ground on inventory levels and a mix of make-to-stock and make-to-order production. However, a compromise is not the solution we seek. A compromise is not always the best solution, if ever.

What the evaporating clouds method does is encourage us to reexamine the problem or issue. We've clearly stated the problem, even written it down, so why do we need to reexamine it? Well, maybe the problem we've defined isn't really the correct problem. Maybe we need to reevaluate the situation and challenge our assumptions.

That's the thing with this problem definition. We think we've clearly stated the problem and identified the conflict, but there are unstated assumptions behind them. In this example, we've stated that the problem is with the delivery lead time and that we need to reduce the delivery lead time. The first question someone might ask is, "why?" Why do you have to reduce the delivery lead time? The answer might be that customers are demanding shorter lead times or your competitors can deliver in a shorter amount of time. That might be true, but let's look at some of the assumptions that haven't been stated yet.

We're assuming that the time from when a customer places an order to the time the customer receives that order is too long. We're also assuming, from the statement of the problem, that to reduce the lead time we either need to keep inventory in stock or wait until the customer orders to produce what they order. If we keep inventory in stock, it can just be picked and shipped, or picked, assembled, and shipped. If we wait until the customer places an order, we can produce only that and not waste time producing something else. To be able to deliver from stock we need to increase inventory levels. If we produce to order, we'll decrease inventory levels. Obviously you can't both increase and decrease inventory levels, so there is an inherent conflict in these two statements.

But look at the assumptions. Start with the first one, the big one: we have to reduce the lead time to satisfy customers; maybe so, but maybe not. Maybe it's not the length of the lead time, but something else. Maybe our lead time varies too much and the customer wants more stability. Maybe we just can't meet the promised lead time. Maybe we have a quoted lead time that has no resemblance to the actual time it takes to build, pack, and ship the products. Maybe we're trying to solve the wrong problem!

The evaporating clouds method isn't just stating the problem and writing it down; it also includes this process of bringing out these unstated assumptions, challenging them, and getting to the real heart of the problem. If we undo just one of the statements in our problem, in the diagram, then the whole problem collapses and the conflict disappears. We'll still have a problem we must solve, but it's now more likely to be the real problem: the system problem instead of the localized problem. We'll be looking at the system problem now because, as we reexamine the problem, bring out the assumptions, and challenge them, we'll be asking the questions with our eye on the goal.

The goal is making money by increasing throughput. By taking the problem as originally stated and looking at it from the point of view of achieving the goal, we've focused our efforts on improving the system to increase throughput instead of just "fixing" one part of the system; the delivery lead time. That's the power and benefit of the evaporating clouds method. It'll take some practice, but give it a try and see how you do.

6.5.2 *Current Reality Tree*

Another tool of TOC is the current reality tree. The current reality tree is a type of logic diagram that shows the current state, or how things work right now. The purpose of the current reality tree is to uncover the root cause of whatever is preventing you from achieving your goal. Like the conflict resolution diagram, the current reality tree aids the problem resolution process by clearly stating and documenting how things work right now. At least it states and documents your perception of how things work. Either way, it's a great place to start. It's like a process map, but it's a logic map. You have to know where you are before you can figure out where you're going.

To construct the current reality tree, you usually start with the observed undesirable effects (UDEs) (pronounced "you-deez"). Then work backwards by connecting causes and effects until you arrive at the root cause that produces all the UDEs you started with. Let's use the example from above and start with the UDE that customers aren't satisfied with delivery performance. Figure 6.9 is a simple current reality tree based on this UDE. In this example, the UDE we're starting with is, "customers not satisfied with delivery performance." Two causes of that are the delivery lead time is too long and customers change their orders at the last minute. These causes are themselves UDEs, so we look for what causes these effects. We keep doing this until we get down to the root cause or causes. In this example, we've traced the causes to setups and changeovers being too long, no penalties for last-minute changes to orders, and sales teams being rewarded for sales volume. This provides an excellent starting point for finding solutions to overcome the root causes that have been identified.

6.5.3 *Future Reality Tree*

Similar to the current reality tree, the future reality tree is used to develop and analyze an envisioned future state and the causal relations that will get you there. The initial development of the future reality tree is a starting point. It presents your initial arguments and thoughts on paper, in a logical format that allows for review and discussion. The arguments presented as causes and effects must be analyzed and validated.

Again, this is a starting point. As you work through the analysis, and especially when you start to make changes, you'll probably have to modify your plan. But that's okay; you can't expect your first thought to work without modification. As you work through it, you'll make it even better and stronger. Figure 6.10 shows an example of a future reality tree.

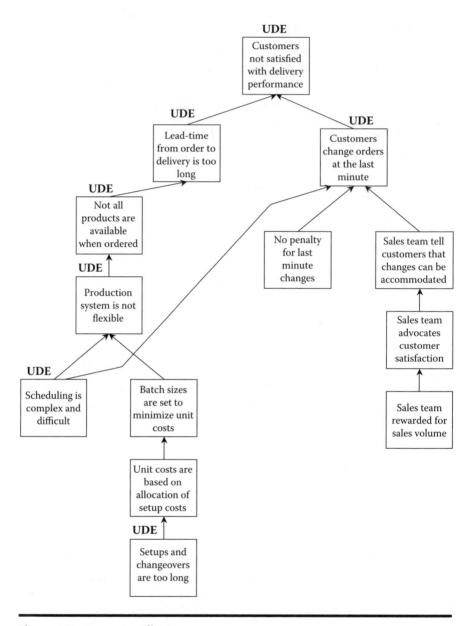

Figure 6.9 Current reality tree.

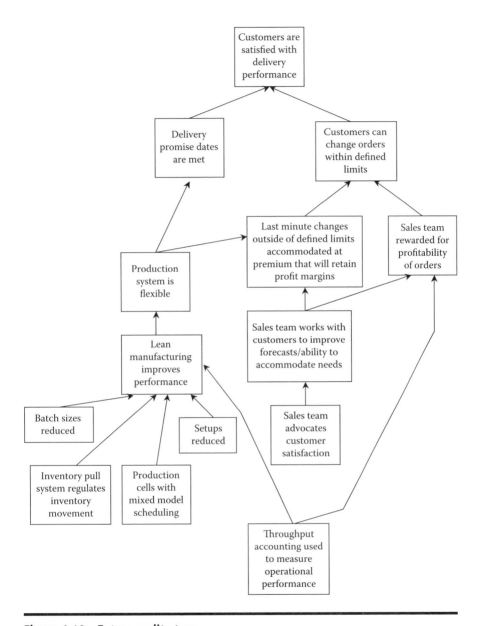

Figure 6.10 Future reality tree.

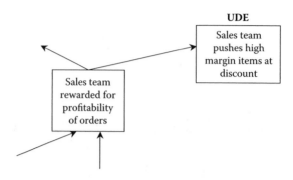

Figure 6.11   Future reality tree with undesirable effects (UDEs).

In the future reality tree, you might also want to include potential negative consequences, or UDEs (Figure 6.11). Whenever you work on a new process or new product, you should try to anticipate potential problems or negative impacts that might arise. For one, this brings more reality into the equation, but it also helps you develop solutions or tactics to reduce or eliminate the problems if they do occur.

These diagrams — evaporating clouds, current reality tree, and future reality tree — are cause-and-effect logic diagrams. They will require a little practice, but they're very useful for analyzing and working through problems and finding solutions. Process maps and value stream maps are also very useful and can be used in conjunction with these logic diagrams. There's no reason not to use all the tools in your toolkit, as long as they are relevant and useful.

Chapter 7

Quality Tools: You've Heard about Them; What Are They?

A number of well-known quality management tools are widely used. You have probably heard of most of them and may even be using many of them. Although they are all very useful tools, many suffer from negative image problems. Many companies have undertaken quality initiatives that have failed, and the quality tools have been blamed. It is not the tools that have failed, but the implementation of them. Numerous factors contribute to these failures, including lack of commitment, lack of education, and misapplication of the tools. My belief is that part of the reason for these failures is a lack of understanding of all of the tools in the toolkit, leading to misuse or inappropriate application of the tools. This chapter will provide a brief overview of some of the more well-known quality tools.

Before we get into the various quality tools and quality systems, we need to spend a few moments talking about quality in general. The question that always arises is, how do you define quality? There are several dimensions to quality, but probably the most important is the idea that, ultimately, quality can be defined only by the customer. That's one of the main principles of the Six Sigma quality system, but it should be adopted by every quality program and be pervasive throughout every organization.

From the customers' point of view, quality is an expectation more than a precise definition. Customers expect the product, or service, to perform in a certain way. They expect certain attributes, a certain level of reliability, a certain result. Your internal definition of quality must tie into and support customers' expectations. If your customer provides you with precise product specifications or specific expectations of a service, you have to provide a product that meets those specifications or a service that delivers the results. Your quality system must be able to support these customer requirements.

Your quality system will use many tools, you will develop a quality policy, and you will develop procedures to deliver the product to the specifications and expectations of the customer. Your internal definition of quality will include the technical specifications and tolerances that are required to meet your customers' requirements. Your quality policy will include the methods that will be employed to ensure that the technical specifications are met, as well as the concepts that are the basis for the policy. The quality policy is the link between the external expectations of the customer and the internal system that is used to meet those expectations. The important thing to remember is that although you define your quality standards and your expectations for quality, they are used to support the delivery of quality products and services as seen from the customer's point of view.

These ideas form the basis of a quality system, and you'll see them repeated throughout the various tools and methods discussed in this chapter. First we'll look at some of the various quality tools, methods, or systems, then we'll talk more about what makes up a quality organization.

7.1 Total Quality Management (TQM)

If you told someone you were going to initiate Total Quality Management (TQM) in your organization, you would probably be laughed at, pitied, or shunned. TQM suffers from a serious image problem. TQM is seen as a fad that is well past its time, and a fad that not only was a failure, but seriously impaired many companies. That's too bad, because TQM is a valid and useful system. The problem has been in the implementation of TQM.

Total Quality Management is a philosophy for managing the quality of an organization, using a wide-ranging and systematic approach. TQM focuses on customers' requirements. It is a strategic approach to managing the organization, using a variety of techniques and process improvement methodologies to embed quality throughout the organization. TQM uses the teaching of a number of quality "gurus." One of the main focuses of

TQM is continuous improvement of processes in the organization. TQM also stresses the use of standardized and documented policies and procedures in all areas of the organization.

Total Quality Management uses many of the techniques discussed throughout this book, especially this chapter. TQM is not made up of a specific set of techniques or teachings. Rather, it is an attempt to develop a quality system that pervades the organization in order to deliver to the customer the product that meets their specifications and needs. TQM evolved from the desire to bring quality to the entire organization. Instead of individual areas or departments making improvements that mostly benefit only them, TQM attempts to integrate quality through the whole system.

The use of standardized procedures and documentation was mentioned. This is one area where TQM may have left a bad taste in many people's mouths. Perhaps there was a little too much emphasis on standardization and documentation. Meticulous documentation of every procedure throughout the plant could become irritating to some people. But documentation is not only important, it's vital.

As with Lean Manufacturing, a quality system stresses the importance of standard work. Each operator should perform the procedure the same way. Although there might not be just one best way to do everything, it's important to find or develop the "best" way to perform every operation in the product delivery process and then to institutionalize it. This standardization provides many benefits. When everyone performs the work the same way, there is a greater level of control of the process and a greater likelihood that the output will be consistent. The measurements that are performed to monitor the process and the output are more valid when the work is standardized. If you read the chapter on Six Sigma (Chapter 5), you know that the reduction of variability is a primary goal. If there is variability in the way that the work is performed, it is very likely that there is variability in the output. And if the work is being performed differently, there is a question of the validity of the measurements used to monitor the process. Are you really measuring the same things, or are you measuring different things? You can't be sure if the process is being performed differently. These things have a profound effect on product quality, both by your internal definition and the customer's definition.

Documentation is important for several reasons. Documentation gets a bad rap when it's being put together, but it is loved once it is completed and being used to train new employees and run the organization. It's not always easy to document all your processes. It's time consuming, for one. It also requires an extreme attention to detail and a willingness to ask a lot of questions and spend the time it takes to clarify the details of the process. To properly document a process, you need a certain set of skills that may be different from the skills that are needed to perform or manage the process.

Document control is another documentation issue that is important but often difficult to manage. Conditions change and processes are improved, and the documentation must be updated to reflect those changes and improvements. A document control system must be put in place to keep up with these changes. The new process must be documented and all the copies of the documentation must be updated. When there are many or constant changes and many processes to keep track of, the challenges can be daunting. However, as is usual, the rewards are well worth the effort if the system operates as it is supposed to.

Another element of TQM is employee commitment or involvement. This is another area that may have contributed to some of the failures of TQM implementations. As you know, you can't force employees to embrace a new management system or a new philosophy. The commitment has to flow from the top down and must be evident in the everyday actions of management. If you just announce a program and expect everyone to follow it, you are doomed to failure. Sure, you can force some changes on people, and they might pretend to go along for a while, but over the long term, if everyone is not committed to the program, it will fail. The failure could be slow and gradual, with the program just sort of fading away. Or it could be quite spectacular, as you lose customers, people get fired, and general disaster takes over.

It doesn't have to be that way. It doesn't matter if it's TQM, Six Sigma, Lean, ERP (enterprise resource planning), or any of the other tools in your toolkit, you can earn your employees' commitment, confidence, and support. You can't just expect these things from your employees; you have to earn them. Just because you're paying someone doesn't mean they're committed to the same ideals as you. And just because you pay them, it doesn't mean you can control them. Of course, you can generally make them miserable by forcing them to do certain things and work them plenty of hours, but that doesn't make for a quality organization. The most productive and efficient employees are those who enjoy their work and their working environment. We've talked a lot about customer satisfaction, but employee satisfaction is at least equally important.

So how do you get your employees to become committed to their work, to a quality system, and to the company? How do you earn their confidence and gain their support? The first answer is simple: you have to be committed. All levels of management, from the very top of the organization, must be committed to the program. They must be committed for the long term, they must be committed when things aren't working as planned, and they must be committed when people make mistakes and there are setbacks. They must be willing to commit resources, time, and support. No program is going to be wonderful from day one. There will be struggles and obstacles; mistakes will be made. But if management believes in the program and is committed to it, this can all be overcome. When the employees see this commitment over a period of time, they will start to see that this is something you really want to achieve. When you support your employees by providing the resources they need, backing them up when they're struggling, and allowing them to make mistakes, they will support you in return. When they see you putting in the extra time and effort to make the program work, rather than sitting back and expecting them to make it work, they will put in the extra time and effort too.

You will gain their confidence when you stand up for them, when you make firm decisions, and when you show confidence in them. You can't make decisions that make everyone happy, and you don't want to. But you have to make decisions, often tough ones, and you have to stand behind them. Nothing annoys employees more than a boss who can't make a decision in a reasonable amount of time. You might need more information or time to think about it, but you can't keep putting off decisions if you want to accomplish anything and gain the confidence of your staff. And if you make a mistake, tell them you've made a mistake and you're changing your mind. They'll accept that, and respect it, better than if you try to cover up or run around the mistake.

You will gain the support of your employees when you support them. Listen to their ideas and accept those ideas that are good, or at least reasonable. Acknowledge their ideas, contributions, and accomplishments. Reward them for their accomplishments and let them learn from their mistakes. If you punish them for mistakes, they'll stop trying anything new and you'll see little progress. Help them learn from mistakes, and don't consider mistakes as failures. Mistakes are valuable learning experiences and growth experiences. Help your employees prevent the same mistakes from happing again, and move on; don't dwell on them.

This element of employee involvement is also reflected in Lean Manufacturing. You can see that many of the ideas and tools discussed in this book share elements and overlap in many areas. There is a reason for that. These tools and the elements of them are very useful and valuable. In the evolution of management systems and improvement methodologies, the useful aspects are kept and the parts that don't work so well are dropped. Some of the tools are divergent from the others, meaning they address different areas or take a different approach to a problem, but they share a common root in the evolution of business systems and philosophies.

TQM attempts to address the organizational structure and the effects of the structure on the organization. Every organization has a different culture and set of values. Some organizational structures work better in different cultures and for different types of organizations. Innovative companies often don't perform well with top-down, hierarchical organizational structures. The structure of the organization should be evaluated in a strategic context, not just as a way to place people on an organizational chart. TQM also attempts to address the integration throughout the organization. Silos and independent department structures often hinder the flow of information and stifle progress. Information must be shared and must be available to whoever needs it whenever they need it. Integration and communication are important for innovation, progress, and improvement.

Finally, TQM advocates a structured and systematic formula for improvement. There must be a clear plan and methodology for quality and process improvement activities. A shotgun approach may result in isolated improvements in different areas, but a structured and strategic approach will result in companywide, planned improvements that provide greater benefits to the organization as a whole.

Total Quality Management is a valid approach to quality and the adoption of a quality system. The basis for it and the ideas behind it are perfectly acceptable. The TQM approach shares many elements of other programs and, if implemented properly, can provide the benefits and improvements you need. It is a good starting point for some of the other tools in the toolkit.

7.2 Statistical Process Control

I include statistical process control (SPC) because, although SPC is widely used, it is often absent in the small manufacturing shop. If you're manufacturing something and are not using SPC techniques, start now! SPC involves a family of statistical techniques, including charts, graphs, diagrams, and worksheets, that help you measure, control, and analyze your operations.

If you read the chapter on Six Sigma (Chapter 5), you will remember that one of the objects of Six Sigma is to reduce the variation in your processes. SPC techniques are used to measure, control, analyze, and monitor the variations in your processes. You can't reduce the variation if you don't identify it. You need to know how much variation is in the process, if the process is capable, if the process is in control, and the type of variation, and you need to monitor the process to make sure it stays in control (once you know it is in control).

To perform statistical analysis, you need to measure the output of your processes. If you are manufacturing something, you should be measuring the output. This means that whatever you're making has some specifications, and you should be doing some sort of measuring or monitoring to determine if you are meeting those specifications. In Chapter 5 we used an example of a process of drilling holes in boards. The specifications state that the diameter of the holes should be a certain size (3.22 mm), with a tolerance of ±.04 mm. At some point you should be measuring the completed boards to determine whether they are acceptable. You will need to make some initial measurements so that you can determine your starting point; then you will take ongoing measurements to determine how you're doing on a continuous basis.

Before going any further, we need to talk a little more about variation. In Chapter 5 there is a fairly thorough discussion of data distribution, particularly the normal distribution. I don't want to repeat that here, so you can go back and review that if you need to. If you haven't read Chapter 5 yet, you can just read through the section on the normal distribution (Section 5.2) before you go on from here. What I want to talk about here are the causes of variation.

When discussing the normal distribution, we talked about sets of data, or the results of measurements we've taken from a particular process. In the drilling example we looked at a month's worth of data. We measured the output of the drilling process by measuring the diameter of every hole drilled on every board during a month. That left us with 16,000 data points that we could analyze (because we drilled a total of 16,000 holes).

Statistical process control recognizes that there are two types of variation in every process. Special, or assignable, causes can be traced to, or assigned to, a specific thing. In the drilling example, assignable causes may be an improperly calibrated machine or a worn drill bit. The variation in the diameter of the holes can be traced back to a specific, or assignable, cause. The improper calibration might cause the drill bits to wobble or to drill too fast or too slow, affecting the size of the hole. Worn drill bits won't cut cleanly or properly, which could result in different size holes.

The other type of variation comes from chance causes. Chance causes are reasons for variation that are just inherent to the system and cannot be controlled. In our drilling example, no matter how precisely the machine is calibrated and adjusted and no matter how often you replace the drill bits, there will always be some small amount of variation in the diameter of the holes that are drilled.

One of the objects of SPC, and the quality system in general, is to eliminate the assignable causes and minimize the chance causes of variation. One thing you need to do is make sure that what you think are chance causes are not actually assignable causes. Perhaps you've calibrated and adjusted the drilling machine and replaced the drill bits and all the holes are within .04 mm of the target 3.22 mm. You might think that this .04 mm variation is a chance cause. But what if you discover this variation is due to drill bits that have some variation in them because you didn't specify to your vendor the tolerance you expected. This variation is actually an assignable cause that you can reduce by working with your vendor on the specs for the cutting tool.

If no assignable causes are present, only chance causes, the process is said to be stable. In a stable process, any variation will occur in a consistent and predictable pattern. In a stable process, the data will tend to be normally distributed. When assignable causes are introduced, the graphed pattern will look a little different. That's because the results are no longer predictable and stable. It is important to find and eliminate the assignable causes of variation. Besides reducing the variation so you can meet the product specifications, you need to know how a process performs when it is stable so that you know what results you should expect. When, or if, you start to see results that don't conform to those expectations, you will know to start looking for those assignable causes.

The causes of variation tend to be classified as originating from five particular areas:

- Materials
- Machines
- Methods
- Environment
- People

Variation can originate from any one of these areas, or from any combination of them. For materials, there might be natural variation such as the quantity of iron ore in a ton of rock or the size and flavor in fruits and vegetables. For purchased parts, the vendor's processes and quality systems will affect the amount of variability of the materials you receive. Under machines, the design of the machine, its age and maintenance history, and how much it is run will affect the variability of the output. Variation resulting from methods means variation that is introduced or inherent in the process of producing the product or providing the service. One reason for developing standard work procedures is to reduce the amount of variation originating from methods. The environment might affect the variation in a number of ways. Temperature, humidity, atmospheric pressure, light, dust, or pollution, can have various effects depending on the product and process. When providing a service, your physical environment, such as the size of the space that's used or the layout, can have an effect. And then there are people. No matter how much you standardize a process, there will always be a people component. People introduce variability because they do things differently, they are in different moods when they come to work, they are different sizes and shapes, they have different work ethics, and so forth. All of these can affect what people produce and can result in variation.

If you understand these different origins of variation, you can more easily uncover them and then reduce or eliminate them. And remember to find the assignable causes of variation and reduce or eliminate these causes. It might be possible to reduce chance causes too, but there will always be some amount of variation that's just inherent to any process. One example of how chance causes are reduced is in the grading of food products. By using only one particular grade of product, instead of anything you can get your hands on, you automatically reduce some variation.

7.2.1 Control Charts

I mentioned above that when no assignable are causes present, a process performs in a consistent manner and we know what to expect from the process. The question then becomes, how do you know when or if the process isn't performing as expected? Of course, if the output is dramatically different than normal (for example, no holes at all after moving through the drilling department) it will be noticed right away. But variation is not always so dramatic. The shifts or differences in the output are often very subtle. The holes might slowly increase in size over time, and that increase might not be noticeable unless the results aren't being carefully monitored. Changes in output might fluctuate from one direction to the other, which makes detection more difficult. Fortunately, there are tools available to help you monitor your processes, and they're pretty easy to use.

One of these tools, one that is commonly used, is the control chart. Actually, there are several varieties of control chart, but we won't get into all the different permutations (maybe more than one, though). Control charts are pretty simple, really. Control charts are used to monitor a process by recording information and plotting the results. Figure 7.1 provides an example of a control chart; we'll discuss the different parts of the chart and how to develop and use it.

Let's start at the top of the chart. You may want to include more information than what is shown here. Record whatever information is useful to you. You might want to include the item that is being processed and its part number. If there is more than one processing step in the drilling department, you will need a separate, labeled chart for each step. You will also need a separate, labeled chart for each machine. If you process different items at the station, you might want to include any characteristics of the measurements that are unique to each item, such as unit of measure. Include any information that the person who will be recording on the chart needs to know to correctly and completely fill it out. Just don't get out of control and include too much information or anything that's unnecessary.

Drilling department						Average and range chart										
Date	Monday, October 4															
Time	7:15	7:30	7:45	8:00	8:15	8:30	8:45	9:00	9:15	9:30	9:45	10:00	10:15	10:30	10:45
Sample 1	3.19	3.27	3.27	3.21	3.26	3.19										
Sample 2	3.25	3.19	3.25	3.21	3.23	3.25										
Sample 3	3.23	3.25	3.22	3.23	3.24	3.23										
Sample 4	3.20	3.23	3.23	3.23	3.21	3.20										
Sample 5	3.26	3.24	3.20	3.20	3.25	3.26										
Sum	16.13	16.18	16.17	16.08	16.19	16.13										
Average	3.23	3.24	3.23	3.22	3.24	3.23										
Range	0.07	0.08	0.07	3.03	0.05	0.07										
Notes	1st run															

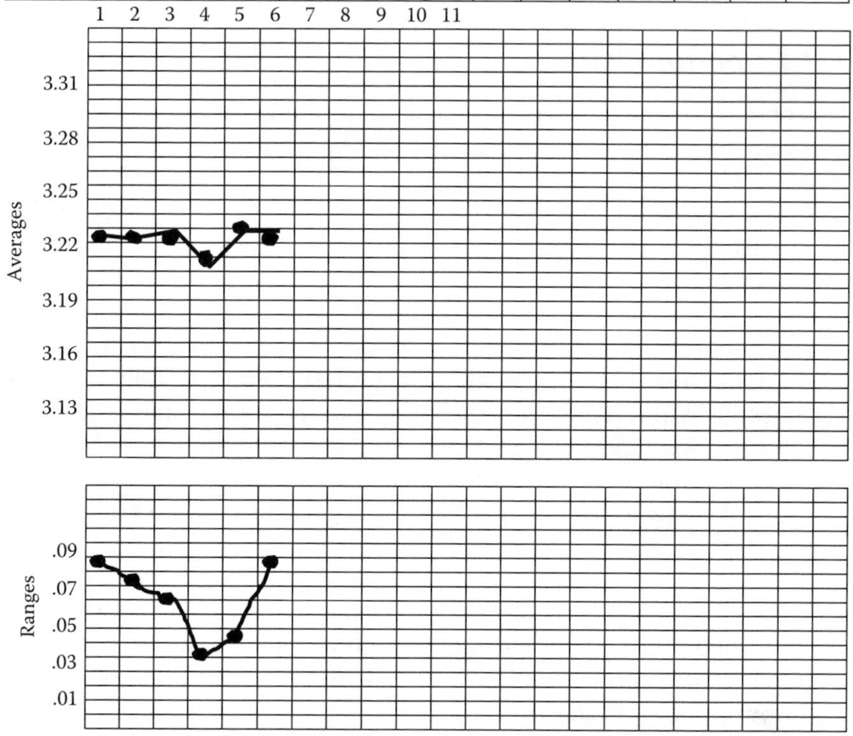

Figure 7.1 Control chart.

Next, you show the actual data that is collected. Fill in the date and the times that the measurements are taken. Throughout the day you will sample the output of whichever process you're evaluating. You could measure every single item that you process, but it's much more likely that you will be sampling. In this example we're taking samples every 15 minutes. Choosing the number of samples to measure for your process is beyond our scope here, but for our example we're selecting five items to sample. So every 15 minutes we will measure five of the items being processed through this step. The measurements are recorded on the sheet as shown in the example. Total the measurements for all the samples taken at a given time and enter that number in the sum row. Then calculate the average (the sum divided by the number of samples, which is five in our example) and enter that number. Next, enter the range. The range is the difference between the highest and lowest sample measurement. In our example, for 8:30, the highest measurement is 3.26 and the lowest is 3.19. Therefore, the range is .07 (3.26 − 3.19 = .07).

Next you're going to plot the data on the graph area of the chart. When you're up and running with SPC, you'll have the control charts set up already, with everything except the data you collect. The scale of the graph will be set already and the numbers on the left that show the lines for the data points will be determined already. When you first set up the charts for a process, you'll have to do some testing and sampling to determine what those numbers are, but in our example, they're already set up. You can see that the numbers run several hundredths of millimeters above and below the specification for this process (3.22 mm diameter), and yours will show a similar range of values. Because this example is an average and range chart, we will plot the average from each set of samples.

Our first set of samples is from 7:15, and the average measurement from the five items we sampled is 3.23 mm. So you just plot that point (3.23 mm) in the column for the first sample set. After each set of samples is measured, you simply plot the average on the graph. Then play connect-the-dots to get a better view of the plot points. Now do the same thing for the range; plot the range for each set of measurements and connect the dots.

When you've done all that you will have a chart that looks like the one in Figure 7.1. It looks interesting, but it doesn't really tell you anything, does it? That's because there's something missing. The plotted measurements don't relate to anything. You must compare the collected data to something, to provide a perspective on how you're doing. You need to see how the data compares to the control limits.

A process is considered in control if it is performing consistently within defined limits. These limits are calculated from a set of measurements taken from the process when you know it is stable. The upper limit is known as the upper control limit (UCL) and the lower limit is the lower control limit (LCL). Pretty original, huh? The UCL and LCL define the limits of a process that is in control. Measures of completed items that are greater than the UCL or less than the LCL indicate the process either is not in control or may be getting out of control. On an ongoing basis, you will measure items that come off the line and determine if they fall within the upper and lower limits. The process is in control if it continues to fall within these limits. As you monitor the process you will be able to see if you start to fall outside the limits. If this happens, you know you need to make an adjustment to the process to get it back in control. Maybe it's time to recalibrate, change the cutting tool, or take some other action.

Figure 7.2 shows what the chart looks like once we add the upper and lower control limits to both the average and range graphs. Now we have enough information on our chart to tell us something about how we're doing. As you can see from this example, the measurements we've taken so far fall within our control limits, so the process is in control. That's good. That means we can continue processing normally; no special action is required.

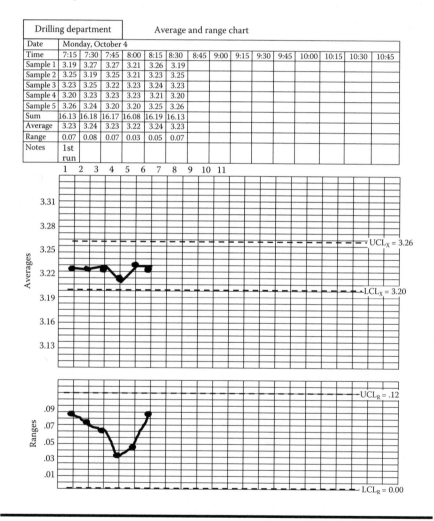

Figure 7.2 Control chart with control limits.

The control chart helps you monitor the process. You can tell if the process is in control, meaning that it is performing within the control limits you have calculated. But just because the process is in control and operating within the limits, doesn't mean everything is working the way it should be. Besides being in control, the process must be capable.

Your process can be in control but not capable. So you need to determine the capability of your process; this includes the particular machine or operation as well as the process as a whole. Let's talk about machine capability first. You need to know whether the machine or operation has the ability to produce the item to the specifications. If the specification is for holes with a diameter of 3.22 ± .04 mm, can the machine actually drill holes of this size?

A machine has a range within which it is normally able to operate. You need to determine what that range is and whether that range is within your specification range. The first step in calculating the capability of the machine is to find and eliminate all assignable causes so that the machine operation is stable. The next step is to measure a number of items processed through the machine or operation and calculate the average, range, and upper and lower control limits. This is just what we discussed above when developing the control charts. The difference here is that you examine parts or items that are produced consecutively, rather than a small number of samples taken at regular intervals. You examine all the items processed for a given period of time, or a certain predetermined number of items. The number of items isn't vital, but there must be enough of them to give you an accurate picture of how the machine is working.

Once you have collected this data, construct an average and range graph, as described above. Include the upper and lower control limits for both the average and the range, just as above. To analyze the capability, you must perform an extra calculation: the upper and lower limits for individuals. This calculation is similar to the upper and lower control limits.* Now compare these individual limits with the given specifications or specification limits for the part and operation you are measuring. If both the upper and lower individual limits are within the specification limits, the machine is considered capable. This means that if there are no assignable causes of variation present — only chance causes — the machine has the ability to produce parts that meet the specifications.

---

* The formulas for the upper and lower individual limits (UL_x and LL_x) are

$$UL_x = \bar{X} + \frac{3}{d_2}\bar{R} \qquad LL_x = \bar{X} + \frac{3}{d_2}\bar{R}$$

Drilling department			Average and range chart											
Date	Monday, October 4													
Sample 1	3.25	3.26	3.28	3.21	3.22	3.22	3.25	3.24	3.24	3.25	3.22	3.25	3.26	3.24	3.23
Sample 2	3.28	3.21	3.29	3.24	3.25	3.28	3.21	3.22	3.26	3.21	3.25	3.26	3.22	3.26	3.25
Sample 3	3.28	3.22	3.25	3.23	3.24	3.26	3.27	3.22	3.21	3.26	3.28	3.26	3.25	3.26	3.29
Sample 4	3.28	3.22	3.24	3.24	3.28	3.25	3.22	3.23	3.25	3.23	3.26	3.21	3.21	3.27	3.24
Sample 5	3.27	3.23	3.28	3.22	3.21	3.26	3.26	3.26	3.25	3.23	3.24	3.25	3.28	3.27	3.24
Sum	16.36	16.14	16.34	16.14	16.2	16.27	16.21	16.17	16.21	16.18	16.25	16.23	16.22	16.3	16.25
Average	3.27	3.23	3.27	3.23	3.24	3.25	3.24	3.23	3.24	3.24	3.25	3.25	3.24	3.26	3.25
Range	0.03	0.05	0.05	0.03	0.07	0.06	0.06	0.04	0.05	0.05	0.06	0.05	0.07	0.03	0.06
Notes															

Date	Monday, October 4													
Sample 1	3.21	3.28	3.28	3.22	3.28	3.21									
Sample 2	3.22	3.28	3.25	3.27	3.21	3.22									
Sample 3	3.29	3.23	3.26	3.24	3.22	3.29									
Sample 4	3.24	3.26	3.23	3.21	3.23	3.24									
Sample 5	3.22	3.28	3.23	3.27	3.24	3.22									
Sum	16.18	16.33	16.25	16.21	16.18	16.18									
Average	3.24	3.27	3.25	3.24	3.24	3.24									
Range	0.08	0.05	0.05	0.06	0.07	0.08									
Notes															

Figure 7.3   Data for capability analysis.

This difference between individual limits and control limits warrants a quick explanation. The control limits you calculate for use with the control charts that you will be using on an ongoing basis actually take into account several different sources of variation. For a regular, ongoing operation, a number of factors will affect the variation in output. Different machine operators will be performing the work, different batches of materials will be used, maintenance activities will occur between samples collected, and other factors will affect the process. In the capability analysis, these factors are removed, for the most part, because the data that is collected comes from a very small, discrete time period. Presumably, the same operator will run the parts, the same batch of materials will be used, no maintenance activity will occur, and all other factors will be the same for all the parts for which the data is collected. Because the capability analysis concentrates on only one attribute — the individual machine being studied — the calculations are slightly different. A more thorough explanation of this difference between individual and control limits can be found in *SPC Simplified: Practical Steps to Quality.**

Let's continue with our drilling example. Figure 7.3 shows the results of a set of 100 samples collected from one machine for the capability analysis.

---

* Robert T. Amsden, Howard E. Butler, and Davida M. Amsden, Quality Resources, 1998.

The average = 3.25 mm
The upper individual limit = 3.31 mm
The lower individual limit = 3.19 mm
The specifications are 3.22 mm ± .04 mm, or
Upper specification limit = 3.26 mm
Lower specification limit = 3.18 mm

The graph in Figure 7.4 demonstrates that the drilling machine is not capable. The lower individual limit is within the lower specification limit, but the upper individual limit is above the upper specification limit, and the average is above the specification target.

The drilling process is in control, but it is not capable. This is valuable information; maybe not what you want to see, but valuable nonetheless. Generally, you will use customer requirements to determine the capability of the process, because the customer determines the specifications. In the example we've been using, the specifications do not come directly from the customer but are derived from customer requirements. The products in the example are various stereo equipment components. The customer has an expectation of how the product should perform, and the detailed component specifications come from the design and engineering team's efforts to meet this expectation. The capability analysis has told us that we are not going to be able to meet the customer's requirements a fair amount of the time because we cannot meet the design specifications.

If the process is not capable, you need to do something to fix it. Perhaps the machine has never been calibrated correctly. Whatever the problem, you know the process doesn't work as you want it to and you know you need to fix it. Of course, you may know this intuitively, but using SPC gives you hard data that makes your analysis and problem solving more effective. Instead of guessing what the problem might be and possibly "fixing" things that don't need to be fixed, you can begin to focus on the source, or root cause, of the problem. You have eliminated assignable causes of error. You know that because the process is in control. You have discovered that the process isn't capable. Now you can determine why it isn't capable. You need to find out why the drilling machine can't drill holes to your specifications. You need to find the reason and change the system so that it can meet the specifications. You can fiddle with it, make adjustments, and blame people all you want, but that won't accomplish anything. Until you make a fundamental change to the system, to the machine, it will continue to operate as it does. So what are you waiting for?

We must discuss a couple more things about capability before moving on. The first is machine versus process capability. We've just discussed machine capability. We were concerned with one machine, or one oper-

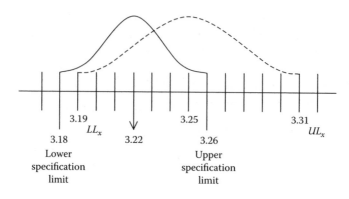

3.19
LL_x
3.18
3.22
3.25
3.26
3.31
UL_x

Lower
specification
limit

Upper
specification
limit

Figure 7.4 Individual and specification limits.

ation, and whether or not it could produce items that meet specifications. That's very important. Just as important is the capability of the entire process of producing items, either parts or finished goods. A process is a series of operations. Each separate operation has its own range of variation. When you look at all the operations, the individual variations can compound. This complicates things. Also, if you remember, when we analyzed the machine capability we minimized or reduced the variation that can be found with the materials, the machine operator, and the environment. To determine the capability of the process we need to allow these variations when performing the analysis. You can do this by collecting data over a longer period of time. Collect data that includes different batches of materials, parts produced by different operators, and when environmental conditions are different. It doesn't do much good to have capable machines if the process isn't capable.

As you learn more about capability analysis you will hear about the capability index. Quality folks have their own secret language, so instead of "capability index," you'll probably hear people say, "C_{pk}" (pronounced "see-pee-kay"). Quality people are precise (which is a good thing), and C_{pk} is a specific capability index. Another one is C_p, but the C_{pk} is probably the more common measurement used. Both these indexes measure the actual performance (the capability) against the specifications. A C_{pk} above 1.0 is considered capable, whereas a C_{pk} below 1.0 is not capable.

Something you might have noticed with the capability example is that the range of the actual results of the process is wider than the range of the specifications, but also that the average of the results was greater than the specification target. The specification target is 3.22 mm, whereas the average of the actual results is 3.25 mm. The C_{pk} incorporates this difference between the specification target and the average from the actual results. The C_p does not take this into account. This is one reason the C_{pk}

is more popular, because it is important to know whether or not the specifications and actual results are centered at the same point.

7.2.2 Principles of Statistical Process Control

Understanding the following six basic principles of SPC will help you understand its usefulness and necessity:*

1. No two things are exactly alike.
2. Variation in a product or process can be measured.
3. Things vary according to a definite pattern.
4. Whenever many things of the same kind are measured, a large percentage of the measurements will tend to cluster around the middle.
5. It is possible to determine the shape of the distribution curve for parts produced by any process.
6. Variations due to assignable causes tend to distort the normal distribution curve.

We've discussed most of these principles, so they should be relatively clear. Remember them as you examine your processes and you will have an easier time finding the sources of your problems. You will also save a lot of time and energy by not looking for problems in processes that are simply experiencing normal variation.

A number of techniques fall under the umbrella of SPC. Although we have reviewed some of them already, the following list provides a review, with short definitions for each:

- *Histogram* — A type of graphical representation of data. Also known as a frequency distribution, it presents the results of measurements at specific points in time. From our example, the diameters of the holes in the drilling operation can be collected and shown in a histogram.
- *Control charts* — A combination of tables and graphs used to record the results of measurements over time. There are two general types of control charts: variables charts and attribute charts. Depending on the type of chart, they include data on when the measurements were taken, the results of the measurements, summary calculations, and graphs of the results. The graphs include the control limits.

---

* *SPC Simplified: Practical Steps to Quality*, Robert T. Amsden, Howard E. Butler, and Davida M. Amsden, Quality Resources, 1998.

▪ *Checksheets* — Various types of forms used to collect data about nonvariable attributes. Nonvariable attributes are results that you want to analyze but cannot measure (for example, if you want to analyze types of defects to find out which type is most prevalent). From our example, let's say you want to analyze the types and number of defects at the drilling operation. Some of the defects you might want to analyze are diameters out of spec, no hole drilled, and holes in the wrong location.

▪ *Scatter diagrams* — Also known as scatter plots, they are used to determine if there is a relationship between two variables (for example, time of day and number of defects). A correlation between the variables does not mean there is a cause-and-effect relationship, but it may mean that you can predict the result of one of the variables based on the result of the other. Graphing the points (results) may allow you to see a relationship.

▪ *Regression analysis* — The mathematical relationship between two variables; the mathematical representation of a scatter diagram where there is an actual cause-and-effect relationship between the variables.

▪ *Capability analysis* — Measures and studies the variation of a process. It is used to determine whether a process is capable; that is, can it produce items within the defined specifications? The capability of a machine or operation or of a process can be determined. A capability index (C_p or C_{pk}) can be calculated.

▪ *Analysis of variance (ANOVA)* — Used to determine if there is statistical significance in a set of data. Statistical significance means you can use the results to predict the results of future tests. ANOVA can help you determine if a sample is representative of the population.

▪ *Design of experiments (DOE)* — Tests the relationships among multiple variables at the same time. It is used to reduce error in experiments or tests, therefore increasing the useful information obtained from fewer experiments or tests.

7.3 ISO 9000

Many companies state that they are ISO 9001 certified, ISO 9002 certified, ISO 14000, or some other ISO or QS something. Some of your suppliers and customers may be in this group. If you've wondered what these certifications are and what they mean, you are in luck. This section will give you a brief overview of the ISO certifications.

As with the other tools in the toolkit, you need to examine why you might want to become certified to one of the ISO standards. Some organizations go through the process for the wrong reasons, diluting its effectiveness and possibly wasting money. Some companies may find that their customers require them to become certified if they want to remain their suppliers. Even if you are told you must become certified to remain, or become, a supplier, you should examine the reasons behind the requirement. What I'm saying is, you shouldn't just try to do the minimum required to get certified. You should look at the benefits of operating your company in a manner that complies with the certification requirements.

ISO 9000 is a family of quality standards. They are an internationally accepted set of standards, or guidelines, that demonstrate a level of control over an organization's quality management system. ISO 9000, the quality standards that we're interested in, was adopted by the International Organization for Standardization in 1987. In 1994 the standards were revised, and now more than a hundred countries have adopted the ISO 9000 set of standards. The standards were updated (again) in the year 2000 and are now referred to as ISO 9000:2000. If you are familiar with the previous version (1994), you will notice that this version includes substantial changes. ISO certification provides your customers a certain level of confidence in the quality and consistency of your products. That is the goal of the standards: to help organizations consistently deliver quality products. ISO certification doesn't certify that your products or services are "quality products," nor does it guarantee quality. It simply means that you have met the requirements of certification. The implication is that if you have met the requirements you will produce quality products, but that's not always the case. See the statements above about why you might explore certification.

To become ISO certified requires an audit by an outside registrar. The registrar is an accredited organization that will audit your company for compliance with the ISO requirements. The audit will consist of two parts. One part consists of the registrar comparing your company's documented quality system with the published standards. The other part consists of the auditors performing an on-site inspection and review to ensure you are actually performing to your documented procedures.

The ISO 9000 standards were developed with eight quality management principles in mind. These principles, which are similar to the Baldrige Award criteria (see Section 7.4), do not make up the standards, but were used as a guide to their development. They are

1. *Customer focus* — Understand and meet customer expectations.
2. *Leadership* — Top management sets the tone.
3. *Involvement of people* — Involve the people, or employees.
4. *Process approach* — Adopt a process approach to management.

5. *Systems approach to management* — Adopt a systems approach to management.
6. *Continual improvement* — Continuously improve your processes and products.
7. *Fact-based decision making* — Adopt fact-based decision-making processes.
8. *Supplier relationships* — Adopt mutually beneficial relationships with suppliers.

The ISO 9000:2000 family of standards consists of three documents: ISO 9000:2000, ISO 9001:2000, and ISO 9004:2000. One document contains the actual standards, and the other two provide the concepts behind quality management systems, the philosophy behind the standards, vocabulary definitions, and tips on making improvements to management systems.

The ISO 9000:2000 document is titled "Quality Management Systems — Fundamentals and Vocabulary." This document provides a basic understanding of the concepts behind quality management systems and the fundamental aspects of such a system. It also contains a number of definitions relating to quality, management, and other terms relating to management systems.

The ISO 9001:2000 document contains the actual standards, or the requirements that must be met in order to become certified. The title of this document is "Quality Management Systems — Requirements." These standards encompass everything from the scope of the quality system to documentation requirements, management responsibility, management of resources, customer focus, production and purchasing processes, design and planning processes, process monitoring and measurements, and more.

The final document, ISO 9004:2000, titled "Quality Management Systems — Guidance for Performance Improvement," provides tips on methods of improving quality management systems. This document provides helpful information on how to use the requirements from the 9001:2000 document to improve the performance of the organization. Remember, the intent behind the ISO 9000 standards is to help organizations improve their performance. It takes a lot of time, effort, money, and resources to implement a quality system that meets the standards of ISO 9000. Why would you do it if it did not improve the performance of your business?

The U.S. automotive industry developed its own quality system standards, QS 9000, from the ISO 9000 standards. The QS 9000 standards incorporate the ISO 9000 standards, with some additions. Another standard you may have heard of is ISO 14001. This standard relates to environmental issues in product development.

7.3.1 ISO 9000:2000 Requirements

As stated, the ISO 9001:2000 document outlines the standards or requirements you must meet to become certified. These requirements are not specific actions that must be taken, but are principles that must be adopted by the organization. You must interpret them so that the specific actions fit your organization; its culture, operations, and the products or services that you produce or provide. The requirements are broken down into separate areas and are listed as numbered paragraphs. I have listed them below, along with brief descriptions:

1. Scope
   1.1 General
   1.2 Application — Some of the requirements can be excluded if they are not part of the organization's business or operations.
2. Normative References — Definitions of terminology used. Because some terms might be used differently by different organizations, terms that are used differently by your organization and ISO must be defined to avoid confusion.
3. Terms and Definitions
4. Quality Management System
   4.1 General Requirements — Identifies the processes that must be addressed by the quality management system. This includes all appropriate internal processes and any external processes that must be controlled by the organization (such as outsourced operations).
   4.2 Documentation Requirements
      4.2.1 Management System — Documents required by the management system.
      4.2.2 Quality Manual — Contents of the quality manual.
      4.2.3 Control of Documents — The controls required for documents.
      4.2.4 Control of Records — Twenty-six documents are addressed in the standards. Six issues regarding the control of records must be addressed:
      A. Identification
      B. Protection
      C. Retention
      D. Storage
      E. Retrieval
      F. Disposition

5. Management Responsibility
   5.1 Management Commitment — Requirements for leadership by top management.
   5.2 Customer Focus — Requirements for top management to focus the organization on customer requirements.
   5.3 Quality Policy — Requirement of a documented and controlled quality policy, defined and supported by top management.
   5.4 Planning — Plans for meeting quality objectives.
   5.5 Responsibility, Authority, and Communication — Definition and documentation of organizational structure and the responsibilities and authority within that structure; includes relationships and methods of communication.
   5.6 Management Review — Requirements for management review activities, including relevancy of review activities, records, and actions taken.
6. Resource Management
   6.1 Provisions of Resources — Requirement for top management to provide the resources necessary to meet the quality system objectives.
   6.2 Competency — Requirements for staff competency and the records to demonstrate that competency.
   6.3 Infrastructure — Determination of the infrastructure needed to meet the quality system objectives, including top management providing the needed infrastructure resources.
   6.4 Work Environment — Requirements for the management of the work environment so that the quality system objectives are met.
7. Product Realization
   7.1 Planning of Product Realization — The plans for how to meet customer requirements.
   7.2 Customer Related Processes — Requirements for identifying customer requirements, reviewing those requirements, and communicating with customers.
   7.3 Design and Development — Defines the process of design and development, including identification of the inputs and outputs of the design and development process. The design and development activities are broken down into separate stages.
      7.3.1 Design and Development Planning — Identifies responsibility and authority for design and development activities.
      7.3.2 Design and Development Inputs — Identification of design inputs and records management requirements.

7.3.3 Design and Development Outputs — Identification of design outputs and records management of outputs.

7.3.4 Design and Development Review — Identification of review activities and records of review findings.

7.3.5 Design and Development Verification — Verification of designs that are developed, including testing and comparison with existing designs.

7.3.6 Design and Development Validation — Validation that design outputs meet input or requirements.

7.3.7 Control of Design and Development Changes — Identification, verification, and validation of design changes, including records of changes.

7.4 Purchasing — Requirements for the management and control of procurement processes.

7.5 Production and Service Provision — Requirements for the actual production of the product or the provision of the service; includes management of the production process, controls, and any necessary testing.

7.6 Control of Monitoring and Measuring Devices — Requirements for calibration of testing and measuring equipment.

8. Measurement, Analysis, and Improvement

8.1 General — Requirements for measuring and evaluating the overall quality system.

8.2 Monitoring and Measurement — Requirements for monitoring and measuring customer satisfaction, internal processes, and products; includes methods for performing internal audits.

8.3 Control of Nonconforming Products — Determines processes for identifying and controlling nonconforming products (that is, defects).

8.4 Analysis of Data — Requirements for analyzing data that is collected throughout the quality management system.

8.5 Improvement — Identification of continuous improvement activities, processes for corrective action, and processes for preventive action. Defines how improvement opportunities are identified and how improvement activities are managed and controlled.

You can see that the ISO certification requirements are extensive. Hopefully, you can also see the benefits of managing your company using the guidelines and requirements of the system. Even if you do not plan to seek certification, you can use these guidelines and requirements to develop an effective quality system that can improve the performance of your organization.

7.4 The Malcolm Baldrige National Quality Award

The Malcolm Baldrige National Quality Award (MBNQA) recognizes business excellence. It is given to a company that meets the award criteria and uses them to improve their processes and satisfy their customers. Many states have their own statewide quality awards that are patterned after the Baldrige criteria. Organizations are generally encouraged to apply for their state's award before moving up to the Baldrige Award. Once you explore the criteria further, you will probably understand why. The criteria are quite comprehensive and difficult to fully adhere to, at least when you first start to use them. Working through the application process at the state level will give you valuable feedback that you can use to make the improvements to help bring you up to the next level.

The award criteria are often used as a model for increasing the overall performance of an organization. The award is based on a TQM approach to operating and improving an organization. You may also notice similarities to the ISO 9000 requirements. The Baldrige criteria are not designed so that organizations can win an award. They are a comprehensive set of criteria designed to be used as a framework for managing an organization, to improve that organization's performance. It is worth noting that companies that win the Baldrige Award tend to outperform non-Baldrige winners, as measured by market value.

As a quick aside, here is a bit of trivia. For a long time I wondered who Malcolm Baldrige was. I knew of the award and had a vague understanding of the overall concept behind it, but I had never heard of Malcolm Baldrige. It may not be important to organizations that are using the award criteria, but it is nice to know a little about the person for whom the award is named. Malcolm Baldrige was President Ronald Reagan's Secretary of Commerce during the time the award and the award criteria were developed. Mr. Baldrige died just before the U.S. Senate enacted the legislation creating the award, and the award was named for him.

The award is given in five categories: manufacturing, service, small business, educational institutions, and healthcare organizations. (A new category for nonprofit and government organizations has been created, and it is expected that such organizations will be eligible to apply for the award beginning in 2006.) There are three separate sets of criteria, for business, education, and healthcare. The differences in the criteria are small and reflect the different environments of the categories. We'll be discussing the business criteria.

Use of the award criteria does not, of course, guarantee success or increased profitability for your business. However, the criteria are useful and powerful as a management model. Some studies have shown that the stock prices of MBNQA winners beat the overall market. Of course, winning the award is a lot different than just using the criteria. The intent is for organizations to use the criteria to improve their performance. Applying for the award means that you are confident in your organization's performance and are willing to have a team of experts evaluate you and provide valuable feedback. Winners of the award prove that they are meeting the criteria and using them to excel in the marketplace.

7.4.1 The Award Criteria

The MBNQA criteria are divided into seven interrelated categories: leadership; strategic planning; customer and market focus; measurement, analysis, and knowledge management; human resources focus; process management; and business results. These seven categories are subdivided into 19 criteria items. The 19 criteria items are further divided into a total of 32 "areas to address," which serve to clarify the criteria items. The areas to address include one or more questions to which the applicant must respond. These questions further clarify the criteria items. Each category has a point value. The first six criteria are referred to as "process" items, and scoring is based on four dimensions: approach, deployment, learning, and integration. Approach refers to how the applicant addresses or meets the criteria item requirements. Deployment refers to how well the applicant applies the approach throughout the organization. Learning implies that the organization has a process improvement focus, beyond just establishing an approach and deploying. Finally, integration implies that all parts of the organization integrate their processes to the benefit of the organization as a whole. Ideally, the requirements of the criteria should be addressed and met (approach), they should be established throughout the organization (deployment), process improvement should be emphasized (learning), and processes and initiatives should be integrated throughout the organization (integration).

The seven categories of the MBNQA are as follows:*

1. Leadership — This category relates to the organization's senior management, all the way to the top. It requires that top management is personally involved in providing leadership to the organization.

    1.1 Organizational leadership — This criteria relates to several areas of senior leadership of the organization. The first is how senior leaders guide the organization: how they deploy organizational values, how they balance value creation for customers and stakeholders, and how the leadership system

---

* From the 2004 Malcolm Baldrige National Quality Award criteria, available from the Web site: www.baldrige.nist.gov.

communications function. This criteria also addresses organizational governance and performance review, including review of the senior leaders' performance.

    1.2 Social responsibility — This refers to the organization's responsibilities to the public. Included are the impacts of the organization's products and services on society, the ethical behavior of the organization, and the support of key communities.

2. Strategic planning — This category relates to the strategic direction of the organization and the key elements of the strategic plan. The effectiveness of translating the strategic plan into organizational performance is part of this category.

    2.1 Strategy development — This category addresses how the organization establishes its strategic objectives. This includes the strategy development process and how key factors are addressed. It also requires that the applicant state the key strategic objectives and the timetable for achieving them. The strategic objectives should address both long- and short-term challenges and opportunities and the needs of all key stakeholders.

    2.2 Strategy deployment — After the strategy is developed, it needs to be deployed throughout the organization. This criterion addresses the conversion of strategic objectives into action plans. Included are performance measures and projections of future performance based on these performance measures.

3. Customer and market focus — In this category, the organization's methods of listening and responding to its customers are examined.

    3.1 Customer and market knowledge — This category addresses how the organization determines the requirements of customers and markets. It includes market segmentation, customer and potential customer analysis, and how the organization continues to learn so that its business needs continue to be met.

    3.2 Customer relationships and satisfaction — This criterion relates to how the organization builds and maintains relations with its customers. It addresses how relationship building affects how customers are acquired, satisfied, and retained. This criterion also requires a statement of how the organization determines and defines customer satisfaction.

4. Measurement, analysis, and knowledge management — The effectiveness of management's use of data to support customer-focused performance is looked at in this category. Also included is the organization's analysis of its own performance.

    4.1 Measurement and analysis of organizational performance — This criterion addresses how the organization develops and uses performance data in all parts and levels of the organization. Performance measurement and analysis are included.

4.2 Information and knowledge management — This category addresses the quality and availability of data that is needed, including how knowledge assets are built and managed.

5. Human resources focus — Here, the workforce is examined for its development and ability to align with the organization's performance objectives. It examines the organization's attempts to build and maintain an environment that encourages employee excellence, participation, and growth.

   5.1 Work systems — This criterion addresses how work and jobs enable the employees and the organization to achieve high performance. It includes the compensation system and career development system. Also addressed are the organization and management of work, workforce diversity as it relates to performance, performance measurement systems, and hiring practices.

   5.2 Employee learning and motivation — This criterion relates to employee education, training, development, motivation, and career development.

   5.3 Employee well-being and satisfaction — This addresses the work environment, including emergency preparedness and business continuity. It includes support of employees, productivity, and employee satisfaction.

6. Process management — This category looks at the management of the organization's processes. Examined are process design, effectiveness of process management, and process improvement systems.

   6.1 Value creation process — This relates to how customer value is identified and created. Customer value refers to those features or attributes of your product or service that the customer finds valuable. In Lean Manufacturing, value is considered those things the customer is willing to pay for, as opposed to your costs, which the customer is not willing to pay for (such as the cost of performing physical inventories).

   6.2 Support processes — These are all the processes that support the value creation process. The criterion addresses such areas as how the organization identifies the support processes, the requirements of those processes, and the design and management of those processes.

7. Business results — Because the award is designed to recognize business success, this category examines performance of the organization in each of the other six criteria areas.

   7.1 Customer-focused results — This addresses customer satisfaction and the results of the value creation process (Categories 3 and 6); it includes measurements and analysis of these areas.

7.2 Product and service results — This addresses the performance of your products and services as perceived by the customer and includes comparisons with your competitors' performance (Categories 2, 3, and 6).

7.3 Financial and market results — This addresses the financial performance and results of the organization; it includes market segmentation data and comparisons with competitors. (This is the real test, isn't it?)

7.4 Human resources results — This addresses the performance of the work systems, employee satisfaction and performance, and learning and employee development (Category 5).

7.5 Organizational effectiveness results — These results refer to the internal performance measures. Measures include efficiency, productivity, performance to plan, supplier performance, and other internal measures (Categories 2 and 4).

7.6 Governance and social responsibility results — This refers to how senior management runs the organization; it includes regulatory compliance, ethical conduct, and fiscal liability (Category 1).

As you can see from these categories, the criteria encompass all aspects of the business. You will also notice that the first category examines the leadership of the organization, all the way to the top (the CEO, general manager, etc., depending on the organization). The criteria are quite comprehensive, but they do not dictate how the organization meets them. Every organization will do so in different ways. The award criteria provide the framework. Each organization has quite a bit of leeway in the methods it uses to meet the requirements of the criteria and to achieve results.

7.4.2 Scoring

The point values for the categories and subcategories are shown in Table 7.1. As you can see, the heaviest weight is on results. That's the true measure of organizational performance — whether it can produce results. You can also see that all of the other categories tie directly to business results. This is an important concept: the relationship between performance and results, and the interrelatedness of all the areas of the organization. If you don't score well in a particular category, not only will the results tied to that category not score well, but it will also negatively affect the scores for other categories. If you don't perform well in human

Table 7.1 MBNQA Point Values

Category			Point Value
1 Leadership			**120**
	1.1	Organizational leadership	70
	1.2	Social responsibility	50
2 Strategic planning			**85**
	2.1	Strategy development	40
	2.2	Strategy deployment	45
3 Customer and market focus			**85**
	3.1	Customer and market knowledge	40
	3.2	Customer relationships and satisfaction	45
4 Measurement, analysis, and knowledge management			**90**
	4.1	Measurement of organizational performance	45
	4.2	Analysis of organizational performance	45
5 Human resources focus			**85**
	5.1	Work systems	35
	5.2	Employee learning and motivation	25
	5.3	Employee well-being and satisfaction	25
6 Process management			**85**
	6.1	Value creation processes	50
	6.2	Support processes	35
7 Business results			**450**
	7.1	Customer-focused results	75
	7.2	Product and service results	75
	7.3	Financial and market results	75
	7.4	Human resources results	75
	7.5	Organizational effectiveness results	75
	7.6	Governance and social responsibility results	75
Total			1,000

resources focus, it's likely that most of the other categories will be negatively affected and overall results will be lower. Think about it. If work systems, employee learning and motivation, and employee well-being and satisfaction don't perform well, how can you expect to see great results anywhere in the organization? The same idea holds true for all categories.

If the support process (the processes that support the core processes of the organization) don't perform well, the rest of the organization's performance is likely to suffer, as will the results. This is one of the things that make the Baldrige criteria such a valuable tool to improve organizational performance. It places heavy emphasis on results, and results can't be achieved without top performance from all the interrelated functions within the organization. As you analyze and improve performance in one area, you will see the weaknesses in other areas.

7.4.3 Application and Review Process

Now that you understand the criteria, how do you go about applying for either your state's version of the award or the national Baldrige Award? First, you might want to evaluate whether you are ready to apply for the award. You should take an honest look at your company. Do you need to bring up your performance level first? You can complete a self-assessment to see how you measure up against the criteria. If you apply, during the selection process the examiners will review your application in great detail, compare your answers to the criteria, and follow up with a site visit if warranted. So be as honest and critical as you can during the self-assessment. The best organizations can reach scores of about 70 percent. Well-run organizations at the beginning stages of development of comprehensive management systems might score at the 20–25 percent level.

If you perform the self assessment and think you're ready to step up and apply for an award, here's what to expect from the application process, the examination review process, the site visit (if you get that far), and the report and feedback stage. The application itself consists of three parts. First is an eligibility form, then an application form. Finally, there is the report, in which you must provide an overview of the organization and your responses to the award criteria. The business overview, known as the organizational profile, consists of two sections (P.1 and P.2) that provide (1) an organizational description and (2) organizational challenges.

In the application you will respond, point by point, to every criterion and the areas to address in all seven categories. You must explain how you are meeting the requirements of the criteria. For example, Criterion 1.1 is organizational leadership. You must explain how top management provides leadership to the organization; how they are involved in creating and monitoring the environment that facilitates excellence throughout the organization. You must document the processes and systems top management uses to lead the organization, monitor performance, and promote their espoused values. How does top management gather information? What is their evaluation and decision-making process? How do they disseminate their values and decisions through the organization to reach all levels? To score highly in this area, you must demonstrate that you have an effective system of organizational leadership.

Once your application is complete and accepted, it will be assigned to a team of examiners. The examiners are volunteers who dedicate a considerable amount of time to reviewing and responding to applications. *Dedication* is a good term to use with the examiners. Besides the time commitment to the review process, they undertake several days of training to prepare them for the review. The examiners are experienced professionals with a desire to give something back to the business community. The examination process begins with an individual review of the application by each team member. Each member makes comments for each criteria item. The comments will either acknowledge that the requirements are being met or indicate that there are gaps between the requirements and the organization's performance. Each examiner will also score each criteria item during the individual review.

You might think that the organization could present an application that shows they are meeting all the requirements. So far the organization is just documenting the processes and systems it has in place to meet the requirements. Well, besides the fact that if the company scores high enough the examiners will visit the site to verify statements in the application, it's not as easy as all that. The examiners have a thorough understanding of the requirements, which are quite detailed when you consider all the areas to address. The applicant might think the requirements are being met, but the examiner might see a gap between the requirements and the organization's stated system to meet the requirement.

After each team member reviews the application and makes comments, the team members come together for a consensus review (if the applicant scores high enough to warrant further review). The examiners will find different things during their individual review, so during the consensus review the individual comments will be consolidated, reworded, pared down, and prioritized. Also, because the individual examiners likely scored the items differently, the team will decide on a score for each criterion.

If the applicant scores high enough in the consensus review, the team will conduct a site visit at the applicant's location. During the site visit, the team will verify that the applicant is meeting the requirements of the criteria, as documented in the application, and will try to determine if the gaps they have identified actually exist or if the application just didn't document clearly that the requirement is being met. Sometimes the team will not be able to verify that a requirement is being met, as the applicant claims, and sometimes they will discover that a requirement is being met although the application wasn't clear on that point.

After the site visit, the team's findings are reviewed by a group of judges. The judges review the comments and report of the team of examiners and make the final determination of the applicant's score and whether they will receive an award.

Whatever the results, probably the most valuable thing the applicant receives is the feedback and report from the judges. The report presents the comments and findings of the examination team in a format the applicant can use to further improve performance. (I say further improve because to get to this stage the organization has to have improved their performance significantly from some point in the past.) The applicant has established a management system, documented that system through the application process, allowed a team of experts to review the system as compared to the award criteria, and received a report on its performance in meeting the criteria requirements. Going through this process and receiving the feedback is extremely valuable. It's even more valuable when the organization takes the feedback and uses it to improve its performance. Remember, the goal is not to win the award, but to use the award criteria to enhance and improve the performance of the organization. Don't forget that the award recognizes business excellence, and excellence includes results.

7.5 The Quality Organization

We've discussed a variety of quality systems and quality tools (TQM, SPC, ISO 9000, and the Baldrige Award criteria), and in Chapter 5 you learned about Six Sigma, but what have you really learned about quality? What is the role of the quality manager or quality professional? What about the quality department? Many organizations don't have a specific quality department, but many still do. Is one right and one wrong? Do you have a clear idea of what quality is and where and how it fits in your organization? I probably haven't given you enough to allow you to make a clear statement regarding quality, but you should be developing a pretty good idea of what a quality system might look like. In the rest of this chapter I will explore more tools and techniques of quality and try to clear up some of the confusion you might have in terms of what quality is and how it fits in your company.

7.5.1 Defining Quality

First let's try to define quality. What is quality, exactly? The *APICS Dictionary* (11th edition) defines quality as, "Conformance to requirements or fitness for use." It goes on to state that, "Quality can be defined through five principal approaches: (1) Transcendent quality is an ideal, a condition of excellence. (2) Product-based quality is based on a product attribute. (3) User-based quality is fitness for use. (4) Manufacturing-based quality is conformance to requirements. (5) Value-based quality is the degree of excellence at an acceptable price. Also, quality has two major components: (1) quality of conformance — quality is defined by the absence of defects, and (2) quality of design — quality is measured by the degree of customer satisfaction with a product's characteristics and features." That's a lengthy description, but is it enough?

According to the American Society for Quality (ASQ), "quality is

- Based on customers' perceptions of a product's design and how well the design matches the original specifications.
- The ability of a product and service to satisfy stated or implied needs.
- Achieved by conforming to established requirements within an organization."*

In Chapter 5, in the discussion of Six Sigma, I stated that the customer ultimately defines quality. Who's right, and can we really come up with a definitive definition of quality? Well, they're all right to some extent, and no, we probably can't come up with an all-encompassing, definitive definition of quality that will satisfy everyone. What you can do, though, is understand quality as it pertains to your organization, and develop a quality system that allows you to perform at your best and meet or exceed your customers' needs.

My definition of quality includes a service component. If you don't provide absolutely exceptional service to your customers (and your suppliers and employees), you can't have true quality. Most definitions of quality refer to the product or service your company provides. Does it meet specifications or customer requirements? Does it meet regulatory, safety, or health standards? Is it "high quality," which usually means sturdy or high performance? To me that's only part of the definition. The service component that I include in my definition of quality relates to the customer experience. Even if you provide a service instead of produce a product, the customer experience component of the service needs to be included.

---

* Quoted from the ASQ Web site: www.asq.org.

Think a little about how you do business and how you treat and interact with your customers, suppliers, and employees. Think about the entire experience of your customers. How do they find you? Are you easy to find? What kind of first impression do you think you make? If your potential customers find you through the Internet, what is their first impression? How does your Web site look and function? Does the look and feel of the site reflect your company? Is the site easy to navigate, is it consistent from page to page, and is there enough information? Do you provide a phone number? Many people still want to speak with a person, even if they find you through the Internet. Have you ever been frustrated with a Web site? That's not a good first impression, and it detracts from the customer experience.

What about the experience of an existing customer? What sort of experience do they have with the ordering process? With some companies, it seems like you practically have to beg someone to take your order. Does your company have options for placing an order? Some people want to speak to someone face to face; some people just want to send a purchase order; and some people want to do everything online. Many companies are moving to Internet-based ordering, but not all of your customers are ready for that. Sure, it might be more cost-effective and productive, and that's perfectly valid, but if your customers aren't ready, how does that affect their experience? You may need to provide them with the tools and training to wean themselves off of the current ordering methods. That would certainly enhance their experience, wouldn't it? If you helped them and worked with them for the benefit of both of you, wouldn't that be a wonderful experience? That's quality.

How about the production and fulfillment processes? Can your customers easily find out the status of their orders? Do you deliver precisely when your customers want delivery? Are the items packaged perfectly for the customers' needs? What happens when the product arrives at a customer's door or when the customer arrives to pick it up? Are the drivers or warehouse people friendly and knowledgeable? Do they look like representatives of your product and your company? What happens if there is an error? How is the error resolution process and experience?

All of this, and much more, is part of the customer experience, which is a component of quality. You can't call yourself a quality organization if anyone who interacts with you has a negative experience. You have to go overboard with the service. You have to make the experience so delightful that people go out of their way to find you and come back to you. When you get to that point, you're on your way to quality.

7.5.2 Quality Strategy

Now that you have some definitions of quality and some ideas of what quality is, you need to think about how you create and deliver that quality. You need a quality strategy. The quality strategy links the strategic plans of the organization with the quality policies and functions.

The strategic plan defines the goals and targets toward which the company strives. This is the direction the company wants to take. The company strategy includes the markets that the company will compete in, the target market share, growth objectives, profit objectives, and the products, services, and other resources that will be needed to fulfill the strategic goals. The quality strategy can be incorporated into the strategic plan or can be subsidiary to it. The quality strategy integrates the quality processes and functions that will be designed and managed to fulfill the strategic plan. Without a clear strategy for integrating, deploying, and delivering quality, you cannot achieve the level of quality that you need to compete in today's global marketplace.

7.5.3 Quality Policies

The quality strategy links the company's strategic plan with the quality processes and functions. The quality policies define and document the principles and ideals of the quality mission of the organization. The strategy is what you want to achieve; the policies define the methods you will use to get there and the principles that drive the policies.

The quality policies provide the guidance needed to develop the processes and procedures that are followed to produce and deliver your products and services. Quality policies must be distributed throughout the organization and shared with all business partners, including employees, suppliers, and customers. The quality policy should include your organization's definition of quality and the methods that will be used to achieve that quality. These policies must be clear, simple, and understandable. They also must be embraced by senior management if they are to be followed and enforced throughout the organization.

The quality policies must be consistent with and linked with the other organizational policies. When developing the policies, you should solicit input from all areas and all levels of the organization. It is not realistic to think that a quality department or quality manager can develop effective quality policies. The organization as a whole develops quality policies, derived from the quality strategy and driven by top management and overall company strategy and policy. The quality policy needs to include appropriate performance measures, or metrics. Metrics will be discussed in more detail in Section 7.6.

7.5.4 Quality Functions

The traditional view of quality includes a dedicated quality department with a dedicated staff of quality professionals. This traditional framework usually assumes and emphasizes the need for a dedicated and specialized team of inspectors, auditors, and quality enforcers. The modern view of quality removes the barriers and assigns the responsibility for quality to everyone in the organization. Every person in the organization impacts quality, so everyone should be responsible and accountable for quality. Of course, the responsibility and accountability are limited to those areas over which each individual has control.

This doesn't mean that there is no room for a dedicated quality department, with a quality manager and specialized support staff. It just means that their roles will change to fit with the modern view. The quality manager might be a high-level executive with responsibility for quality throughout the organization. This manager would be responsible for developing the quality strategy and policy, finding people with any necessary specialized skills that are needed to maintain quality, organizing any training that is required to develop the workforce as quality practitioners, and developing the performance targets and measurements.

Even in the modern quality organization, you need dedicated and specialized quality professionals. Although the responsibility for quality is spread throughout the organization, some jobs should be attended to by trained specialists. Probably the biggest need for specialists comes in the areas known as quality engineering, reliability engineering, and metrology. In the Six Sigma organization, green belts, black belts, and master black belts are the specialists in the statistical methods, project management, and team-building areas.

Quality engineering, reliability engineering, and metrology are responsible for quality as it relates to product design, testing and test equipment, analysis and failure prediction, product reliability, and measurement and calibration of test equipment. In the traditional quality view, these functions would act independently and might lead to confrontations and adversarial relationships. In the modern quality organization, the reporting hierarchy is not as important. The emphasis is to utilize the expertise of these specialists to design, develop, produce, and maintain the products and services of the organization as well as the processes and procedures that are used to produce them.

Dedicated quality control and quality assurance personnel, auditors, and inspectors are irrelevant in the modern quality organization. These tasks are performed by the line workers as part of their regular duties. The role of the quality department, if there is one, is to train and develop the workforce to be able to perform these duties as part of their duties.

Of course, regulatory requirements may require separate inspection by dedicated personnel, and if so, that's fine. Where this isn't required, it should be abolished if at all possible. Many organizations will have difficulty making a switch from the traditional quality hierarchy to the modern quality organization. That's understandable, because change is often difficult. But you should consider the benefits of training all your employees in quality principles and techniques and assigning responsibility and accountability for quality to all areas and all levels of the organization.

7.5.5 Organizational Environment

The structure and environment of the organization will affect quality, especially as defined by me (which includes the customer experience component). A hierarchical organizational structure, a bureaucratic organization, or a departmentalized organization will be very different from an entrepreneurial organization, a matrix-structured organization, or a decentralized organization. This isn't to say that one structure or another will provide a higher level of quality, just that the structure will have an affect on quality that must be considered.

A hierarchical, bureaucratic organization may lack flexibility and may adhere to the dedicated quality department philosophy. That doesn't mean that high levels of quality can't be achieved and maintained. An entrepreneurial, matrix-structured organization might have a looser or more carefree attitude that assumes that quality will be embraced by everyone. High levels of quality can be achieved in this environment. The point is that the structure and environment of the organization must be taken into account when developing the quality strategy, quality policies, and the assignment of responsibility and accountability for quality. The quality functions must work for your company, your environment. They must fit with your corporate strategy and your corporate structure. If you try to fit incompatible organizational and quality structures together you won't achieve the quality levels that are needed to compete in the marketplace.

7.5.6 Quality Resources

As with any other function, quality management requires adequate resources. The three categories of resources needed for quality are people, tools and equipment, and money and support. People are the most important resource, but they need the proper tools and equipment to do their jobs, and the quality system needs budgetary and management support. We've discussed different scenarios for roles of responsibility for quality. Either way that the quality functions are managed, either in a dedicated quality department or spread throughout the organization, employees need certain skills and abilities.

The people must be technically proficient in the products you produce and the processes used to produce them. This is pretty much a general requirement for all the employees who work in your plant, but the level of proficiency needs to be raised, or the areas of knowledge broadened, to raise or maintain the level of quality. Technical proficiency means the employees understand the products, their uses and functions, markets, and the needs they fulfill. They need to understand the production process, the planning system, and production methods. All of these things affect quality and they must be understood by the people responsible for quality.

The people must have an understanding of and proficiency in quality tools and techniques. They might need to understand how to complete control charts and why they are used; they might need to understand statistical methods, measurement techniques, or the use of testing equipment. They need to have a clear understanding of the quality policies and the quality mission, including the importance of quality to the organization and its success. The people also need soft skills such as team building and team dynamics skills, facilitation and conflict resolution skills, and problem-solving skills.

These skills can be acquired by hiring people who already have them. They can be acquired by utilizing outside experts and consultants. And they can be acquired through education and training of existing employees. A combination of methods might be the most appropriate, but again, the method of acquiring the necessary skills needs to fit in with your organizational environment and your goals and strategy.

Quality personnel may require specialized tools and equipment, such as those needed for testing and measuring. Laboratory space, dedicated testing or administrative space, or training space and equipment might be necessary.

All this takes money and top management support. Top management must recognize the need to hire and train personnel, purchase and maintain tools and equipment, and acquire or allocate space and facilities. The quality functions must be included in the budgets. Other possible budget items to consider include travel expenses and activities related to and membership in professional associations.

7.5.7 Cost of Quality

I'd be remiss if I didn't mention "the cost of quality." We've discussed the need for resources to effectively manage quality, but are these resources the cost of quality? I'd argue that, no, they're not. They are investments in resources needed to produce the products; we've just identified them as "quality" resources. A cost implies that you don't get any benefit from the expenditure. An investment implies that a return is generated that is greater than the amount invested.

I agree with the philosophy of quality guru Phil Crosby, who asserts that quality is free, but there is a cost associated with the lack of quality in an organization. The costs associated with a lack of quality are manifest when products are produced that don't meet requirements. These products must be reworked, scrapped, or returned by the customer. If systems can't be developed that ensure the production of products that meets specifications, testing and inspection procedures must be adopted. If nonconforming products reach the customer, costs increase exponentially. Customer confidence and relationships are affected; delivery, return, and replacement costs are incurred; and rework or scrap costs are added. Add to that the costs of investigating the failure and taking the steps to prevent it from recurring. Many times, blame is assigned and people are fired, which increases costs.

Developing and investing in an effective quality system will prevent these costs. Many of the tools discussed throughout this book can be used in an effective quality system, including Lean Manufacturing, Six Sigma, theory of constraints, SPC, and all the quality tools in this chapter. The idea is to identify your customers' needs; produce the products efficiently, effectively, and without defects; eliminate waste; and deliver products to your customers when and where they want them. That's pretty much everything we're talking about in this book.

7.5.8 More Quality Tools

Quality management encompasses many tools and techniques. This section provides a brief description of some of them, many of which you will recognize.

7.5.8.1 The PDCA Cycle

The PDCA cycle (Figure 7.5), or plan-do-check-act, is primarily a process improvement tool. You can see that it is a closed-loop system, meaning that it never ends. That's continuous process improvement, or kaizen. In the first stage (plan), the problem that is being addressed is studied and analyzed. You develop a plan for improvement. In the next stage (do), the plan that was developed is executed. In the check phase, you review and analyze the results of the action that was taken. Did things go as planned? Were the results as expected? Were there any unanticipated effects? In the final phase (act), you take action to address the findings

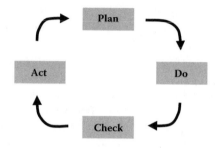

Figure 7.5   PDCA cycle.

of the previous phase. If everything went as planned, or if corrections were made to address issues that arose, then the new process is adopted as the new norm.

7.5.8.2 Process Maps, Flowcharts, and Value Stream Maps

Process maps, flowcharts, and value stream maps are used to define a process, analyze it, and identify areas for improvement. A flowchart is a diagram of the process. A process map includes definitions of the work process and more process specifics, including the how's and why's. Lean Manufacturing uses value stream mapping.

7.5.8.3 Ishikawa Diagram

Also known as a cause-and-effect diagram or a fishbone diagram, the Ishikawa diagram (Figure 7.6) is used to identify the cause of a problem. It is similar to the five why's method of root cause analysis in that it attempts to identify the reasons for a result. Also, because it is a graphical tool, it adds another dimension to the problem-solving process, which can be very useful. For complex problems, the main stem may have many branches.

7.5.8.4 Pareto Analysis

Pareto analysis, also known as the 80/20 rule, is well known in inventory management, especially regarding cycle counting. The 80/20 rule states that 80 percent of the results are attributable to 20 percent of the sources. This is a general rule that holds true in many situations and is a useful analysis tool.

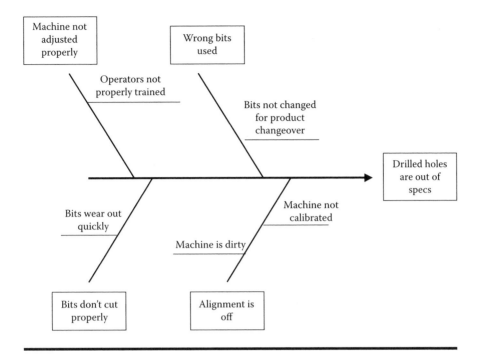

Figure 7.6   Ishikawa diagram.

In inventory management, it is often found that 80 percent of the total inventory value (in dollars) is attributed to only 20 percent of the total number of stock numbers. In quality, it is often found that 80 percent of the variation in a process is due to 20 percent of the possible causes of variation. In general, it is useful to assume that a large portion of the observed effects are due to only a small portion of the possible causes. This assumption allows you to focus your efforts on the most pressing issues by prioritizing them.

Table 7.2 shows a simple example of a Pareto analysis based on the causes of defects in the drilling operation as identified in the Ishikawa diagram (Figure 7.6). In this example, we are using just the four primary areas of causes, but in reality you would use all the specific causes (such as "machine not calibrated") if you could trace them to specific numbers of defects. If you cannot determine the number of defects attributed to specific causes, just use the broader areas.

This example shows us that greater than 79 percent of the defects are due to 25 percent of the causes of defects. That's pretty close to 80/20, and it allows us to see where we need to focus our efforts. If we can eliminate the top cause we will eliminate 79 percent of the defects. On the other hand, if we had not performed this analysis and started inves-

Table 7.2 Pareto Analysis, Defects

Cause of Defects	% of Cause	Number of Defects	% of Defects
Bits don't cut properly	25%	8,100.00	79.3%
Alignment is off	25%	850.00	8.3%
Machine not adjusted properly	25%	750.00	7.3%
Wrong bits used	25%	510.00	5.0%
Total	100%	10,210.00	100.0%

Table 7.3 Pareto Analysis, Inventory

				Cumulative % of Total	
Rank	Item No.	Annual Dollar Value	Cumulative Value	Total Dollars	Total Items
1	7	$64,025.00	$64,025.00	49.7%	10%
2	3	$37,810.00	$101,835.00	79.0%	20%
3	9	$6,240.00	$108,075.00	83.9%	30%
4	2	$6,000.00	$114,075.00	88.5%	40%
5	6	$5,490.00	$119,565.00	92.8%	50%
6	4	$2,880.00	$122,445.00	95.0%	60%
7	5	$2,000.00	$124,445.00	96.6%	70%
8	1	$1,920.00	$126,365.00	98.0%	80%
9	8	$1,440.00	$127,805.00	99.2%	90%
10	10	$1,080.00	$128,885.00	100.0%	100%

tigating why the wrong bits are used, we would have been able to eliminate only 5 percent of the defects. Where would you like to begin your improvement efforts?

An example of Pareto analysis of inventory value is copied here from Chapter 2 (Table 7.3). This example shows that 79 percent of total inventory value (annualized) is found in 20 percent of the inventory items (two out of ten stock numbers). This is useful in cycle counting to determine where to focus your control efforts. Spend your time and effort worrying about the two items that represent the biggest portion of the inventory value. Spend less time worrying about the one item that represents only eight-tenths of a percent of the value.

The 80/20 rule holds true in many different situations: 80 percent of human resources problems are due to 20 percent of employees, 80 percent of total sales volume comes from 20 percent of the customers. You may find other examples where Pareto analysis can be used to prioritize efforts to improve performance. Try it the next time you start to look at a problem or initiate an improvement activity.

7.5.8.5 Activity Network Diagram

If you are familiar with project management tools or if you took an operations research course in college, you have probably seen PERT (program evaluation and review technique) or CPM (critical path method) diagrams. The activity network diagram (Figure 7.7) is similar to these, but it is a simplified method. You may want to use an activity network diagram when you don't need or have the time or expertise required to create a PERT diagram. It also works well as a preliminary analysis or the beginning of a process map.

The activity network diagram shows which tasks can be performed at the same time (in parallel) and which tasks must be completed before the next one is started (in series). In this example (Figure 7.7), Tasks 2 and 3 can be started and worked on at the same time. Task 4 must be completed before Tasks 5 and 6 are started, and so on. Identify each task, the time required, and the person or team performing the task. This will highlight any obvious conflicts or constraints that need to be dealt with before proceeding with the project.

7.5.8.6 Brainstorming

Brainstorming is an activity that is often misunderstood or improperly conducted. People often say things such as, "let's brainstorm on that." What they really mean is that they want to discuss and evaluate various proposals or options. That's not brainstorming. Brainstorming is a specific technique with clearly identified steps.

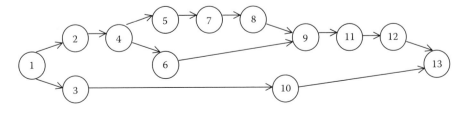

Figure 7.7   Activity network diagram.

The first step is familiar to many people, at least in part. It is the creative idea stage, and it can be the most difficult step to complete properly. The topic to be worked on is identified (usually a problem that needs to be solved) and written down. The first step of brainstorming involves coming up with ideas to solve the problem. But it is much more than that. Brainstorming is a group activity. All members of the group are encouraged to participate and contribute ideas. A good facilitator will make the process work properly. At this stage there are no bad ideas. No comments or changes to an idea are even allowed at this stage, and this is where the facilitator's skills are tested. Free thinking and creativity are encouraged here, and ideas and excitement build. Sometimes, as the ideas are written down, new ideas are built on previous ones. Finally, when no more ideas are forthcoming, this stage ends.

The next step is to categorize and group the ideas. In this stage, ideas can be combined or revised to clarify them. The result from this stage is a list of ideas that could be turned into action items. No action is taken yet. The next step is to narrow down the list to those ideas the team thinks should be examined further. This can be done by voting, either by secret ballot or voice vote. This step might take several rounds if the list cannot be narrowed down sufficiently the first time through. The final result is a list of ideas that can be prioritized and acted on to begin the process improvement activities.

The greatest difficulty in performing brainstorming successfully is the first stage. Without a good facilitator or a disciplined or experienced team, the process can break down quickly. If not controlled, people may start to make judgments or inappropriate comments that will quickly put an end to the creative process and flow of ideas. In the beginning, no idea is too outrageous. Throughout the process, ideas should not be judged, because it will discourage further brainstorming efforts. Later stages will provide an opportunity to eliminate or give low priority to ideas not considered worthwhile for pursuit. Don't sabotage a great opportunity by making judgments about someone's ideas.

7.5.8.7 Affinity Diagram

An affinity diagram is similar to brainstorming or an extension of a brainstorming session. Ideas are generated and written on notecards, and the notecards are placed face up on a table and arranged into groups. During the idea generation and grouping activity, no judgments are made, as in brainstorming. The grouping may take a few rounds to complete, because there may be some disagreement on where an idea should be placed. When agreement is reached and all the ideas are placed into groups, a new discussion might begin, with new ideas generated. The affinity diagram is a good tool for projects that aren't yet clear or are complex and need to be organized before work can begin.

7.5.8.8 Interrelationship Digraph

An interrelationship digraph is like an extension or an evolutionary step of the affinity diagram. In brainstorming and affinity diagramming, ideas are generated and grouped so that they can be prioritized and used to start improving a process. Brainstorming generates the list of ideas and the affinity diagrams show them in a more graphic or visual format. This is a good starting point, but as you know, just about any idea that you start to take action on will have an effect somewhere else. The interrelationship digraph is a tool to show where the effect might have an impact and which ideas or tasks are related. Anticipating where effects might be felt and visualizing relationships can help you avoid or deal with potential problems. See Figure 7.8 for an example of an interrelationship digraph.

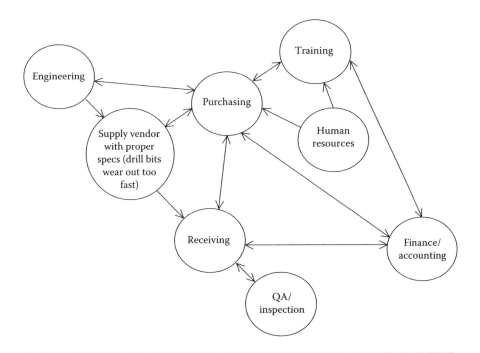

Figure 7.8   Interrelationship digraph.

7.5.8.9 Force Field Analysis

Before undertaking a major project or one that will significantly affect your organization, you should perform an assessment of the strengths and obstacles that will help or hinder your efforts. Major initiatives, such as a Lean Manufacturing or ERP implementation, will have significant impacts on all areas of the organization. You need to realistically assess where you stand right now, what things will help you with the undertaking, and what things are obstacles to success. One form of assessment is a force field analysis. (Shields are down. Reroute power through the auxiliary power grid! No, not that kind of force field.)

In the force field analysis, a team will identify as many strengths and obstacles as it can. The list need not be exhaustive, but it should include the key areas that are important to the success or failure of the project. Then each item on the list is assigned a weight along a scale (Figure 7.9) that runs from negative to positive; for example, ±1 to 5 or ±1 to 10. Because the weights are relative, the scale shouldn't be delimited too fine (meaning, don't use a scale of ±1 to 100). The team members will determine the weights for each item. Because there will probably be some disagreement, a consensus or an average might be used. Instead of just listing the items and the weights, the force field analysis uses a graphic approach. (Notice that many of the tools we've been discussing use graphic representations rather than just text.)

Many more tools can be found under the umbrella of "quality" and with process improvement activities. These are just a few of the more popular ones. You can use them immediately to help you with your next project or activity. Give one or two of them a try and see which ones might work best for you.

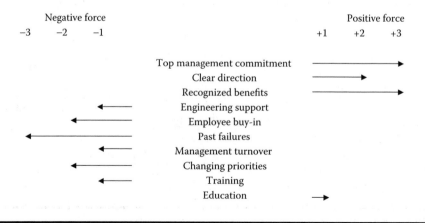

Figure 7.9   Force field analysis.

7.6 Metrics

Performance measures, or metrics, are an important tool that should be utilized in every organization. ISO 9000 certification and the Baldrige criteria both require that performance measurements be used to make improvements. Most organizations use some sort of performance measures for their employees or their processes, but they are sometimes more arbitrary than actual performance measurement systems.

Many times, organizations collect data for the purpose of justifying a decision that has already been made. They collect data that supports their decision and try to call this performance measurement. Far too often, supervisors and managers document behavior they don't like simply as a way to fire someone; then they try to justify the firing as not meeting performance. This is not performance measurement. A formal performance measurement system will capture data and report results, but a true performance measurement system takes this data and results and uses them for two purposes: to provide feedback that can be used to make decisions and to guide the behavior of individuals so that the company's goals are achieved.

The difficulties are deciding what to measure and finding out how to measure it. One question that must be asked is, why am I measuring this particular item? If you can't justify a measurement based on what feedback it will provide or what behavior it will guide, it probably isn't a necessary measure. The feedback that is received through the metrics is very valuable for evaluating the results of activities. Results over time, or results before and after some change has been made, can be very enlightening. A formal performance measurement system removes the guesswork that is often used in evaluations. The success or failure of a change can be quickly recognized, and the reasons can often be discovered based on the results.

The question of what to measure shouldn't be taken lightly. You can't measure everything and you don't want to; that would just give you a lot of unnecessary information. You need to concentrate your measurement activities on those areas that will help you achieve your goals. Your performance measures must support your strategic plan and the strategic goals of your organization. You need to determine the key factors for success and develop measures for these factors. You need measures that will help you determine if you are making progress toward your goals.

As I mentioned, performance measures should provide feedback that helps you make decisions (organizational measures) or guide behavior (individual measures). Organizational measures might measure a department, a particular group or area, or a machine. For example, defect rate, fill rate, and space utilization are organizational measures. Individual measures assess the performance of an individual or worker.

Most companies love to do this — measure an individual's performance. But, as mentioned above, these measures often are used simply to punish people. How many days were they out sick? How many mistakes did they make? Did they meet their sales quotas? This might be important information, but you have to step back and look at whether the numbers serve any real purpose besides punishment. To guide behavior you need three things. You need to define the results you are trying to achieve, measure the progress toward those results, and provide the tools and support that will let your employees achieve the results. You can't just tell people to "improve, or else" and expect great things from them. They need to know what they need to do, how to do it, and why they're doing it. They need to know where their efforts fit in and how they're doing along their way.

MANAGEMENT OF THE BUSINESS

Chapter 8

Project Management: You're Gonna Do It; Do It Right

Just about any book or article on project management begins by defining what a project is, so I won't talk much about that. Pretty much any undertaking that is not a part of your normal operations is a project. Just about anything with a defined beginning and ending is a project. Selecting a new vendor, performing a kaizen event, installing new equipment, building a new facility, restructuring your logistics system; these are all projects. You may not call them projects or think of them as projects, but they are projects. You are constantly involved in projects, so why not get better at working on them?

Project management has its own defined body of knowledge with standardized methodologies to better ensure the completion of successful projects. The Project Management Institute (PMI) has developed a widely recognized set of standards for project management. It also administers the project management professional (PMP) certification in project management. Although many organizations and private companies use their own project management systems, the PMI standards are available to anyone and are widely recognized in industry.

8.1 Project Phases

In most operations, activities take place in two distinct phases — planning and execution. Projects are no different, but the planning and execution phases are generally divided into either four or five phases. These project phases are

1. Initiating
2. Planning
3. Executing
4. Controlling
5. Closing

Some people separate executing and controlling into separate phases, resulting in five phases; others group them together, resulting in four phases. Whether you consider execution and control as separate or not isn't all that important, as long as you recognize that separate activities are associated with each of them. Control needs to occur concurrently with project execution, but discussing them as separate phases clarifies the activities of each (Figure 8.1). Let's talk a little bit about what goes on in each phase.

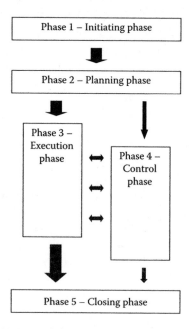

Figure 8.1   Project phases.

8.1.1 Phase 1 — Initiating Phase

Before you actually start doing anything, you need to determine what exactly you are going to do and why. This phase might also be called the "what are you going to do and why" phase. If you don't start with a clear idea of what you are trying to accomplish, you will likely not accomplish much of anything. Begin with a clear and concise statement of the project. Define the goals of the project and what you expect to achieve with it. Justify the project — the need for it and the benefits that will be derived from it. One way to start is to identify either the problem that you are trying to solve or the opportunity you are planning to take advantage of. This allows you to clearly state the goals and the justification. Let's say we state the problem as, "information requirements are not adequately supported with our existing technology." The project will be to implement an enterprise resource planning (ERP) system. The justification might sound something like, "an enterprisewide information system will allow the company to leverage enhanced information and information flows to improve companywide performance, resulting in increased customer satisfaction and profitability."

When you define your goals, make sure you are really sure about what you are saying. For example, is your goal to install an ERP system, or is your goal to implement a software solution that will enable the organization to improve performance? You can install an ERP system and get no benefit from it, so carefully word your stated goals. Your objectives, on the other hand, can be more action oriented. An objective might be to install the necessary hardware and software components of the ERP system. You will have a list of objectives but only one or a few goals. The bigger the project, the longer the list of objectives you will most likely have.

At this point you can begin to identify the resources you will need to complete the project. This will be a preliminary list based on the nature of the project; it won't be too specific at this point. You won't have a project plan and might not even have a project manager yet, so the resource estimate will be a guess based on past experience more than anything else. Still, it does provide a starting point and might help with the initial justification.

You also need to declare any assumptions you are making; for example, you will be able to find and acquire any expertise needed for the project, or you will be able to free up people as needed to complete the project successfully. All these assumptions need to be spelled out. It will be an important document to refer to in the future, as the project progresses and obstacles are encountered. If a project is justified and initiated based on assumptions that turn out to be untrue, it will affect any decision that might need to be made about whether to continue the project or abandon it.

Risks and potential risks also need to be identified. Assuming that everything will be hunky-dory (even if you state that assumption) is not a good way to begin a project, especially for a large or complex problem. For the ERP project, one of the risks might be that a technological breakthrough might hit the market halfway through the project. This risk should be stated, along with the action or decision that will need to be taken should the risk bear out. In this example, a decision to continue implementation of the current technology or abandon it and move to the new technology will need to be made. You might also weight the potential risks and estimate their likelihood, as in failure modes and effects analysis (FMEA) (see Chapter 5).

Project charter is a term that is often used for the document that formally authorizes the project. The project charter can include all the points addressed here, and it might be used as a contract if the project is for an outside customer.

8.1.2 Phase 2 — Planning Phase

If Phase 1 is the "what are you going to do and why" phase, Phase 2 is the "how are you going to do it" phase. The planning phase is much more detailed than the initiation phase. This phase is quite extensive, but if performed properly it can make the difference between success and failure.

All the activities, sliced and diced and divided and subdivided into as much detail as is necessary, need to be listed. Probably the most popular method of defining and listing all the activities is the work breakdown structure (WBS) system. The WBS is a hierarchical structure consisting of individual work packages. Each work package is clearly defined and responsibility is assigned to a specific individual. We'll talk about the WBS in more detail later.

The WBS will facilitate the estimates of time and money that must be included in the project plans. Each work package will have a definite time estimate and a "cost" estimate. I don't like the term "cost," but that's what most people will use and what you will most likely see in project management literature, so I will use it here.

The WBS will also facilitate the scheduling of the project. Each work package has a time estimate, and the hierarchical nature of the structure helps to develop the sequence of work and activities. Some work will be able to be performed at the same time as other work (in parallel), and some work will have to be sequential (in series). The WBS and the scheduling of activities will go hand in hand, but they are distinctly different activities. See Figure 8.2 for an example of a project schedule in the activity network diagram format (see Chapter 7). Each work package in this diagram is numbered and the estimated time to complete the work is included.

When the schedule is complete, the critical path needs to be identified. The critical path is the sequence of activities with the longest estimated completion time. The critical path is important because delays in any of the work packages along this path will affect the time of completion for the entire project. Because most plans, budgets, and promises are based on completing a project in a certain period of time, the work along the critical path is, well, critical. Figure 8.3 highlights the critical path of our sample project.

Many people might think that writing a formal project proposal at this point is unnecessary. It can be necessary if the project has not received final approval, but it is useful as a reference and control document regardless. The proposal should include the issue being addressed by the project, the goals and objectives, and the plan and schedule that have been developed. Resource requirements and costs will also be included.

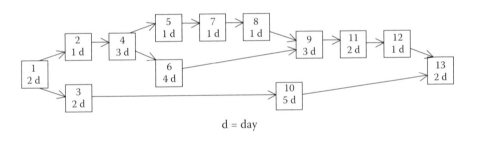

d = day

Figure 8.2   Project sequence.

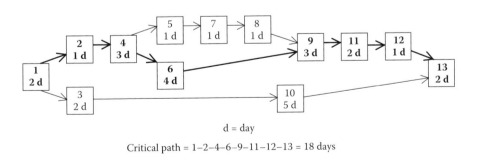

d = day

Critical path = 1–2–4–6–9–11–12–13 = 18 days

Figure 8.3   Critical path.

If a final decision to proceed has not been made, this document will contain all the ingredients needed to make that decision.

8.1.3 Phase 3 — Execution Phase

As they might refer to it at Nike, this is the "just do it" phase. The planning and justification have been done and approval has been given. It is time to put the plan into action. Some of the first things to do in this phase might have already begun or been done on a preliminary basis. If you don't already have one, you need to assign a project manager to the project. This is, of course, a key position. The project manager can make or break the project. With all other resources allocated and all the plans completed and approved, the project manager is the one resource that will probably have the most impact. The project manager will need to be a person of varied skills, but the term manager is a bit misleading. The project manager will manage the project, but he or she will also lead the project and the project teams.

Project managers need a diverse set of skills to complete a project successfully. One important skill is leadership, because the project manager is also the project leader. Leadership skills are an identifiable set of skills. Although some people seem to be more natural leaders than others, leadership skills can be learned. Many strong leaders have a military background, most likely due to the great leadership training programs the military has developed over thousands of years. Regardless of where it comes from, your project managers need to have this skill. You should seriously consider a leadership training program to develop future leaders in your organization.

Other skills the project manager needs are communication skills (both written and verbal), team-building skills, negotiation skills, and problem-solving skills. The project manager will need to communicate with all levels within the organization, from the project sponsor (senior management) to every project team member. These people will possess a wide range of positions, experience, skills, ages, and perceptions — all factors affecting how they communicate and perceive communications. The project manager must be comfortable and competent in writing reports, leading informal meetings, and making formal presentations. He or she might need to become involved in negotiations with suppliers, department managers and supervisors, and senior management to acquire the necessary resources to complete the project. The project manager has to have problem-solving skills because problems and conflicts will arise. It's inevitable. The project manager must be able to overcome obstacles and solve problems. The project manager will have help with all these areas, but he or she is the final authority and has the responsibility. Choose your project manager wisely.

All the personnel resources needed for the project must be determined, acquired, assigned, and scheduled. Some personnel may not be needed right away, but the skills needed and the people who have those skills and will be assigned to the project need to be allocated. For those people who will be needed in the future, preparations should be made. If they need to be freed of other responsibilities to work on the project, their current work assignments will have to be adjusted and their supervisors and colleagues will need to be a part of the process. Don't wait until the last minute to pull someone off their normal duties and assign them to a project. It may not hurt the project, but it will hurt their normal work department and can affect the entire organization.

The project team needs to be assembled and developed. Many project team members might need training to perform their best. Some will need formal team training and some might need specialized training in some aspects of the project. Personnel need to be assigned to teams and their reporting responsibilities need to be clarified. Unless they are assigned to the project full time, there will be inevitable conflicts with their time. These conflicts can be minimized significantly by proper planning and with the proper assignment of responsibility and authority.

The work packages that were developed in the planning phase are now assigned to individuals or teams. Each work package consists of specific tasks to be completed, with defined output or deliverables. The project manager will assign the work packages, as well as the responsibility for completing those packages. This defines exactly what work will be done, when it will be done, and who is responsible for completing it.

The planning, scheduling, and the WBS are not all that different from the planning done on a regular basis in manufacturing. The planning hierarchy (discussed in Chapters 2 and 3) starts at the highest level (sales and operations planning) and makes its way down to the lowest level (materials requirements planning and production sequencing). This isn't much different from starting with the project goals and objectives and developing the WBS into the individual work packages. In the planning hierarchy, you perform capacity checks at every level, to make sure you can actually accomplish the plan with the resources available. Don't you think this is a good idea when developing the project plan, the WBS, and the work packages? There should be a formal mechanism for checking the capacity of the project plan. Determining the resource requirements is one thing, but verifying that the resources are available when you need them is another. *A Guide to the Project Management Body of Knowledge*, known as *PMBOK® Guide*,* states that "resources will be obtained either through staff acquisition or procurement." That seems like a rather big

---

* Project Management Institute, 2000.

assumption. I'd feel better if the availability of resources were more quantifiable.

One rather basic issue that might be overlooked initially is the need for a project office or project control center. For a small project, this might just be the project manager's office, but for larger and more complex projects you will need a dedicated project center. For large projects, the required space and equipment can be substantial. The project manager will be stationed there and a large part of the work will probably be done there. Desks, chairs, and tables, computer equipment, phones, and anything else you use in a regular work area will be needed. Setting this up might be a project in itself.

8.1.4 Phase 4 — Control Phase

This is the "are you sure you're doing what you're supposed to be doing, and how is it going?" phase. The control phase activities will occur at the same time as the execution phase activities, for the most part; but it can be beneficial to view these phases as separate. This helps to clarify the differences between the execution and control activities and helps to strengthen the idea that control is needed (control systems are weak in some organizations, unfortunately).

You must develop a system for managing the project. If your organization is used to working on projects, you probably have some sort of project template. Still, each project is different, and the project manager, project teams, and resources will be different, so the particulars for how the project will be managed need to be developed and defined. One vital part of the management system for the project is the communications system. Communications methods between team members must be established, as well as communications between the project manager and the project teams, project sponsor, and team members and their normal supervisors.

The communications system at your company probably isn't something you think much about, until something bad happens. When something bad happens or some problem surfaces as a result of poor communication, somebody might tell you to fix it. If a fix is put into place (telling people to communicate more isn't a fix), it's probably a Band-Aid solution. PMI's *PMBOK*® *Guide* recognizes project communications management as a distinct management area; communications planning has its own section. Communications planning consists of identifying the information needs and methods of communication for the project. What information is needed, who needs it, when do they need it, how will it be delivered to them, and who will deliver it are all issues that must be addressed. A communications management plan is the result of the communications planning processes.

The controls and control system have to be established. The control system is for the most part a performance measurement system. You must establish and define which tools or methods of control or measurement you will use and which items will be monitored and when. As with other performance measurement systems, the project control system will be used to track and monitor progress, check performance to plan, and identify areas where corrective action is required.

An extremely important component of the project control system is the status report. Have you ever been involved with a project where the boss asked, "what's the status?" or "where are we on that project?" This might indicate a lack of formal status reporting. It might indicate a lack of communication too; but either way, it shows that there is probably room for improvement in management of the project. The status report should be a formal and required document. It doesn't necessarily need to be long or complex, but it does need to accurately communicate the status and whether any corrective action needs to be taken.

One format that is used for both project scheduling and status reporting is the Gantt chart. The Gantt chart is probably the most popular method of showing and sharing project status information. It is easy to put together (most project management software will generate one at the click of a button) and easy to understand. A Gantt chart shows the activities down the left side and the timeline from left to right. In the most basic form, that's all that is necessary. Often, however, other information is provided. Usually the responsible party is listed next to the activity. See Figure 8.4 for an example of a Gantt chart.

Along with the status report, the project schedule needs to be reviewed. This review goes hand in hand with the status report but goes beyond it to review the entire remaining schedule and whether it has been affected

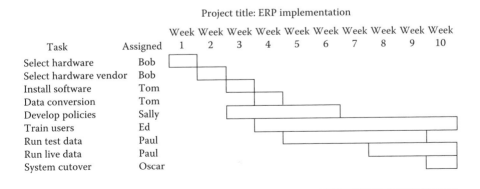

Figure 8.4 Gantt chart.

in any way. By reviewing the schedule, the project manager will be able to determine whether the project is still on schedule. If it is ahead of schedule, that's probably a good thing, but actions still might need to be taken. If downstream activities can be rescheduled forward, that needs to be reviewed. If rescheduling would add costs, it might not be desirable to do it. If, however, the project is behind schedule, actions will need to be taken to get it back on schedule, if that's even possible. If it is not possible, action must be taken to mitigate further problems or delays.

If the schedule review results in any changes to the project, either to work requirements or schedule changes, change orders will need to be issued. Change orders should be formal, written documents; not verbal commands. The change orders will state the change (what the new work or schedule will be) as well as the reason for the change and the impacts of the change. This ensures that the changes are clear and the need is well defined. It also documents the changes for review at the completion of the project.

8.1.5 Phase 5 — Closing Phase

In Hawaii, we say it is "pau hana" time. That means "done work" or time to go home. But you can't just sit down and stop, you have to make sure you are really done and tie up any loose ends. Maybe we should call this the "make sure you're done before you go" phase.

Before the project can be officially closed or considered complete, the client's acceptance must be obtained. The client can be either internal or external. External clients are easy to identify. They are the people who contracted for the project or are paying for it. Internal clients might not be as easy to identify. If the project is the implementation of an ERP system, who is the client? Many people and departments will be using and relying on the new system, so who is the person who must be satisfied and accept the project? It will probably be a senior manager or member of the executive team. In smaller organizations it will be the president or CEO, or possibly the board of directors. The important thing is that the client is clearly identified and that they accept the project before it is closed.

Naturally, the deliverables have to be installed or delivered. The formal installation or delivery is part of the closing phase of the project. This is easy to identify when the deliverable is a physical item that is delivered to a client or customer. It is a bit more esoteric if the deliverable is a recommendation by a consultant or the implementation of a complex software system. For the ERP project, most, if not all, of the system will be up and running before formal closing of the project. It may be hard to say just when the project has delivered its promise.

Formal documentation of the project should be completed as part of the project closing. The documentation should include items relating to the completion or termination of any outside services that were used on the project. Final payments to contractors, return of equipment, compliance with any regulations or contracts, and any other information relating to the project should be included.

Along with the final project documentation, you need to issue a final report. The report will summarize the project and the results. The final report will indicate whether the goals of the project were met, if the client was satisfied, and if the project was completed on time and on budget. A project financial report should be included. This report will not only summarize the actions and results of the project, but will serve as a valuable reference for future projects.

Finally, a post-project audit should be conducted. The post-project audit will identify the things that went well and areas that need improvement for future projects to be successful. Information about things you might want to include in future projects should be collected and added to a project planning template. For those things that didn't work well or needed improvement, include an honest assessment of why they didn't work or why they were a problem. Offer suggestions of actions that can be taken to eliminate the problem in the future or reduce the effects. Any recommendations that will be useful for future projects, project managers, and project teams should be included in the audit.

8.2 Project Scope

Project scope management is a vital activity, but it is not always well understood by people who aren't seasoned project managers. The term *project scope* is tossed around a lot, but it does have a definite meaning. It is also sometimes very hard to control. Controlling the scope of a project takes understanding and discipline, by the project manager, but especially by the client and senior management outside of the project.

8.2.1 Scope Planning

During the development of the project charter, you will define and describe the result, product, or service that will be delivered. *Product scope* is a term that is sometimes used to refer to the description of the final product and its specifications. This clearly defines the finished product, its features, and its functions. It is important that the client and the provider (you) agree to the product scope so that everyone is clear on what will be delivered and what constitutes successful completion of the project.

There will almost certainly be requests for changes to the scope of the project or product. The scope management plan, which comes out of the scope planning process, documents how the scope will be managed and how any scope changes will be handled. The communications process for scope change requests and the process of managing any changes that are approved have to be addressed. Of course, the person who has the authority to make changes in scope needs to be identified. The boundaries of what constitutes a change in scope and who will pay for any changes needs to be addressed too, as much as they possibly can. When is a change small enough that it can be handled without affecting the project, and when does it constitute a change in scope? You should try to identify that point if you can. A small change by itself might not be an issue, but many small changes might add up to a significant change in scope. The scope planning process should address these points and identify procedures for addressing them. Simply acknowledging that changes will be requested and identifying a method for handling them is a huge advance for many organizations. Many organizations don't have a plan, and end up paying the price with a late and over-budget project, or an unsuccessful project.

8.2.2 Scope Definition

Scope definition and the WBS go hand in hand. One could argue that they are the same thing, but the *PMBOK® Guide* recognizes that the WBS is a result of the scope definition process. Basically, scope definition involves dividing and subdividing the project into the smallest possible (and reasonable) pieces.

Scope definition helps improve the estimates for the cost of the project, the completion time, and the resources needed. By breaking down the project into clearly definable work units, it makes estimating easier and more accurate. These work packages serve as a basis for the performance measurement system and the control system. The outputs or results of these work packages will be measured and compared to the plan, and will therefore be the measure of performance and tracking for the project. Although the work itself might be assigned to a team, specific individuals will be given responsibility for the completion of the work and the deliverables. Because the work packets are clearly delineated, this responsibility assignment is easier to make.

8.2.3 Scope Management

"Scope creep" is not a name you call your little brother. It is the unplanned growth or expansion of a project from its original parameters, and scope management is what keeps it in check. Scope management is an active process. Simply monitoring progress and letting the project move along on its own is not enough. As I mentioned, you will receive requests for changes to the product or project. Some changes will be legitimate or required, and some will not. The project itself must be actively managed, but the scope must also be actively managed for it to remain under control.

Active scope management includes scope change management and control. Requests for changes must be analyzed and evaluated. Change requests should be documented, along with the decision that was made. Any changes have to be justified and approved. Changes in scope need to include documentation on the impact of the change, of cost, time, and the deliverables. And, of course, the changes need to be incorporated into the project. Changes might affect all other downstream and parallel activities. Any downstream changes must be coordinated and the affected activities might need to be rescheduled. Performance measures might need adjustment or revision. Scope management is an important activity that should not be taken lightly.

8.3 Resources

A project must have adequate resources to be completed successfully. If you have ever been given a project to manage and been expected to do it as part of your normal duties, without any additional time, money, people, or equipment, you understand what I mean. Unless it's such a small project that it is hardly even considered a project, failure is highly likely. In fact, it might seem like a test or a reason to get rid of you (and it might be). This is poor management. A project needs resources, and the resources needed have to be defined in as much detail as possible. The time to plan the project and define the resources is a needed resource itself.

A project needs a sponsor. The sponsor is a senior-level manager who overseas the project and has ultimate responsibility for its completion and outcome. The sponsor may be the person who approved the project or a member of the team that approved it. The sponsor has the overall fiscal accountability for the project. He or she does not get actively involved in the management of the project but will keep a close watch on its progress. The project manager will keep in close contact with the sponsor. The sponsor will approve any scope changes and resolve any conflicts or problems that arise with the client.

The project manager, as I mentioned, must have a diverse set of skills. Although experience isn't necessarily a requirement, it is likely that the more complex the project, the more experienced the project manager.

The project manager has the day-to-day control of the project. He or she is responsible for coordinating all the resources that are needed to successfully complete the project. The project manager will assign the work packages and the responsibility for completing them, and will coordinate the scheduling of them. The project manager is responsible for completing the project on time and within budget.

The project manager will be responsible for acquiring most of the resources needed to complete the project. Actually, it's probably better to say that he or she is responsible for assembling the resources. The project sponsor is responsible for making sure that the resources are available and that an adequate budget is in place and adequate funding is available. The project manager will have a role in developing the budget, because the budget cannot be completed until the project is broken down into small enough work units to estimate the time, costs, and resource needs.

After the project manager, the first set of resources that needs to be considered is human resources: the people who will perform and supervise the work of the project. Some of the people who perform some of the work won't actually be part of the project team, but someone on the project team will be responsible for their work that is done on behalf of the project. For example, if the project involves the development of a new product, and a prototype has to be built or a test run on the production line is required, the production workers might not be part of the project itself, but the results of their work will be the responsibility of someone on the project team.

Assembling the project team is an important task, and not one the project manager should delegate. Determining the human resources that are needed is part of the project planning process, but the assembly and development of the project team is part of the execution phase. The skills required of the project team are identified in the planning stage, but now the project manager will have to assemble those skills. In some cases, the manager will have to find the skills and acquire or recruit them. The people and teams will have to be assigned; people will be assigned to teams when necessary, and people and teams will be assigned to specific work packages.

In many cases, the project manager will have to negotiate to acquire the necessary skills and desired personnel. The best people are usually hard to get. Their functional manager is probably reluctant to give them up to work on a project, especially if the assignment is for an extended period of time or a full-time project assignment. Some difficult negotiations might require that the project sponsor get involved. Assignments of personnel to a project when the assignment adversely affects a functional department but positively affects the company as a whole will probably have to be made at a senior management level.

Sometimes, project personnel will be acquired from outside the company. There are a variety of reasons and situations when this is either appropriate or necessary. If unskilled labor is needed, an outside staffing agency is a cost-effective way to get the needed labor quickly. When the organization doesn't have the needed expertise available, independent consultants or contractors are a ready source. A capital improvement project will probably require an outside construction contractor, and a contract manufacturer might be utilized for test production or the initial production of a new product.

Regardless of where the project team originates, they must be coordinated, and they will need team development and training. Even experienced team members will need training and development, because each assignment is different and each team has different dynamics and characteristics. Team-building activities might be appropriate, especially if the team will be working together for a considerable amount of time, their task is complex or stressful, or they need to produce results quickly. It is usually beneficial to spend the time on team-building exercises instead of just throwing the team together and letting them develop on their own.

The personnel performance measurement system and any compensation and reward systems have to be developed and communicated with the project team. People need to know what is expected of them. They need to know how they will be measured and compensated. As with any measurement system, the project performance measures have to be tied to the desired outcome of the project. The measurements should evaluate the status and performance toward the project's goals, and they should help guide the behavior of the individuals performing the work.

Another important project resource is money. The project budget is an important document and its development is an important activity. The project budget will include the estimates of the expenditures and costs associated with the project and the system for capturing and reporting these costs and expenditures as they occur. Costs include any items and materials that need to be purchased, payroll expenses, and overhead and other miscellaneous costs. Some organizations will have to develop new systems for capturing and reporting project costs. If projects aren't a normal part of your business, your financial systems probably aren't set up to capture and assign costs to the project. Materials and outside contractor costs are relatively easy to assign to a project. The difficulty usually lies with the internal personnel and overhead costs, and allocating and reporting them as project costs.

Remember that when you're developing the project budget, the costs are estimates. Cost control and cost management deal with the expending of funds for the project and the issues that arise when actual figures don't match the budget. You need to have a plan or process in place to handle variance between actual and budget. Some variance will be small and within the authority of the person responsible for that particular item or the project manager. Other variances will be more significant, requiring authorization beyond the project manager. Of course, if the expense or cost is below the estimate, that's great. The problems generally occur when the actual is above the estimate.

Although you must report and explain all variances, the extent of the variance might require different actions, including additional budget decisions. A cost change request form and cost change documentation system might need to be put in place, depending on the size and complexity of the project. One decision you may need to make is whether the extent of a cost variance, and its effects on other parts of the project, is significant enough to modify or rework the budget.

I'm sure many people wouldn't agree with me, but I believe that if changes are significant enough, or the effects on other areas are significant enough, a new budget should be produced instead of sticking with the original budget and measuring against it. I'm not just talking about project budgets either, but regular departmental and organizational budgets. If your original budget reflects information and assumptions that turn out to be significantly off, why continue to use it? I can hear the defenses and excuses already. Okay, if there are legal or regulatory requirements I can understand keeping it, but that doesn't mean you have to use a meaningless document for internal purposes. For heaven's sake, if you just acquired an incredible new customer and sales will double over your original estimate, won't the rest of the budget be affected significantly also? Why not revise and reissue the budget so that it is realistic? If you must, keep the original for comparison and posterity, but don't use it for planning. If sales double, materials costs, payroll costs, and many other costs will probably need to increase significantly. You might even need to make capital improvements and buy capital equipment. Prepare a new budget and use it. Some people will say that there's no difference between a modified budget and a new one, but there's a good chance that with a modified budget, performance measurements will be based on original estimates, which aren't really appropriate.

You need to assemble, schedule, or acquire all of the equipment, tools, and materials you will need to complete the project, but the final resource we're going to discuss is time. Time is the one thing you can't get back if you lose it. Money, people, and materials can all be replaced, but time can't be. You can add more time, but you can't replace time you've used or lost. There are two ways to schedule a project, or any other activity. You can determine when the completion is required and work backwards to calculate the start time, and the start and completion times for all the individual activities that are required. Or you can calculate the completion date or time by adding up the duration times of all the activities that fall on the critical path.

The time estimates for each work package will be determined when the WBS is being constructed. Some activities can be performed in parallel, whereas others must be performed in series. Activity sequencing is the process of assembling all the individual work packages so that all the work is performed in the order required to complete the project by the due date. On complex projects, the activity sequencing can be quite involved, because many activities will depend on some other activity being completed. Sometimes paths will cross or work completed at one point won't be needed by another activity until much later. Even on a small project, activity sequencing is important to ensure that all the work is completed when it needs to be for downstream activities to begin.

The critical path is the sequence of activities with the longest completion time. It is called the critical path because the activities in this sequence have to be started and completed on time for the project to be completed on time. If any activity on this path takes longer than estimated, the completion date of the project is in danger. The project manager will spend considerable time and energy ensuring that the activities on the critical path are completed on schedule. If any of the critical path activities slips, one or more activities downstream will need to be compressed. This might require a reallocation of resources from non-critical path activities. However, this can get tricky. Just because an activity isn't on the critical path doesn't mean it can be delayed indefinitely. There is a range of time within which the other activities must be completed to ensure completion of the project on time.

8.4 Risk Management

Project risk management is another important activity that is often overlooked or misunderstood. In the risk management process, you identify and analyze risks or potential problems and prepare a plan for responding to or dealing with them. All sorts of things can potentially go wrong on a project. The more complex the project or the activities that make up the project, the more things can go wrong. It is important to identify all the risks, but it is also important not to get carried away. There is a risk that everyone assigned to the project could get sick at the same time, but because it is not very likely, you probably don't want to list that as a risk. That might fall under contingency planning, but not under risk management as we're defining it here.

After the risks have been identified, they need to be analyzed. There are two parts to the analysis. One part assesses the impact of the risk or the effect on the project if the risk materializes. The impact can be identified as a point on a scale, a severity scale. If a problem that has been identified actually occurs, but the effect on the project is small, it would be rated low on the severity scale. If a problem occurs that would stop the project in its tracks, it would rate very high on the severity scale. A numerical scale can help to quantify the risks. Usually, the risks are then prioritized in descending order of severity. You need to worry more about the bigger risks than the ones that have little impact.

The second part of the risk analysis is to determine the probability that the identified risk or problem will occur. This is pretty straightforward, but it is not always easy. You need a way to determine whether a problem is 70 percent likely to occur or 30 percent. It might be easier to use a limited number of ranges, say 0 to 25 percent, 26 to 50 percent, and so on. Then each item can be placed in the appropriate bucket. If you use these buckets, you will need to determine the numerical score to assign to each bucket. For example, the 0 to 20 percent bucket might equal 0.1, which is the midpoint of the bucket.

Once you determine the severity scale and the probabilities, you can develop a risk matrix and calculate the risk score for each identified risk (Figure 8.5). The risk matrix calculates the risk score for each identified risk. For a risk that has a very low severity and a low probability (0.1), the risk score is 0.1. For a risk that has a very high severity (5) and a high probability (0.9), the risk score is 4.5. The risk scores allow you to quantify the risks and compare them with each other. Again, you might want to prioritize the risks based on the risk scores. The items with the highest score are the items on which you want to focus your efforts. You want to do what you can to prevent them from happening or develop plans to deal with them if they do happen.

Along with identifying the risks and performing the risk analysis, you need to develop a plan or a process for addressing the impacts and effects when the problems materialize. You'll have a different response to a low-severity item than a very high one. You might want to develop a set of options for responding to the problems and a process to determine which option will be selected. You will want to identify the person who will be

			Probability			
		0.1	0.3	0.5	0.7	0.9
Severity	Very low (1)	0.1	0.3	0.5	0.7	0.9
	Low (2)	0.2	0.6	1.0	1.4	1.8
	Medium (3)	0.3	0.9	1.5	2.1	2.7
	High (4)	0.4	1.2	2.0	2.8	3.6
	Very high (5)	0.5	1.5	2.5	3.5	4.5

Figure 8.5   Risk matrix.

responsible for responding to the problems. For the very low severity problems, the person responsible for the activity might be the appropriate person to respond. For high-severity items, the project manager might be the person responsible, and for very high severity someone beyond the project manager might have to step in to address the issue. The response might also be affected by the cost associated with the problem — either the cost to fix the problem or the cost of the impact on the rest of the project. The cost might be included in the severity scale or it might be addressed separately.

Finally, you need to track the risks. Some risks might have a warning indicator that alerts you to an impending problem; others might not. Either way, you should track any actions and decisions that are made to avoid, mitigate, or address the problems when they occur. This documentation will let you go back to review the problem and the actions taken. This can help you if similar problems recur on future projects and in the event of a dispute with the client or an outside legal or regulatory agency. You probably want to document the effects on downstream activities and the project as a whole, and this will feed into the communications system for disbursement.

8.5 Miscellaneous Issues

At some point it will be time to deliver something, either to another person or team working on the project or to the client. The logistics of this delivery need to be worked out, preferably before the delivery is to take place. An official handoff should probably occur, with documentation. The person delivering needs to document and ensure that the task or product is complete and ready to be delivered. The person accepting delivery must be satisfied that the task is complete or the product is acceptable. The time, location, and any necessary equipment needed to deliver must be arranged. Sometimes these activities will be simple, such as checking off or signing off that a task has been completed (for example, all duplicate stock numbers have been purged). Other times, the task might be more complex. If a new prototype of a complex product, or a hazardous material, has to be delivered halfway across the world, the logistics and planning necessary to complete the task might be a project in itself.

There will be various reporting requirements for the project. The project sponsor and senior management will want to see regular status reports as well as interim reports if an issue warrants it. The reporting requirements fall under the communications system and should be well planned for. The results of performance measures need to be reported. Some performance measures only need to be reviewed by the project manager, but some important measures will be included in the regular status reports. Any performance that is below standard needs to be addressed, and the actions taken or to be taken should be included in the report. Don't just report a problem; report the actions taken to resolve the problem.

Progress needs to be reported regularly. A progress report will probably be a part of the status report but may be issued at other times if progress is either well ahead of schedule or behind schedule. A percent complete report is a common progress measure, as is completion versus planned schedule. Work in process might be another component of the progress report. Some sort of estimate of the future progress or status is probably appropriate too. Letting management know the status of risks, any potential new risks, or whether everything looks good for completion on time is always a good idea.

8.6 Critical Chain Project Management

Dr. Eli Goldratt, the developer of the theory of constraints (TOC), has extended the principles of constraints management from the shop floor to project management. Chapter 6 presents TOC in some detail, but one of the most familiar aspects of TOC is drum-buffer-rope production scheduling. To review, the drum is the constraining resource, which sets the pace of production. The rope ties the drum to the first step in the production process, so that this gatekeeping operation operates at the same pace as the drum. The buffer is a stock of inventory that is held in front of the constraining operation (the drum) so that there is no interruption in the constraining operation. Because the output of the entire system depends on the constraint, the constraint must never stop running because of a lack of parts to run.

In his book, *Critical Chain*, Dr. Goldratt presents the adaptation of TOC to project management. Specifically, the area of buffer management has been adapted to project time management. One of the hardest things to estimate and manage in a project is the time. The time estimates for each individual work package involve a measure of uncertainty. The estimated completion time for the project might be more of a pronouncement from management than a calculated or realistic estimate. A certain amount of safety time will be included with all estimates. All of these issues work against an accurate and realistic estimate of the time requirements for a project.

That's not all. You must consider other factors when developing the project schedule and managing the project. Generally, there are no incentives to finish a work package early. Dr. Goldratt even suggests that early finishes aren't even reported. Not only are there no incentives to finish early, there are actually disincentives, especially for people and organizations that work on projects regularly. If your organization rarely undertakes projects, at least formally, completing a step early, and reporting it, probably isn't a big deal. But what if you regularly work on projects and have to work with many of the same people when you work on them? If you complete your work early and report it, there will be increased pressure on everyone else to either complete their work early or shorten their estimates for future work or projects.

Furthermore, even if one step is finished early, the likelihood of the next step starting earlier than planned is small. It seems to be human nature to put off starting a task until the last possible minute. What are the chances that you will start your work early if the step prior to yours is finished early? More than likely you will start your work when you originally planned to start or when you need to start so that you're done when you're scheduled to be done. Why start early if there is no benefit to it?

This brings us back to the time estimates for the project. It is suggested that naïve or inexperienced people will provide time estimates that are "accurate," or that have little safety included. Those with more experience will provide estimates that give them a high probability of completing the work within the estimated time. This means that there is significant safety time included in the estimates. And the more uncertainty there is in a particular step or work package, the more safety time will be added. If there are no incentives to completing work early, there are certainly penalties to completing the work late. You will certainly hear about it if you are late. The person responsible for the next step will have something to say about it, even if they don't start their work right away (as mentioned above), and the project manager will want to know why. All the work downstream will be affected because they will have to start later than planned or try to make up for the delay. If the work falls on the critical path, the fate of the entire project is at stake. Delays usually cost money, often considerable amounts of money, and often include penalties assessed by the client. It's no wonder that a fair amount of safety is included in time estimates.

Looking at these points I just brought up, you can see that early completion of a task is generally meaningless to the project, but lateness is passed on to the next step. This means that the protection that is built into the project is wasted. The safety time included in the time estimates for the work packages works the same as inventory in the production process. It is protection against uncertainty. But unless the resource is a constraint, the inventory buffer is unnecessary. It's waste. Unless the work package is on the critical path, the time buffer is unnecessary and it's wasted.

Where does this leave us? Time estimates for work packages include buffers to protect against uncertainty (meaning problems that will delay the work). Most of the buffers are wasted and they're unnecessary unless they are on the critical path. The conclusion would be to remove or eliminate the time buffers for all the work that isn't on the critical path and do something about the buffers that are on the critical path so they're not wasted. Well, that sounds pretty simple, but how do you go about doing that?

For one thing, there will still be uncertainty in the project. This means that something will happen that will cause a delay or otherwise affect the project. You need a buffer to protect against this uncertainty; but, as we've seen, the individual work packages aren't the place to put the buffers. If you took all the safety time out of the individual work packages and added it all together into one buffer package, you could call it the project buffer. Look at Figure 8.6 to see the difference.

Of course, your next question is, what do I do with that project buffer? Did I just move it, or am I going to do something with it? Am I going to use it differently? We are going to use this project buffer differently. This first step is just to figure out exactly how much buffer we have. By taking it out of the individual steps and adding it all together we can see how much we have. If you came up with your time estimates in the regular way, when you remove the buffers, as we've done, there still might be some left in the individual steps. Either the person revising the estimate still wants to keep some protection or the original estimate has an unknown amount of uncertainty to it, so it's not clear how much safety is really there. But that's okay for now; we've already made an improvement.

I said earlier that the work on the critical path needs to be protected with time buffers, but not the work that's not on the critical path. That's true, but we do need to protect the critical path from delays on other paths if those delays would affect the critical path. To do that we'll take the safety out of the individual steps that aren't on the critical path and

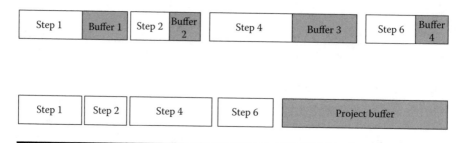

Figure 8.6 Buffer comparison.

put it into feeder buffers that will be used to protect the critical path. When we take the safety out of the individual steps and add it up, we're not going to need all that time, so we can just throw some of it away. If we keep just half of it, we'll still have a good size feeder buffer. See Figure 8.7 for what our project and buffers look like now.

The feeder buffers are needed only where the feeder activities intersect the critical path. This is similar to how we protect constraining resources in production with inventory buffers. The feeder buffers protect the critical path, but there could be a delay on a feeder path that's greater than the time in the feeder buffer. The project should still be on time because the project is still protected by the project buffer.

Notice from the section heading that we called this method of time estimating, scheduling, and management of the project "critical chain." You might then be wondering why it's called critical chain and what the difference is between critical chain and the critical path. You'll be glad to know that the difference is not that great, but the concept is a little different than what you may be used to.

We defined the critical path as the sequence of activities with the longest completion time. That's true, but it might not be a complete answer. Take a look at the critical path again (Figure 8.8), but notice the highlighted steps: 5, 6, and 10. Step 6 is on the critical path, but Steps 5 and 10 are not. But let's say all three of these steps are assigned to the same person to complete. This person is a resource that can't work on all three steps at the same time. That means that this person is a constrained resource. It could be a person with a special skill who is assigned to work on several steps, or it could be a machine or some other type of resource. Because these steps cannot be completed in parallel (because they depend on the same resource), they must be completed in series. That would make the critical "path" jump around from the originally

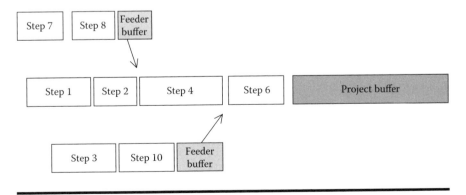

Figure 8.7 Feeder buffers.

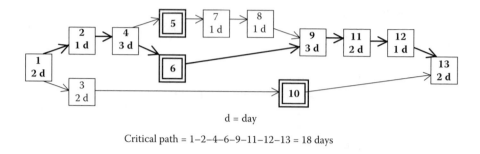

d = day

Critical path = 1–2–4–6–9–11–12–13 = 18 days

Figure 8.8   Critical path.

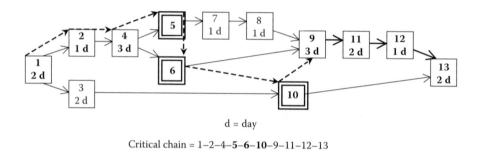

d = day

Critical chain = 1–2–4–**5**–6–**10**–9–11–12–13

Figure 8.9   Critical chain.

defined path to the non-critical paths and back again. Because this isn't the same as the original definition of critical path, the term *critical chain* was coined.

If critical path is the sequence of activities with the longest completion time, we need a new definition for critical chain. The definition of critical chain is the longest chain of dependent activities (Figure 8.9). I said that feeder buffers are needed where the feeder activities intersect the critical path, to protect the critical path. I said that the critical path is the constraining resource, and that's why we have to protect it. Now we have a critical chain that we have to protect. We have to protect the constraining resource. We protected the critical path with feeder buffers (time buffers where the non-critical path intersects the critical path). We protect the constraining resource with the same type of feeder buffers. The buffers protect the constraint, so they're placed in front of the constraint. Some of the feeder buffers might change location — those where the constraining resource is on the non-critical path. Take a look at Figure 8.10 to see what we have now.

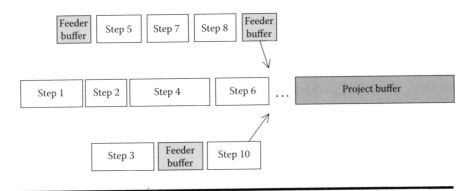

Figure 8.10 New feeder buffers.

We've laid out a wonderful framework and built our arguments step by step, but now you might have some fundamental questions, such as, what does this really mean and how does it affect a project? Let's try to tie it all together now.

Let's start with some questions. How accurate are the time estimates for individual work packages? How much uncertainty is there in these time estimates? How often do projects fail to meet their deadline or original planned completion date? How much extra time is included in the time estimates to act as protection? You can see what these questions are getting at. Projects often miss their deadlines, and even if they do, a lot of extra time is built into the project schedule. If time is money, completing projects on time, or early, saves money. That's what critical chain project management helps you do — complete projects on time and save money by offering a new way to schedule and manage the project.

Change is never easy, and changing time estimates by eliminating the protection inherent in most estimates won't be easy. But look at the advantages. The time estimates for the work packages will be more aligned with the actual time needed to complete the work, instead of including extra safety time. All the safety time will be "packaged" together into the feeder buffers and the project buffer. Maybe at first you will keep all that extra time and put it in the buffers. But as you complete projects and see that you have some of that safety time left over (meaning you completed the project early), you'll start to cut out some of it. Eventually, your projects will have only the work content time and time buffers that are just big enough to protect against the amount of uncertainty that you calculate. Doesn't that sound great? Give it a try.

Chapter 9

ERP: Information at the Speed of Light

There is a good chance that you have implemented an enterprise resource planning (ERP) system already, are considering implementing an ERP system, or are enhancing, upgrading, or replacing your ERP system. In this chapter we'll discuss what an ERP system is, what it encompasses, and how to select, implement, and maintain it.

The subtitle of this chapter is "Information at the Speed of Light." Whether they are good and useful or garbage, data and information will be flying around at the speed of an electron. This can be good or this can be bad. It depends on you. Data accuracy is more important than ever. With today's technology you can promise a customer a product and a delivery time while you're standing in their warehouse. But if the information you're using is defective, your promises are worthless. Technology is great, if it is used properly.

With inventory, as the lead time decreases, the amount of inventory in the system goes down. The inventory starts to flow, rather than move in batches. With information, the same sort of thing happens. As your system becomes more integrated and electronic, and less paper based, the information flows continuously. Instead of stacks of paper and folders moving across your desk, information constantly bombards you. Reports can be sorted through and responded to in reasonable time (okay, not always reasonable), but constant information flow is hard to ignore and becomes overwhelming. How to handle all that information — how to summarize it, sort through it, and manage it — is a consideration that is often overlooked. Keep this in mind as you work through your ERP decision-making and implementation process.

9.1 What Is ERP and Why Would I Want It?

There is no one definition for what is included in an ERP system, although many people will tell you there is. Generally an ERP system refers to a computer-based system that integrates most, or all, functional areas of a business. Depending on the size of the company and the software purchased, the ERP system may include modules for different areas. For a small manufacturer, the ERP system may include modules for the accounting and financial areas, sales, purchasing, receiving, all inventory data, production planning and scheduling, and warehousing and shipping. Larger companies, or small distributors, may have a separate warehouse management system. Other companies may include modules for engineering, quality, capital equipment, employee tracking, safety, customer relationship management, and others.

What it boils down to is an information system. That terminology is almost outdated these days, but an ERP system is a sort of grand information system. Most people think of a computer system when they think of an information system, but it doesn't have to be computer based. Before computers, organizations had information systems. An information system is simply a defined method of communicating and sharing information. An ERP system is a sophisticated and technologically advanced information system. The idea is to integrate the information flow among the different people who need it, use a single database for the company, and standardize the information among the users. Many companies, even those with ERP systems, still share, or pass along, information very inefficiently. Without an integrated system, the receiving department could receive a shipment, but no one else may know about it for days afterward. The receiving paperwork might have to be reviewed by a supervisor, sent to the inventory department for input, sent to accounts payable for payment processing, and sent to accounting for posting. The data could be changed or modified along the way, accidentally or purposely; some data may end up in the wrong place; and it may take various amounts of time to be processed at each stop. With an integrated system, the receiving data can be input directly at the receiving dock (or simply verified from an advance notification) and be available immediately to everyone who needs it.

One of the main reasons companies buy ERP systems is an expectation of greater efficiencies and improved information. You need to ask yourself up front, why do I want to implement an ERP system? How will an ERP system improve my company's performance? Why do I need an ERP system? What are the expected benefits of implementing the system, and do they outweigh the costs of implementation? You need to ask these questions and answer them honestly. You need to be ready for the system and the process of implementation, or you need to wait until you are ready. Poor selection and implementation of an ERP system can seriously damage your company (and your career). The costs are large. Hardware, software, employee time, consultants, possible disruptions; the costs add up quickly. Careful analysis and a lot of work are required. Don't let the look and sizzle of a fancy system cloud your thinking.

But then, the possible benefits are enormous. A good implementation can start to show benefits even before you buy anything (we'll discuss how later). The efficiencies, improvements in information and information flow, flexibility, and enhanced planning and execution tools can improve efficiency and productivity, which show benefits far beyond the costs of implementation. As with anything, done wrong, it will be expensive and useless. Done right, it will be a useful and beneficial tool. A useful and beneficial <u>tool</u>. An ERP system is simply a tool to help you process, analyze, and use information to run your business. An ERP system is not an answer to your problems. If you have problems running your business now, you will probably have problems running your business after you implement an ERP system. An ERP system, like any other tool, can help you, but it can't solve your problems. Hopefully, this chapter will guide you down the right path.

9.2 History and Evolution

Why should you care about the history of ERP? Knowing the history of ERP may not be of great importance to you and may not help you select and implement a system, but it may help you understand ERP, how it fits in with business strategy, and how you can use it for your benefit.

Early on in the computer age, BOM processors were developed so that companies could automate the task of sorting through all of their bills of materials (BOMs) to help the inventory planners. For those of us who are accustomed to computers and automation, the idea of manually working through a stack of BOMs to calculate materials needs is unthinkable. As computers became smaller, faster, and easier to use (by the standards of the day), BOM processors soon evolved into materials requirements planning systems.

The materials requirements plan (MRP) was quite an advance. Besides just calculating material quantities needed, it introduced timing of the materials needs. The MRP became a materials ordering tool and a scheduling tool. Due dates and lead times were incorporated into the equation. Besides working out what was needed, how many you had, and how many more were needed, you could now determine when they were needed. Materials requirements planning is concerned with materials; raw materials and manufactured components. The expected completion dates of the manufactured parts and the arrival dates for the purchased parts are monitored along with the need dates for these parts. Priorities can be set and schedules developed based on this information.

Continuing advances added functionality and processing capabilities. Capacity planning was an important enhancement, and other higher level planning tools were added. Execution control and monitoring and integration with the financial system brought us to the closed-loop system, or MRP II. Additional enhancements and functionality further integrated the business, to bring us to what is commonly and generically known as ERP. ERP, depending on the system, attempts to integrate the entire organization. All the core manufacturing planning and execution tools, the financial system, sales ordering, human resources, payroll, employee monitoring, and more combine in one integrated system designed to increase the effectiveness of the organization. Whether any system actually accomplishes this goal is a matter of debate, but the tools are widely available and being enhanced every day. With the level of sophistication of these systems, the users need an equivalent level of sophistication, in their knowledge and use of the systems and how the system does and should work.

What's next? It's hard to say. Computing power is still increasing exponentially, and ERP systems become more advanced and sophisticated every day. New enhancements are added, functionality is added, and companies are becoming more integrated with each other. ERP is sure to evolve into a more powerful and sophisticated tool, extending the tools up and down the supply chain. Internet applications with a central database available to all supply chain partners are a possibility. You need to be prepared for the evolution. You can do that by preparing your business, and your information systems, to perform at their best.

9.3 Modules and Functions

ERP is a generic term that generally means a computer system with modules for several areas of the business. For a manufacturing company, an ERP system usually consists of, at a minimum, the core manufacturing, planning, and inventory functions and the financial module. It is hard to discuss ERP systems in generic terms. Although the different vendors' packages share many commonalties, there are also many differences.

However, some of the more common modules and functions are discussed in this section.

The inventory functions are a good place to start. Sometimes, very basic inventory functions are included with the financial module. These are generally not sufficient, except for the smallest companies. Other systems include the inventory functions with the core manufacturing system, and some systems include inventory as a separate module. Regardless, the inventory functions are essential for any manufacturing company. The inventory functions allow you to define and manage your products. Stock numbers, item descriptions, on-hand information, inventory locations, materials costs, and vendor information are some of the basics found in the inventory functions. The purchasing functions may be included with the inventory module. These functions would include identifying vendors, defining vendor lead times, and lot sizing information.

Again, I want to emphasize that every system will be different, and the core manufacturing and planning functions may vary considerably, although many of the systems are quite similar. The manufacturing functions may include the BOM functions, routings, machine and workcenter definitions, work order processing, master production scheduling, materials requirements planning, and labor tracking and reporting. The sophistication and capabilities of the manufacturing module will vary depending on the industries and company sizes targeted by the vendor. Of course, the inventory and manufacturing modules must integrate seamlessly. The manufacturing module will remove materials and components from inventory and place finished goods into inventory. Inventory may be allocated or issued, it may move between inventory locations, and it will be received and sold; this information must flow freely between the modules. If there is a separate planning module, this information must integrate with the planning tools too.

Much of the planning functionality may be included with the main manufacturing module. Sales forecasting is a basic planning function. The sales and marketing department is responsible for the sales forecast, and they should be responsible for entering, maintaining, and updating the forecast in the system. The sales forecast may be within the master production scheduling module or the sales order processing module. Regardless of where it is in the system, it will be integrated with the planning tools. Don't let the sales and marketing department pass on responsibility for maintaining the forecast in the system just because it may be included in a module other than sales order processing. If there is a separate planning module, it may include more advanced planning tools, such as sales and operations planning, simulations, or advanced planning and scheduling. Sales and operations planning isn't necessarily a more "advanced" tool, but it is not as widely used and integrated into

ERP systems as master production scheduling and materials requirements planning.

When developing the sales and operations plan, master production schedule, and MRP, you must perform capacity checks to ensure the feasibility of the plans. If you don't have the capacity to produce what is planned, the plan is worthless. Capacity planning may be included with the planning functions, or there may be a separate capacity planning or scheduling module. The planning functions will assume you have an infinite amount of capacity. That is, the master production schedule and the MRP will be calculated and developed regardless of how much actual capacity you have. It will calculate how many items you need to make based on what the demand is (forecast or customer orders) and what you already have (on hand and on order). On the other hand, finite loading and finite scheduling take into account the actual amount of capacity you have available (in other words, it will use the data you have entered into the system). The capacity needed to produce an order is compared to the capacity available, or the capacity already scheduled, to develop a schedule that does not overload your machines or workcenters. As a small manufacturer, you probably don't need the level of detail and sophistication that finite scheduling promises. I mention it here because you may come across it during your ERP search and implementation and you should at least understand what it is.

The module that most companies probably look at first, and one of the most important, is the financial module. The financial module should include all the financial and accounting functions, beginning with the general ledger. It will also most likely include accounts payable and receivable, subsidiary ledgers, capital equipment, maybe payroll, and possibly budgeting. The financial module may be the most integrated module, because it must interface with every other module. Just about every transaction has a financial impact and must be recorded and posted to some account or ledger. The financial module should be able to reconcile transactions and roll up the subsidiary ledgers into the general ledger. Various levels of automation may be allowed for the posting of transactions, and your company will have to decide how much automation will be allowed and which transactions can be automated. You probably don't want transactions to post automatically if you don't have sufficient controls in place and a high level of confidence in your internal systems.

A sales order processing module is standard. Without sales you don't have a business; but you need to manage your sales orders. At the very basic level, the sales order module is used to track and monitor customer orders, but sales order processing can be much more than that. The sales order module is vital to the integrated system. The forecast and actual customer orders, as reflected in sales order processing, affect the planning systems. Firm customer orders affect the scheduling system. The sales and marketing department utilizes the sales order processing system to track customer orders and to measure and report on customers and customer activity. Sales orders integrate directly with warehouse systems for pick ticket generation, inventory allocation, fill rate analysis, and delivery scheduling. Sales orders integrate with the financial system, so that upon order fulfillment, invoices can be generated and the related accounting transactions can be processed and posted.

Some systems will have, if needed by your company, a sales configuration processor. The sales configuration processor is used when there are multiple options available for particular products. Automobiles, industrial machinery, even jewelry, are some industries that may benefit from a sales configuration processor. For jewelry, you may start with a basic ring model, then have the ability to custom configure it by choosing the size, metal (gold, silver, platinum), stone (diamond, ruby, emerald), stone size, head (the part the stone sits in), and engraving (inside, outside, style). If you have a large number of options or possible combinations, such as in an automobile or industrial machinery, a sales configurator is almost a necessity. The configurator interfaces with the BOM processor and the inventory and planning systems.

Related to sales order processing is customer relationship management (CRM). Some CRM programs are stand alone, or separate from the ERP system, but many ERP systems include it as a module or option. CRM is concerned with monitoring, tracking, and reporting on customers and potential customers. Sales promotions are often tied into or utilize the CRM system. Frequent, loyal, or high-dollar customers are often provided customized promotions or discounts based on the information that has been gathered about them through the CRM system. Information such as the customer's birthday or anniversary can be tracked through the CRM. The system can be programmed to use this information to automatically send a prompt to an employee for a card to be sent to the customer, or an e-mail can be sent directly to the customer. The idea is to better track and keep in touch with your customers so that you can provide them a higher level of service. Of concern during an ERP implementation is how or if the CRM module affects the other parts of the system. There may be no direct impact, but if there is it should be known up front and proper procedures put in place to account for it. For example, maybe a free gift is sent during the holiday season to every customer who has spent a certain dollar amount in the past year. How would that affect the planning, scheduling, distribution, or inventory systems?

A human resources module may be included or optional with your ERP system. The human resources module may integrate with the other modules through employee job classifications, work hours, and pay rates.

Employees may be assigned different skills classifications, and these may be tied to workcenters for assignment and scheduling. Employee pay rates can be tied to workcenters or production processes and rolled up in product costs. Employee efficiency rates may be assigned and utilized for production scheduling. Payroll functionality may be integrated into the human resources module, as well as time and attendance monitoring functions. Many other human resources functions, such as tracking employee training, may be included in this module.

Depending on the products you manufacture, you may need an engineering module. You will probably need this if you have highly engineered products, complex products, or products where changes are made frequently. The engineering module may include functionality for testing, revisions and version tracking, materials substitutions, schematics or drawings, or formulas. Even small companies with relatively simple products may need an engineering module. I worked for a food manufacturer, a commercial bakery, that needed to develop, enter, and monitor the formulas for the cookie and cracker products they produced. These formulas constituted the BOM for the bulk product and became a part of the BOM for the finished goods, along with the packaging. This just illustrates that you will need different functionality depending on your products and your company.

A warehouse management system (WMS) may be the primary component of the ERP system for a distribution company, but it may also be included in ERP systems for manufacturing companies. A WMS goes beyond the functionality found in most ERP systems. Most ERP systems include basic inventory location, including bin location, functionality, as well as pick ticket generation, inventory allocation, and invoice generation. A WMS provides the greater functionality and more powerful processing needed in larger warehouse or distribution environments. Pick scheduling, warehouse zoning, employee classification and scheduling, order consolidation, dispatching, and delivery planning and scheduling are functions that may be included. Warehouse space planning and layout, random stocking locations, automated identification and sorting, and metrics are other possibilities. As with all systems, and the other modules, there should be a business justification for a WMS.

Finally, I want to include E-commerce. E-commerce may not be considered by some to be part of the ERP system, or a module of the system, but it should be viewed this way. The E-commerce system should be integrated with the rest of your systems. Many organizations forget this in their excitement to begin an E-commerce initiative. The E-commerce system is often implemented without much regard to its integration and impact on the functioning of the rest of the system. This is a mistake that can cause huge problems. Many companies think they can create a Web site and start selling through the Internet without needing to thoroughly analyze this action. They may think sales will be negligible, at least in the beginning, so it won't have much effect on the regular operations. But if someone must manually check the orders that come through the Web site, then manually enter them into the "regular" ordering system, this has an impact. Any time you have two or more electronic systems that don't integrate, you have a system that is prone to problems and errors. It is far better to spend the time and money to plan ahead, perform some analysis, and integrate your systems prior to implementation, than to expect to do it later, when you have different systems up and running.

9.4 Selection

This section discusses issues regarding the selection of an ERP system. Note that Section 9.6 lists software as Step 12 in a 16-step implementation process, meaning that system selection is not the first thing you should do. However, because many organizations tend to think they need to find and select the system before they can implement it (rather than the selection process as part of the implementation), I will discuss it here as a step that is separate from implementation.

Before you select an ERP system you need to analyze your business and your business processes. If you think buying and installing an ERP system, or any other system, will magically transform your business, you're just plain wrong. The only reason to implement an ERP system is because it fits into your business strategy and will help you run your business and achieve your goals. You shouldn't implement an ERP system just because you think you need one, you want one, your friends all have one, you got a great deal on one, or your want to move from a paper-based system to an electronic system. If you can't justify the need for an ERP system through a thorough business analysis, including cost/benefit and return on investment, you shouldn't buy one.

You need to analyze your current business processes in relation to your company's strategic plan and mission. If your current processes do not support your strategy or you are not achieving your goals, you need to determine why. What is preventing you from achieving your performance goals? Are policies and procedures out of line with your strategy? Are equipment, machinery, or facilities obstacles to improvement and growth? Are employees not properly trained, unmotivated, and unproductive, or are you experiencing labor problems? Are customers dissatisfied with your service? Are lead times too long? Is quality too low? Are suppliers not meeting your standards? If you have any of these problems or issues, you need to examine their root causes and make changes to your current operations.

As you examine your business and your processes, you will identify areas for improvement. Make the improvements immediately. Don't wait until you have looked at every detail or every part of the organization; make improvements as you identify the need for them. A series of small improvements starting now is better than a big improvement later. Too often that big improvement never comes. Then you have no improvement at all and have probably gone backwards while waiting.

Once you've performed your business analysis and process improvements, you may be ready to justify the need for an ERP system to help bring you to the next level of performance. Or you may find that you don't need an ERP system at this time. If so, don't try to justify the need for a system because you think it would help. If you've gone through the time and trouble to perform the business analysis, you should rely on the findings. If through the analysis you find you don't need a new ERP system at this time but you may in the future, don't buy one now; wait until you do need it. Why might you not need an ERP system or not justify the purchase at this time? If you are currently undertaking other initiatives and you don't have the resources to add an ERP implementation, you probably should wait. If you don't have the expertise in house and would have to rely totally on outside expertise, you might want to wait until you are able to bring some of the expertise on board. If you don't have adequate cash flow to pay for the system, you are experiencing substantial growth, or you are making substantial changes to your business environment, you may want to delay an ERP system implementation. Perform a thorough analysis, see what you find out, and go from there.

When you get to the point where you start looking for an ERP system, things can get interesting, and you're going to have to control yourself. It is easy to get carried away with slick presentations, lots of extra features, and enhanced functionality. When you start to look for systems, you need to send out requests for proposals or a set of requirements. You need to list your business requirements; in other words, the system functionality and features you need to run your business the way you currently run it now that you have performed your process improvement activities. Don't try to fit your business to an ERP system's functionality. Find an ERP system that includes the features and functions you need to run *your* business. That said, you may have to make accommodations for the standard nature of many systems' functions and features. As unique as you think your company is, it is very similar to many other manufacturers, whether or not they make similar items to yours. Once you've listed all the features and functions you need to run your business, along with some nice-to-haves, you must submit the proposal to potential vendors.

You need to see how well each vendor's system fits your requirements and how they answer your request. Be wary of vendors who tell you their system fits your needs exactly (possible, but unlikely), they can customize their system to fit your requirements (not a good idea), or they don't have a system right now that fits your needs but are coming out with one that does very soon (yeah, right). Remember, this is an important investment and it requires all the diligence and scrutiny of any substantial investment.

Before you can submit your requirements, you need to find potential vendors. You can take a number of avenues. This is one area where professional networking and business associations start to pay off. You are not the only person who is looking for, or who has looked for, an ERP system. Ask around and see where other people began their search. If you know people in a similar business or industry, you might ask them what system they are using or how they found their list of potential vendors. Some associations print lists of software vendors along with the products and features they offer. They are a good start. Many consultants or consulting companies either keep lists of software vendors with the products they offer or are familiar with a number of software systems or vendors. Some consultants specialize in software selection services. There are some Web sites dedicated to listing and comparing features of different software products, including ERP systems. You should do a thorough search, possibly using multiple sources, to find your initial list of potential systems and vendors. If you don't have a good list to start with, you won't find the right system for you.

Your initial list of potential systems or vendors may include from six to twelve systems that you want to investigate further. Review and investigate the systems on this list with enough detail and effort to narrow it down to three or four. The systems on the initial list will have most of the features you need, but in various combinations. The three or four systems you narrow it down to will have most of the features and functions you need and an interface and usability you find attractive. You may also eliminate some systems based on price (either too high or too low), although your price range might be fairly wide at this stage. The three or four systems you hone in on are the systems you want to examine more closely through a product demonstration.

A number of issues may surface when it comes to the product demo. The first concerns the data that the vendor uses to present the demo. If the vendor uses its own data to present a prepackaged demo, you won't get a good feel for how the system will work in your business. It is far better for the vendor to give you a demo using your own data. This will allow you to see how the system will work in your business, and you will be more comfortable with the demo because it will be more understandable to you. If vendors tell you they can't use your data or that it isn't necessary, be suspicious. Why can't they use your data or why don't they want to? They will not use everything; just a certain amount of select data that allows them to perform the demo to your satisfaction. Will their system not work with your data? Do they not want to take the time to provide a customized demo? Do they not want to take the time to understand your business enough to give you the demo that you need and deserve? If they can't or won't use your data and can't give you a truly reasonable explanation, you might want to remove them from your list of potentials.

Now, is the vendor giving you a sales pitch or a product demo? You don't want a sales pitch. You've done a lot of work up to this point, researching your needs and the features and functions of the potential software. You're ready for a demonstration of how the system works, with your data. You don't need, or want, someone trying to sell you something. If they've made it this far in your selection process, they should be confident that they can give you what you need without pressure tactics. Who is presenting the demo? Are they the technical people, meaning the hardware and software people? Or are they the business and manufacturing experts that can answer your questions and guide you through the demo? Are they dedicated product demonstrators, or are they the system implementers who will be working closely with you if you purchase and implement their system? You may not be comfortable working with an ever-changing cast of characters. If the people who give the demo are never seen by you again, how can you be assured that any answers they give you will be the same as the ones you get when it is time to implement? When it is time for the implementation, you need to be comfortable and confident in the team with whom you will be working. The demo is a good time to meet with, talk to, and become familiar with the implementation team, or at least part of it.

Another demo issue is standardization or customization. Are you being given a demonstration of their standard product, or have they customized any parts of the system to show you what you want to see? Are you going to be able to use their standard product to run your business, or is customization necessary? If their standard product cannot be used, why not? How did they get this far in your selection process if their standard product isn't adequate? You want to see how their standard product works, specifically with your company's data. If their standard product has deficiencies, you want to know what they are and whether or not they are a showstopper. Don't let them show you what you want to see, only to find out later that you need customization or need to spend more time and money for their system to work for you.

Can you work through the selection process yourself or do you need outside help? The selection process is a major undertaking. If you haven't been through the process before, don't have anyone in the company who has been through the process, don't have the resources necessary to do it all yourself, or want some additional expertise to help you through, you may want to get outside help. This is not uncommon, but it does require extra work on your part. In addition to working through the software selection process, you need to work through the consultant selection process. Unless you have a trusted outsider available to you, you need to find a consultant. The benefits may be well worth the effort, but it is still extra work.

9.5 Education and Training

Education and training are two different things. Education answers the question, why? Training answers the question, how? You need to perform both education and training. However, the needs are different for each. The methods, the topics, the settings, the people involved, and the amount of time will be different for each. Training involves learning, using, and practicing how to operate the system. It includes learning how to manipulate the physical aspects of the system, such as entering information, generating reports, posting transactions, and navigating through the system. Education involves learning about why the system is designed the way it is, for your particular system and for ERP systems in general. Education includes concepts and how they are applicable to your business and business in general. Education will lead to questions, analysis, improvement activities, and performance expectations.

Training will be heavily weighted to the employees who enter and process transactions on a daily basis. Education will be heavily weighted to supervisors and managers who review system performance, perform analysis, and hold decision-making positions. There will be some overlap of education and training; some people will require both. In smaller organizations, the overlap will be greater than in larger organizations. In smaller organizations, some people will perform daily processing tasks as well as higher level analysis and evaluation. Of course, this is true not just for companies undertaking ERP implementations.

Budgeting for education and training is a requirement that is often overlooked. Education and training require the company to use, or expend, resources. These resources include cash outlays, employee time, equipment, and facilities. Time, equipment, and facilities are often viewed as fixed costs or resources that are already available and thus don't need to be included in a budget. You have to remember that employees' time is very valuable. Any time spent in education or training is time away from their normal duties that needs to be "made up" somehow; the work backs up, other employees perform their tasks (as well as their own), temporary labor is brought in, or the employees have to work overtime. Overtime for hourly employees is a cash outlay and overtime for salaried employees affects morale and, therefore, productivity, efficiency, and quality.

Budgeting for education and training for an ERP implementation needs to include all resources involved over the timeframe of the implementation. The timeframe includes the initial stages of business analysis and process improvement, through implementation, and on into evaluation of the systems performance. Budgeting often fails to include all costs over the life of the project, often for political reasons. Budgets are usually prepared for a specific time period, and if the project runs over multiple time periods, the budget is broken up into pieces that fit neatly into the predetermined windows. This often distorts the total cost picture and is often done purposely. Because of established budgeting procedures and the ingrained thinking that goes along with it, games are played with the numbers to give a false impression of the total cost or impact of the project. It is important to gain a clear and concise picture of the total costs of the ERP implementation, and this includes the total cost for education and training. The performance of the system needs to be evaluated once it is up and running. This evaluation should include the impact and results of the education and training, and the costs of the education and training will be part of the calculations.

What it boils down to is that the education and training portion of the implementation needs to be budgeted, so that resources can be allocated and the education and training can take place when they are required. Unbudgeted tasks are unlikely to take place or will take place at an insufficient level to be effective. Education and training are requirements of the ERP implementation process. Required tasks need resources allocated to them. Allocation of resources tends to occur when the resources are budgeted or planned. Therefore, education and training must be budgeted.

So, the big question becomes, what education and training are needed? The education and training needs will vary for every implementation and every company. However, at the most simplistic level, the employees need to know how to use the system to do their jobs and they need to know why certain things are done the way they are. After that it begins to get more complicated. Many factors are involved, such as whether the employees are familiar with ERP systems already, the extent of their knowledge of ERP systems, their knowledge of the company's systems, processes, and business practices, and their level of experience with education and training activities.

There are any number of ways to determine the educational and training needs of a company for an ERP implementation. However, formal analysis and planning for this aspect of the implementation are not common. What usually happens is that some cursory planning is done at the system evaluation and selection stage and much of it is left up to the ERP vendor. The thinking is that because vendors have more experience implementing systems, they should have a good idea of the education and training needs. This is a poor way to plan such important activities and usually results in less than adequate performance.

One method of evaluating and analyzing the learning needs of the organization that I find quite promising and worthwhile is learning requirements planning. Learning requirements planning is a comprehensive and effective method for determining the educational needs of the organization and then filling those needs.

9.5.1 Learning Requirements Planning

Learning requirements planning was developed by Karl Kapp, Bill Latham, and Hester Ford-Latham and presented in their book, *Integrated Learning for ERP Success: A Learning Requirements Planning Approach*. Usually, one of the last things people think about when exploring, buying, or implementing an ERP system is the learning that must go along with the technical installation of the components. Users need to know not only how the system works, but why. Different people need different education and training, different types and different amounts, and different people have different styles of learning. The result is a matrix of needed education and methods of instruction. The learning requirements planning approach offers a highly effective method to identify, plan, and execute the educational requirements of an effective ERP implementation. At first glance you may be skeptical of the need for a detailed plan for education and training, and of this approach. But you should seriously consider this method if you want to maximize your chances of a successful implementation.

As a manufacturer, you should be familiar with materials requirements planning and BOMs (if not, see Chapter 2). Learning requirements planning is an adaptation of materials requirements planning. Materials requirements planning is used to determine the quantity and timing of the materials required for production. Learning requirements planning is used to determine the amount and timing of learning (education and training) required for an ERP implementation. Learning requirements planning is a systematic, comprehensive, and integrated approach to fulfilling the learning requirements of an important and high-risk project.

As materials requirements planning utilizes the BOM, learning requirements planning develops and utilizes a bill of learning. The bill of learning is a breakdown of higher level objectives into individual learning goals. Figure 9.1 illustrates a bill of learning and shows the comparison to a BOM.

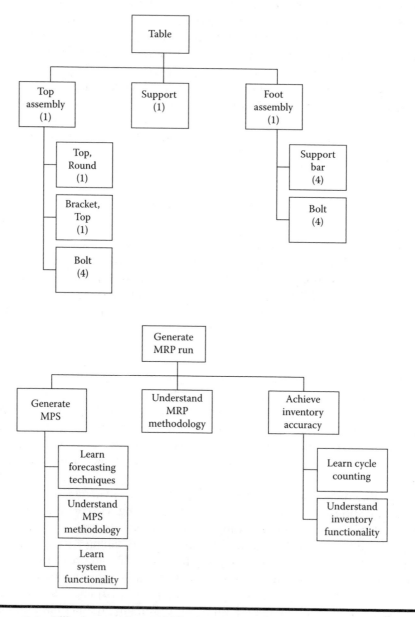

Figure 9.1 Bill of materials and bill of learning.

Just as a BOM identifies the lower level materials and components needed to produce the higher level item, the bill of learning identifies the prerequisite and corequisite learning objectives that must take place to achieve the higher level objective. The learning is time phased to correlate with the needs of the ERP implementation. Learning must occur within certain timeframes to be effective and must be reinforced and practiced. If a lower level learning objective is achieved, but the knowledge or skills gained are not then utilized or reinforced in a certain amount of time, the learning will be lost and forgotten, and relearning will need to be done. It is important for the learning — the education or skills training — to take place at the appropriate time in order to be effective. This is one reason that learning requirements planning is an effective method, because it incorporates the time phasing that is necessary to achieve the desired results of the education and training.

Learning requirements planning consists of a series of identifiable components that need to be performed. As with any other tool, the learning requirements planning model can be scaled and adapted to each individual organization, depending on the size of the company and the resources available. As with the other planning tools discussed throughout this book, all of the steps in the process should be performed, but the extent to which each step is performed can be scaled to the size of the organization. As a small company, you should not attempt the level of detail and scope that larger organizations can perform, but you should equal or exceed their level of commitment.

The first step in the learning requirements planning hierarchy is analysis. In the analysis phase, you identify the learning objectives and develop the bills of learning. This phase corresponds to the business process analysis stage of the ERP implementation, as discussed above. As the business case supporting the need for an ERP system is developed, learning needs can be identified. Part of the business analysis is identifying changes and improvements the company should make prior to implementing an ERP system. To make those changes and improvements, the company must have the requisite knowledge available. If there are any gaps in knowledge, you can identify them and develop learning objectives to close the gaps.

The next phase is diagnosis. In the diagnosis phase, you compare the knowledge that is needed and the knowledge you already have. The difference between the knowledge needed and the knowledge possessed is the gap that must be filled. This is the same as comparing the materials needed to produce a product and the materials you already have, to determine the materials you need to obtain. The analysis identifies what you need; the diagnosis compares what you need with what you have to determine what you need to get.

Next comes the design phase. The design phase encompasses the planning and development of the education and training, including the design of the education and training content, the delivery methods, the timing, and the logistics of providing the education and training. The learning plan will be quite detailed; it must be quite detailed. The types of classes and other delivery methods must be established. You must define whether the instruction will be classroom-based education, hands-on training, informal information sharing, or something else. The individual lessons and lesson plans must be developed. It does little good to plan and schedule a class, only to have an instructor show up without a well-prepared presentation.

Implementation of the learning is tightly integrated with the implementation of the ERP system. Although some educational and training events will take place before the system is even purchased, much of the learning will occur along with the implementation of the system. General education about manufacturing and business systems, ERP systems, and related topics will begin during, or even prior to, the business analysis that precedes the ERP implementation. The bulk of the education and training, however, will occur during the implementation. Any system-specific education and training must be taught while utilizing the system, because it will be tied to the specific functionality of the particular system that has been purchased. Implementation includes the actual learning events (conducting the classroom presentations and training sessions) as well as the selection of the educational and training providers and other implementation team members. In short, implementation is where the actual learning and learning events happen.

Every project or investment should be evaluated to some extent. Evaluation forces you to examine and question whether the effort expended was worthwhile. For an investment, you expect a return. Measuring the return on investment is one type of evaluation. This and other types of evaluation will be used to some extent in evaluating the effectiveness of the education and training. In the learning requirements planning context, evaluation begins with the development of the education and training programs in the design phase. The material that is developed must be evaluated for applicability, appropriateness, completeness, and quality. Evaluation also occurs during and after the learning events. The instructors must continually evaluate whether the knowledge is being absorbed and accepted. After the completion of the learning event, the participants evaluate the instructor, the instruction, the content, and their results. Evaluation includes testing to determine if the students can demonstrate that they achieved the learning objectives, and monitoring to determine if what was learned is actually put into practice as expected. Evaluation is vital but seldom conducted to the level necessary to ensure success. Evaluation leads to feedback, which is used to correct deficiencies and improve the development and implementation of the learning process.

The final component of the learning requirements planning model is continuation. Continuation refers to the fact that learning is not a one-time event. Learning must constantly occur and must continue to grow and change as the organization grows and changes. An ERP system is not just installed and then left to its own devices. As I have tried to reinforce, an ERP system is a tool to enhance and aid the business processes of the organization. As the business environment changes, the ERP system must change along with it and may need to be upgraded or replaced. New learning must accompany these changes, and existing knowledge must be reinforced and refined to remain effective.

9.5.2 Education Providers

One question that must be answered is, who is going to provide the required education and training? Your choices include internal providers (both technical experts and education and training experts), your ERP vendor, external consultants, and other external organizations, such as professional organizations (APICS) and educational institutions (colleges and universities). Most likely, some combination of these providers will be necessary and appropriate. Part of your decision will depend on your timeframe and your budget (another reason for budgeting). If your time horizon is long, you may be able to develop internal resources that can produce the content and execute the learning events.

If your time horizon is short, you may have to rely more on your ERP vendor and outside providers, such as consultants. There are costs and benefits to each choice (isn't there always?). Developing or hiring internal expertise can be costly and time consuming. If the needed skills are not already present they have to be developed, or people with the necessary skills need to be brought on board. Costs involved with developing internal experts include the fees for any classes, travel costs if required, books and other educational materials, and the time spent away from regular duties. Some education may occur after normal business hours. It is important to recognize that this is additional "work" time for the employee. For short periods, employees may be able to perform their regular duties and participate in educational courses after hours. However, this workload cannot be maintained for a long period (physically and mentally), and morale will eventually suffer. Remember, monetary compensation goes only so far.

The other alternative is to hire the necessary expertise. Besides all the costs associated with hiring, bringing in expertise from the outside has other costs. Because you are hiring for skills that you don't already have, you may have to pay higher wages than your current wage structure permits. As confidential as wages and salaries should be, word will get out or speculations will be made. This will affect the morale of the existing employees and can lead to reduced productivity if not handled properly. Some existing employees may feel that the so-called expert is not worth his or her salary; others may believe that they should have been given the opportunity to become the expert. These are valid concerns that need to be addressed up front. You should evaluate the total costs and benefits of developing existing employees versus hiring from outside, and that may not be easy.

The benefits of developing internal expertise, however, either through educating existing employees or hiring new employees with the needed skills, are great. Developing internal experts is a long-term investment in the company. Internal experts are extremely valuable. Not only can they perform the duties required and solve problems, but they can also educate others and perpetuate the knowledge within the organization. One caveat, though, is that these employees must be retained in order to be effective. The fastest way to lose good employees is to not utilize their expertise. The second fastest is to not recognize their contributions and achievements. Please note that compensation is not at the top of the list. If you have developed or hired the expertise needed in your organization, utilize it at every opportunity. Recognize the contributions and reward the achievements. Also, be on the lookout for expertise that you have that you don't know about. Look for the employees who spend their own time and money updating their skills and learning new ones. You might be surprised where you will find them and what they know. Don't take anyone for granted, or you may regret it if they end up employed by your competitor.

What does all this have to do with ERP? ERP implementation requires specialized skill sets that may or may not be currently available within your organization. From the business analysis that must take place prior to an ERP purchase through the skills necessary to operate an installed system, there are many roles requiring expertise that must be filled by qualified individuals. Look at your implementation timeframe, your available resources, and your corporate culture to determine where you can get the education and training providers. Also, don't forget to investigate and evaluate the potential providers. Professional organizations often provide excellent service and real-world experience, but because they are often volunteer organizations, quality and availability may vary from location to location. Colleges and universities may be more consistent but may lack flexibility and experience. Consultants run the gamut from excellent to awful, from cheap to outrageous. You need to find the right fit for your organization. This is an investment like any other and should be analyzed and investigated before a commitment is made.

9.6 Implementation

What is a successful implementation? It is important to answer this question before embarking on the project. If you can't define success, you won't be able to measure it. So what is success in relation to an ERP implementation? Well, go back to why you are considering or implementing the system in the first place. If you are implementing an ERP system, your business analysis justified the need for the system to help you run your business and help your business achieve its goals and grow. So a successful ERP implementation is one in which the installed system meets the expectations for productivity, growth, or return on investment. Anything less cannot be considered successful.

From this definition of success you can see that it is important to define your expectations up front. If you are purchasing a system simply to automate processes, will the benefits of automation be greater than the cost of implementation? You will need to measure the benefits of automation, as well as capture all of the implementation costs, to make the comparison. Is the ERP system expected to improve productivity? If so, by how much must productivity increase in order to justify the expense of implementation? This assumes that during the business analysis you determined you could not make these improvements without using an ERP system. Is that a correct assumption? And again, you must be able to measure your productivity increases and be able to tie those improvements directly to the use of the system to determine whether the implementation is successful. If you define your expectations at the beginning, develop the necessary measurements, and measure your results, you will be light years ahead of many of your colleagues and competitors. The true measure of success is results. The one question you need to answer is, have you achieved your desired or expected results?

9.6.1 The Implementation Team

An ERP implementation is a team effort. Because an ERP system is an enterprisewide, integrated information system, all areas of the organization need to be represented to some extent. To achieve the success you want, you need to assemble a top-notch team for the implementation. This includes everyone from the project manager and team leaders to the trainers and data entry clerks who will provide valuable information during the implementation process. You will be tempted at times to substitute top performers with less than stellar performers on the implementation team. You will want to do this because, at the time, you will think that the top performers are too valuable to take off their jobs to spend time on the ERP project. You will think this and you will try to justify this.

But, don't do it! Remember the justification for purchasing and implementing the system, and remember the definition of success. To be successful, you need the best people working on the implementation. You need to allow these people the time to work on the project, which may mean removing them from their regular duties for some period of time. You need to take this into account and plan for it at the beginning. Don't sacrifice long-term success for short-term benefit.

When selecting the team, you need to consider several variables. The technical skills of each team member are important. They need to know their areas or responsibilities completely. They need to know why you do things the way you do and how you do them. But you need more than technical skills on the team. You need team members who can work effectively on a team and who can contribute to the team. All of the normal team dynamics will come into play over the course of the project, and you need to be prepared for them. Part of this preparation includes carefully selecting the team members and part includes training them in how to work as a team. Team training is another important educational aspect that yields tremendous benefits in relation to the costs incurred. A little time spent training and educating the team members will result in a more effective team and faster results.

The scheduling of the team's activities is a necessary consideration. One thing you must clearly identify is each team member's time commitment. You need to determine whether each person is going to be dedicated full time to the implementation project or if they will participate on a part-time basis. For those employees who will be dedicated full time, arrangements must be made for their normal duties to be performed. This may require spreading their work among several other employees or bringing in outside help for the time they will spend on the implementation (another reason for budgeting of the project). Employees who will be assigned to the team on a part-time basis are a little trickier to handle. The team leader or project manager will need to work with each employee's supervisor or manager to schedule their time as needed. Conflicting demands on an employee's time can create severe problems. Being pulled in two directions can cause stress and negatively affect productivity and morale. The implementation team thus may encounter delays and less-than-desirable performance from the team members. The same effects as well as other disruptions may also be felt within the employee's department. These conflicts, delays, and problems must be overcome immediately. Any delays in the implementation are costly, and declines in productivity and morale are costly in both the short and long term. These obstacles should be anticipated and worked out prior to or at the commencement of the implementation. Sometimes, higher authority must be consulted to alleviate conflicts or potential conflicts.

9.6.2 The Proven Path

Thomas F. Wallace, in his book, *ERP: Making It Happen*, describes the "proven path," which is a series of steps that have been shown to greatly increase the chances of a successful ERP implementation. I would advise anyone who is considering the purchase of an ERP system, or who is in the midst of an implementation, to thoroughly read Mr. Wallace's book. Following is a brief overview of the proven path. The proven path consists of 16 steps:

1. Audit/assessment I
2. First cut education
3. Vision statement
4. Cost/benefit analysis
5. Project organization
6. Performance goals
7. Initial education
8. Defining the sales, logistics, and manufacturing processes
9. Planning and control processes
10. Data management
11. Process improvement
12. Software
13. Pilot and cutover
14. Performance measurements
15. Audit/assessment II
16. Ongoing education

Whew, that's a lot of steps. But you will soon see the importance and necessity of each one of them. You will, of course, be tempted to skip or ignore some of the steps. I would not recommend that. This is called the proven path for a reason. Too many ERP implementations end in failure (remember our definition of success); following these steps in their entirety will greatly increase your chances for success.

Please note that three of the sixteen steps involve education. I think I have stressed the importance of education, and here is more justification. Also note that software is not mentioned until Step 12 of the 16-step process. Too many organizations put software first on their list, then omit many of the other steps listed here. These are the organizations that are least likely to see a successful implementation.

The first step, audit/assessment I, is what I referred to earlier as the business analysis. During this step, you analyze the current condition of the company and ask many questions. An assessment of the company's processes and procedures takes place. The company reviews its strategic plan and strategic direction, analyzes its performance relative to these goals, and investigates methods for closing the gap between performance and expectations. At this stage, process improvement activities may begin or occur. The question of whether an ERP system will help the company achieve its goals is asked and answered at this stage. If the decision is made that an ERP system is necessary and appropriate, the implementation will begin and continue on to the next step.

First cut education involves top management. Key executives and managers who will be critical to the success of the implementation need to be educated in ERP; what it is, how it works, how it is used to help the company, and what it takes to successfully complete the implementation. Depending on the results of the business analysis, education that is not directly related to the ERP implementation may begin. This would be education to immediately aid the operations of the company, such as general production and inventory management education, cycle counting, warehouse management, and other topics. Although not directly tied to the ERP implementation, this education will provide a strong foundation for improvement and for the implementation and education associated with it.

Your company may already have a vision statement, but in relation to an ERP implementation, the vision statement may need to be reviewed and possibly modified. The vision is for the organization after the ERP system is installed and running. If this is different than the current vision, the vision statement should be updated. It is important that the company's vision is documented and distributed to everyone in the company. This gives everyone a clear understanding of where the company is heading, or wishes to head. The strategic plan provides a map to get there, and the tactical and operating plans describe the methods that will be used.

The next step is the cost/benefit analysis. This may have been done during the initial business analysis, but it should still be reviewed at this point after the top management education and the development of the vision statement. Top management may have some insight and input that affects the earlier cost/benefit analysis.

During the project organization step, the management of the ERP implementation project will be designed and planned. The project plan will be fairly detailed at this point, but much of the detail will not be able to come until later. The project manager needs to be selected, as well as most of the key members of the implementation team. A rough timeline for the project can be outlined, and much of the education and training can be planned at this point. Because there will be process improvement activities and the actual selection of the system to follow, many details of the project will be undefined here.

Performance goals are established in the next step. This is where specific goals, and the measurements that accompany them, are defined. Remember, the ERP implementation has been justified by being expected to improve the performance of the company. Specific performance improvements must be defined and measured. What areas are expected to improve, and by how much? If inventory is expected to decrease because of better inventory management by use of the system, by how much, exactly, is the inventory expected to decrease? What specific types of inventory will decrease and in which departments? You need to define these types of specific goals. And if you don't have the measurement systems in place to measure whether the goals are being met, they need to be developed as well.

The next step in the proven path is initial education. In this context, this education refers to ERP education. This is education directed toward the use of the ERP system to run the business. This may require education in the broader topics of production and inventory planning. Some of this general education will take place at this time, but some of it may have already occurred. After the business analysis or audit/assessment I, you may have conducted some education and training and made improvements prior to moving formally into the ERP implementation. At this step, more of the education will be directed specifically toward ERP and the use of an ERP system.

The next two steps define various processes that will be put in place and used once the ERP system is up and running. These two steps may overlap and flow from the first to the second without a clear transition, but they should be viewed as two separate steps. First is defining the sales, logistics, and manufacturing processes. These processes are linked closely to the vision statement that has been developed, because they define how the company will look and operate in order to fulfill the vision. At this level, the processes are broad based and somewhat con-ceptual. From these you will develop the detail of the day-to-day proce-dures. The planning and control processes step flows from the previous step but is a separate step. The planning and control processes include everything from sales and operations planning (see Chapter 3) through master production scheduling (see Chapter 2), materials requirements planning (see Chapter 2), order promising, supplier scheduling, and the quality management system. This is a big step, and an important one. Plan for and budget the resources necessary to complete this step; you will be rewarded.

Following the development and definition of these processes, it is time to turn your attention to data management. This is a good time to mention the "garbage in, garbage out" truism. Remember the subtitle to this chapter, "Information at the Speed of Light"? That refers to both good information and bad information. The information is only as good as the data on which it is based. The data I am referring to here is all the basic data that will be used to populate the system. This includes everything from BOMs, routings, lead times, stock numbers and descriptions, inventory balances, vendor and customer files, customer orders, forecasts, and workcenter and machine capacities. This is when any data cleanup needs to happen. If you have any duplicate stock numbers, nonstandard item descriptions, inaccurate inventory balances, missing routing steps, or any other missing, extra, or incorrect data, you need to clean it up, update it, or fix it immediately. Any errors will be perpetuated throughout the system and will delay or halt the implementation. During testing of the system, if results are not believable or valid due to inaccurate data, acceptance of and confidence in the system will disappear. You may have workarounds in your current system to overcome data errors, but why would you want to implement a brand new (and expensive) system that won't work as designed? The answer is, you don't want to. You want it to work as designed, be implemented as planned, and producing the results desired. That won't happen with inaccurate data.

The next step is process improvement. You may have already done some process improvement prior to the formal implementation, or performed some during some of the previous steps, but this is a good time to make any remaining changes and improvements. It may be difficult to allocate the resources to make process improvements as well as implement the ERP system, but they have to be done. It's better to do it now than to implement a system with poor processes that need to be changed later.

Now, finally, we come to the step you've been waiting for — software. We will cover commercially available, off-the-shelf software here. As a small manufacturer it is unlikely that you have the resources available to develop an integrated ERP system in house. You probably don't want to look for someone to write you a custom-developed package either. As unique as your company is, it is also the same as any other manufacturing company. There are some differences in manufacturing strategy (repetitive versus process) and some unique features in different industries, but when it comes down to it, making jelly and making concrete blocks are strikingly similar. As a result, all ERP packages are very similar. There are some fairly substantial differences in features and functionality, but most of the basic operations and core functions are very similar. So where do you start when you get to the software step? Well, if you've followed the recommended steps up to this point, you should know your business, your processes, and your goals and strategy. You want to find the software package, and vendor, that fits your business and your processes and will help you reach your goals. The vendor plays an important role and should not be discounted. This isn't like a computer game that you buy at your local retailer and install yourself in five minutes. You will be working very closely with the vendor, probably for an extended period of time. The vendor will provide a high level of technical expertise with the physical installation and setup, but they will also provide a substantial level of expertise in the use of the system, in general and specific to your company. Depending on the vendor, they may also be able to provide some level of general production and inventory management education. The point is, an ERP implementation is much more than the installation of hardware and software, and you need a highly reliable and competent vendor with which you can establish a relationship.

The software step includes the selection, purchase, installation, and ongoing maintenance of the ERP package. As we just discussed, the software selection process includes defining your business and finding a vendor who can fulfill your needs. When you have narrowed down your possibilities to three or four choices, you will want to see some demonstrations (do not waste time looking at demonstrations before you have narrowed down the choices). The best demonstrations will use real data from your company so you can better see how your business will be run, not the vendor's demonstration company. The demonstration can also tell you a lot about the vendor. Who is performing the demonstration and what is their level of knowledge and expertise? How is the demonstration performed and does it go smoothly? If there are technical glitches during the demonstration (for example, it won't run with your data, or the person(s) performing the demonstration can't answer your questions), how much confidence can you have that the product will be able to perform after the purchase?

Everything is negotiable, as anyone in purchasing and most people in business can attest. It is no different with the purchase of an ERP system. There are many variables in an ERP purchase, so there are many things to negotiate. The software itself, any hardware or peripherals that may be included, the number of people who will perform the installation and training and all the costs associated with that (travel, time), warranty, consultations, and many others. But, as with any other negotiation, there are productive negotiations and nonproductive ones. Remember, you want an ERP system that will help you run your business and help you grow and reach your goals. The vendor wants a satisfied customer but also wants to make a profit. You need a good, solid relationship with the vendor for the implementation to be successful. Don't negotiate the goodwill just to save a few dollars. It won't work out in the long run.

The installation itself never seems to be easy, no matter how many times the vendor has done it. Computers are wonderful tools, but they just don't seem to work the way we want them to all the time. Even if the operating system and all the hardware is the same, they don't seem to be consistent; one installation will be relatively smooth whereas another simply won't work. Maybe this is just my perception, but be prepared for some glitches during the installation process and be prepared to deal with them.

Ongoing maintenance is included in the software step, but ongoing maintenance is, well, ongoing. Technology changes and your business changes. Your ERP system will change. It also simply needs to be maintained much like your car does. You may need to clean up files, archive data, and perform upgrades. You can't plug it in one day and expect it to stay the same forever. Plan for ongoing maintenance, budget for it, and don't forget to perform it.

The pilot and cutover step is when the real action starts. If you thought the action already started, this may be a surprise. There are different ways to perform the cutover, and you need to determine which is best for you. Cutover refers to moving from your current method of running your business to running the business using the new ERP system. Some people may want to cling to the old way of doing things, but at some point you need to start operating and using the new system completely. How you get there, and how quickly, depends on your choice. Basically, you have two choices: just stop using the old system and start using the new or take a phased approach. Choosing a date and just switching from old to new in one swift cut is risky and dangerous. No matter how much testing you have done, there is no guarantee that the new system will work as expected when you start to rely on it exclusively. Do you want to bet your whole business on that gamble?

A better approach is a phased implementation. Several methods are available. The parallel approach operates the old system and the new system together for a certain length of time. Of course, this means double work for everyone during that time, but it is less risky than the full commitment cutover. In the parallel approach, the old system continues to run the business while the new system is up and running but is not yet used to run the business. Many people expect to get the same results from both the old and new systems when running in parallel, but that doesn't make much sense. If you achieved the same results, why would you spend all the time and money implementing a new system? The truth is, you expect different results from the two systems, although there will be some that are the same. For example, a cash sale will result in a posting to the cash account and a reduction in inventory in both systems. But the old system may require several manual processing steps to get to that result, whereas the new system automatically posts the transaction to all necessary accounts and through all systems. In the old system, that inventory transaction may take several days to be reflected in the inventory records, whereas in the new system the inventory records are updated in real time and may trigger a replenishment notification. In the parallel approach, the old system runs the business while the new system is analyzed for proper functioning.

Another phased approach to implementation is the pilot approach. With the pilot approach, one small part of the business is run using the new system while the remaining parts of the business are run using the old system. One product, product line, or group of items is selected for cutover in the pilot approach. This may result in some overlap and require extra work due to the integrated nature of the ERP system, but it can be used to prove the validity of the new system before cutting over everything. This is a relatively safe approach, but be careful not to extend the pilot too long or accept it too soon. You want to make sure the new system works, but you don't want to get in the trap of holding off the total cutover until everything works perfectly. Nothing will ever work perfectly. If the old system worked perfectly and filled all your needs, you wouldn't be implementing a new system.

Some companies utilize a phased approach that involves implementing one or a few modules of the new system before the rest of them. One method is to implement the financial modules before cutting over the operations. This may have some validity, but it will require extra effort to post all the transactions into the financial system instead of letting them flow from the operations side. Some companies may reverse this and implement the operations first and the financials after. As I mentioned, you need to determine which method of cutover is best for your company. Just remember that you want to implement the new system and rely on it to run your business; you are not implementing a new system just to install some software and do things differently.

After you complete the pilot and cutover, and the system is up and running and helping you run your business, it is time to start measuring the performance of the system. In Step 6, the performance goals were established; now we need to perform the measurements and compare actual results with planned or expected results. If you expected inventory to decrease (specific types of inventory by specific amounts), you can compare any actual change in inventory to what was expected. If inventory has not decreased by the amount expected, you need to find out why. Is there a problem with the system, the operation of the system, poor education or training, or something else? And no matter how tempted you are, you cannot simply measure the difference between actual performance and expected performance and beat someone up because of it. You need to investigate and analyze the causes and take corrective action so that the goals can be achieved. Of course, you may discover that your initial performance goals are unrealistic at the present. Maybe they can be achieved in the long run, but not right now. You may need to modify some of your performance goals, but don't be too hasty. Don't fall into the trap of changing expectations just because you haven't met them yet. Modify those goals that need to be modified and leave those in place that are valid.

Next comes the audit/assessment II step. This is very similar to the business analysis or audit/assessment I that was performed earlier. You have analyzed your business, performed process improvements, implemented the ERP system, and are using the new processes and new system to run your business. It is time to reassess your business. A lot of changes have been made and your evaluation now will be quite different than your previous one. Have additional opportunities presented themselves due to the changes that have been made? If the implementation has not gone smoothly or is not achieving the results that were expected, are there new threats to the company? Is the company's strategy still valid, or do the opportunities and threats suggest a change in strategy? You may need to ask tough questions and make hard decisions at this point. They may be positive: the company has improved to such a point that tough decisions on growth strategies must be made. Or they may be negative: the company has been hurt by an unsuccessful implementation and needs to make some hard choices on direction. Some people won't want to hear this, but maybe it is time to start the next initiative or implement the next tool (see the rest of this book for possible choices).

And now we finally come to the last step: Step 16, ongoing education. Whether it is the ERP implementation or just normal business, ongoing education is vital to any business. Running a business is an ever-changing adventure. Customers' wants and needs change, technology changes, suppliers and customers come and go, economic conditions change, regulations change, and our thoughts and ideas change over time. If your company doesn't keep up with changing technology, changes in the market, and improving tools and techniques, you will find it hard to compete in today's global marketplace. Also, employees come and go or move into different positions. Their skills need to be updated or enhanced and maybe refreshed if skills haven't been used regularly. Every company should invest in ongoing education and training, and it is a necessary step in the ERP implementation process.

9.7 Maintenance

9.7.1 Upgrades

Upgrades are a fact of life when it comes to software and computers. Technological improvements and advances in hardware and systems architecture come at an incredible pace. It is impossible to keep up with all of it, but you will be faced with decisions on upgrading your system, both the hardware and the software. Your software vendor is constantly improving and upgrading the program. They will release version updates for relatively minor corrections and changes, and they will release new versions with more substantial enhancements and improvements. Some new versions will be nearly complete overhauls of their programs. You will need to decide when and if to purchase and install any of these version updates or new versions.

You must address several issues when considering updating your system. One consideration is whether a software upgrade requires an operating system upgrade or hardware upgrade as well. This is not uncommon. It isn't likely for the smaller updates but may be necessary for new versions. New versions of programs often take advantage of advances in operating system or hardware technology. You will also have to pay for these updates and upgrades. You may receive a certain number of updates, or even new versions, for a certain period of time after your initial system purchase, and you should include this in your negotiations. Generally, however, you will have to pay eventually. If you haven't budgeted for this, the decision to upgrade will be even more difficult.

Another consideration is the expected ease of the upgrade installation. If you have customized the system, the upgrade may not go smoothly. Upgrades are generally designed to be performed on the vendor's standard program, without customization. You may have worked with the vendor to customize the program during installation, or you may have customized the system through workarounds or programming that is "tacked on" to the vendor's program. Either way, any customization will hinder the installation of any upgrades to the system.

If you make any changes to the system through upgrading, you will need to provide education and training to any employees affected by the changes. If your implementation has gone, or is going, smoothly, you have well prepared your employees for change. But if you change or upgrade your system frequently, even the most change-oriented employees will begin questioning the need for and wisdom of the changes. There are also costs associated with the additional training, which you may not have budgeted for.

Let's say you have successfully implemented the ERP system and have been using it to run your business for some time. When should you consider going through the whole thing again by purchasing a completely new system? At some point you will have to face this decision, so you should be aware of it now. You may need a new system because of significant growth or other changes in your business. Your products change, your customers change, your markets change, and your business changes. The system you have used to succeed up to this point may not be adequate to allow you to continue this growth and success. Use the same methodology you used to analyze and justify your existing system when exploring the need for a new system. You will have to go through all the same steps, although some of them may be shortened or modified the next time around. Don't skimp, though; go through all the steps if you want the next implementation to be as successful as the last one.

9.7.2 Data Integrity

With any system, ongoing maintenance of data integrity is a requirement. Whether a manual system or an automated one, data integrity is vital to success. With an automated and integrated system, data integrity, or the lack of errors, is especially important. Errors can cascade through the system, wreaking havoc along the way. An error in a BOM can cause planning mistakes, purchasing errors, production stoppages, decreased inventory accuracy, billing errors, accounting errors, missed deliveries, and unfulfilled promises. All system data must be accurate and complete.

Customer orders that are not entered into the system, but are held by the sales team for whatever reason, diminish the integrity of the system. This may cause the order to not be delivered or not be delivered on time, it may cause materials not to be ordered on time, it may cause scheduling and production problems, and it may disrupt the planning cycles.

Purchase orders, materials receipts, engineering changes, vendor lead times, routings, production lead times — everything — must be maintained and kept accurate and up to date. You have spent a lot of time, energy, and money purchasing and implementing the system; don't allow poor data maintenance to derail the project or interfere with its success. Data integrity requires a significant investment by the organization. Communication is vital. If engineering changes a part, that change must be communicated to everyone affected: purchasing, production, planning, accounting, sales, and anyone else who needs to know. If production changes a routing, that needs to be communicated. If sales changes the forecast, that needs to be communicated. But if responsibility and accountability are assigned properly, and performance is measured, it is a doable task. The alternative is to not rely on the system and allow manual workarounds. That isn't what you want.

9.7.3 Employee Turnover

As mentioned above, every organization will have to deal with employee turnover to some extent. Some well-run companies, and some lucky ones, will have little turnover, but others will have a significant level. In relation to ERP, turnover represents a substantial expenditure of resources. Besides the usual costs of hiring and firing, there are the additional educational and training costs directly related to the ERP system. No matter how knowledgeable a new employee is, unless he or she has used the exact same system, there will be a learning curve to get up and running with the ERP system. Almost all systems undergo some sort of modification during the implementation. Some customization probably occurs that makes each implementation of the same software slightly different. Then there are the different versions. Some companies upgrade more frequently than others; both the vendor and the customer. The chances that a new employee was using the same version as you at a previous company are not good. Your rate of employee turnover is a definite consideration when investigating ERP. The additional education and training costs associated with a high turnover rate need to be budgeted so that there are no surprises later. It may be possible that the changes and improvements made as a part of the ERP implementation will help to reduce turnover. Then again, if you do have a high turnover rate, you should investigate why and make some changes regardless of ERP.

Chapter 10

Supply Chain Management: You're Not Alone

You hear a lot about supply chain management these days. Many companies even have managers or executives with titles such as supply chain manager or vice president of supply chain management. But what does supply chain management encompass and what do supply chain managers do? Every company is different, of course, but we will cover some of the basic concepts of the supply chain and supply chain management. Figure 10.1 represents an example of a supply chain.

The first thing I notice when I look at the diagram in Figure 10.1 is that it doesn't look much like a chain. It looks more like a net or a web, and this is just a simplified example. Think about just one item that you produce. Think about all the components, raw materials, intermediaries, distributors, suppliers, customers, and the final consumer. And I bet you are thinking only one direction. How about returns of unwanted items, wrong items, and damaged or defective items? And how about the scrap, packaging, and shipping containers along the way and back to the source? It is a complex system with multiple interdependencies, access points, and transfer points. You need to get your arms around it as much as possible.

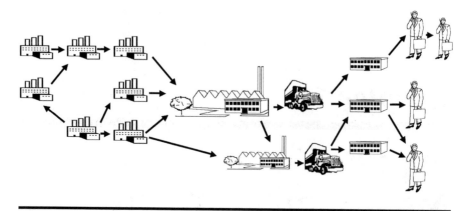

Figure 10.1 Supply chain.

10.1 You Don't Operate in a Vacuum

Globalization. That one word says it all. You are competing in a global marketplace. No matter how small you are and no matter what industry you are in, your supply chain extends around the world. So you had better understand how supply chains work and where you fit in. Whether you are delivering across town or around the world, you need to have a strategy, plan your tactics, and execute. A better understanding of supply chains and supply chain management will help you improve your performance and meet your goals.

The first thing is strategy. Where are you competing in the market, where do you want to compete, and how are you going to go about doing it? If you don't know where you're going, you'll never get there. Too many businesses struggle with day-to-day issues, fight fires, and react to the current situation. No matter how difficult it might be to set aside the time, you need to develop a strategy to set your direction. This will help you with your day-to-day items and your fires, because when you have a strategy, you will have clear direction and a clear path to follow. Any activity that does not support the strategy or that is not on your path isn't necessary. You will be able to prioritize activities and eliminate unnecessary tasks.

Once you have your strategy defined, you need to develop plans that will get you where you want to go. What will your supply and distribution network look like? Where will ownership reside and where will ownership be transferred (ownership of the goods and ownership of the network components and information)? You will need to plan the distribution methods, inventory levels, payment systems, information flows, supply points, and supplier agreements. Okay, I have to say it: failing to plan is planning to fail, or as I learned in the army, prior planning prevents piss poor performance (pardon my French).

Then, of course, you finally have to take action, or execute the plan. However, if you have defined your strategy and fully developed your plans, execution should (*should*) be a lot easier. A small roofing job I once did illustrates how execution is easier if there is a plan in place. Without a plan, the job probably would have taken twice as long, and there is a good chance it wouldn't have achieved the desired result (stopping a leak).

The project was repairing and replacing a roof on a carport. The job included fixing an area of damaged wood, removing the existing roofing material, and installing new roofing material. Because I had never done a job like this before, I could have, and probably should have, done the work differently; but the job was completed satisfactorily (although, as of this writing, it hasn't rained yet). Let's look at what might have happened if I had just started working without doing some planning first.

Without a plan, I might have just climbed up on the roof and starting tearing off the old roofing material. It's a small roof, but it turned out that there were at least four layers of roofing paper plus a layer of felt or tar paper. I would have started tearing up all this material, then looked for someplace to put it. Not too many choices there, so it would have had to go over the edge on one of three sides, onto the grass, the stairway, or the sidewalk. Considering the removal of the old material took at least three days, using manual labor, none of those choices would have been very good. If I had just climbed up and started working, I might not have seen the damaged area, forgotten about it, or not clearly marked it. That could have ended badly for me, because I'm not a light person and the damage was pretty bad. I could have ended up on the ground before the stuff I was tearing off.

If I had just climbed up and started working, I would have had to use whatever tools just happened to be lying around. I know I could have found better tools than the ones I was using, but if I just picked up whatever was handy, I might still be up there. Without a plan, every time I needed something, tools, materials, water (lots of water), I would have had to climb down, drive to the store, come back, and get started again where I left off. Given the cost of gas and the distance to the store, it would have added cost and time to the project. Also, I could have run out of something that was time sensitive. If I had put down some adhesive and run out of roofing paper, I would have had to go to the store, and when I got back the adhesive would have already started to set. Not good. As you can see, just jumping into the execution phase isn't the best place to start. Then why do so many people try to do it?

With this little roofing project, a little planning went a long way. The damage to the wood was an important issue, and fixing it helped the whole project go more smoothly. The damage was close to the middle of the roof, so I cut out the damaged portion, placed a dumpster under the hole, and threw the old material through the hole into the dumpster. Knowing the job wouldn't be completed in one day, I brought in a tarp to cover the roof at night. All tools, materials, and supplies were gathered before the work started. Not wanting a vertical seam in the roof, I purchased enough rolls of roofing paper so a seam would be avoided. Except for the physical labor of tearing off the old material, the job went smoothly and was successful. Without planning before taking action, the results would have been quite different.

The point here is that a supply chain is a complex system. Before you jump in and start trying to find suppliers, buy materials, move items around, and then ship them to your customers, you need to develop some detailed plans. Otherwise, you will end up with a disconnected and inefficient mess of independent operations. An effective supply chain is an integrated system designed to maximize customer satisfaction and profitability. Let's look at some of the components of the supply chain and how they fit together.

10.2 Supply Chain Management

10.2.1 Information Chain

The term *supply chain* implies a one-way flow of materials and products. Raw materials flow to component manufacturers, components flow to finished goods assemblers, and finished goods flow through the distribution system to the final consumer. In Chapter 4, we discussed pushing inventory through the system versus the pull system advocated by Lean Manufacturing. When you push inventory through the system, you rely on forecasts that are usually based on your customers' past ordering patterns. When inventory doesn't move as expected, incentives are offered, which is a method of pushing more inventory downstream through the system.

In an efficient and effective supply chain, information flows freely both upstream and downstream. Forecasts are still likely to be used, but they are not the sole means of determining demand. More information is shared among supply chain partners, and there is more collaboration in determining customer demand and market needs.

In the more highly evolved supply chains, collaboration goes far beyond demand planning. Customer demand and market needs are only one piece of the puzzle. Supply chain partners need to share supply capabilities, capacity availability, resource needs and availability, technology changes, and other information that ensures the supply chain performs at peak effectiveness.

Outsourcing of noncore functions to third parties is common in many organizations. Outsourcing functions such as payroll, human resources, and information technology is common in many smaller companies, but also in many larger organizations. It is often both cheaper and more effective to outsource those functions that are not your core strengths. These third-party service providers become part of the information chain, often branching off in directions that surprise people who are not familiar with the way some of these services work. For example, if a company outsources its accounts receivable function, the customer might remit payment to the third-party provider instead of directly to the supplier. Figure 10.2 shows an information chain.

The value of information is not so much with the information itself, but in the usefulness of the information and the organization's ability to use the information. A one-way flow of information isn't really information; it's data. The usefulness and appropriateness of the data is suspect because one party has little or no input into what data is supplied or when. The value of information is derived from the methods used to gather, analyze, and share it. When partners work together to gather and share information, cooperate in the analysis, and develop improved information delivery methods, the value of the information increases significantly.

Information is great. Information is important. But you must always beware of information overload. More is not always better. If you do not have the capabilities to process and use all the information you receive,

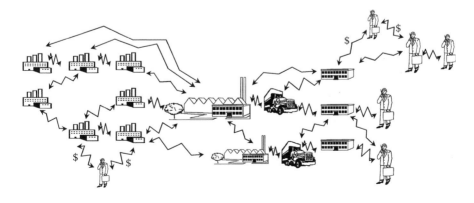

Figure 10.2 Information chain.

receiving it all might be counterproductive. You need to have processes and procedures in place to analyze and utilize information. A good example is the growth of RFID (radio frequency identification) and the volumes of information that can be provided through this technology.

RFID proponents promise real-time inventory status information. Using RFID tags, you can track inventory in real time as it flows through the supply chain. Within the walls of an organization, inventory can be tracked as it moves into, within, and out of the warehouse. This is done passively as the tagged items pass by tag readers at designated points. As the item moves, the reader detects it, and the inventory system is updated with the new information. When items move out of the facility, they can be tracked in the same way throughout the distribution system. As trucks are unloaded, the items can be tracked. When trucks move past designated points along their routes, the items can be tracked.

A lot is promised in the promotion of this technology — promises of improved customer service, increased efficiency, and more and better information. For those who are prepared and have the systems in place to utilize this vast amount of new information, there will be many benefits. The big question I have is, will you be able to use this information? Will you be able to use is effectively? You have to ask yourself, can I use the information I already have effectively? If not, how can you expect to use huge amounts of new information effectively? Or is this going to cause information overload and set you back instead of moving you forward?

The information chain is a vital part of the supply chain. Planning the flow of information is as important as planning the flow of materials. Throughout the remainder of this chapter, consider the value of information as it relates to the topics we discuss.

10.2.2 Value Chain

You cannot have a serious discussion about the supply chain without talking about the value chain. In Chapter 4 (Lean Manufacturing), we discussed the concept of value and the process of value stream mapping. I defined value as aspects of the product that the customer is willing to pay for. In a production process, assembling parts into a finished product adds value because customers are willing to pay for a functional, assembled product rather than a bunch of parts that they would need to assemble themselves. Moving parts and partially processed assemblies between departments is not value added, because customers do not want to pay for the additional costs associated with this movement.

The same concepts apply to value in the supply chain. There are many steps and activities in the supply chain, all working toward getting the product to the customer. Many of the activities add value to the product, and the customer is willing to pay for these activities. However, there are many activities that the customer is not willing to pay for. The more effective supply chain improves those activities that add value and eliminates those that don't. The costs of inventory that is held in the system because of ineffective communication and collaboration between supply chain partners is non-value added, and the customer is not willing to pay for these costs. Activities that are performed by an organization, but which could be performed more efficiently and cost effectively by a third party, are non-value added, and the customer is not willing to pay for this inefficiency.

As I said in relation to Lean Manufacturing, value stream mapping helps you identify the value-added steps in the process, which in turn allows you to remove the waste from the system. We just talked about information and the information chain. Value stream mapping also includes the information flow that is necessary to produce the product or provide the service. Value stream mapping can examine the supply chain as well as the individual organization. You may find, though, that activities that are found to add value in an individual organization are not value-adding activities when seen in the context of the supply chain. See Figure 10.3 for a view of the value chain.

That is the dilemma, then. For the supply chain to maximize its effectiveness, and profitability, some of the individual organizations in the chain should not maximize their performance. Some components should not even exist as individual units. Just as theory of constraints (Chapter 8) shows that it is not desirable to maximize the performance of every machine or workcenter in a plant, it is not desirable to maximize the performance of every component in the supply chain. The system is limited by the capacity of the constraint. This is true in a single organization

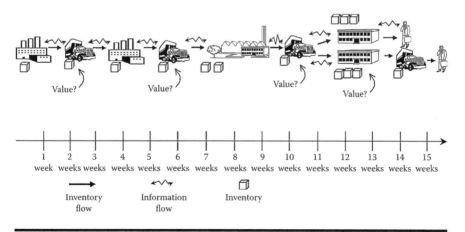

Figure 10.3   Value chain.

as well as in a complex supply chain system. The problem is how to maximize the performance (or throughput) of the supply chain when every individual organization in that chain is trying to maximize its performance (or output).

It might not be easy, but the answer is with collaborative planning and execution and the adoption of supply chain performance measures. This solution is for existing supply chains. For new or developing supply chains, the entire supply chain should be analyzed, planned, and developed as an integrated system. For existing supply chains, it would benefit those organizations with the greater capacity to work with those organizations with less capacity to increase their capacity; that is, of course, if the throughput of the entire system would increase. For planning a new supply chain, or perhaps a new distribution system that is under your control, you will want to plan the capacity and capability of each component so that no constraints hinder the performance of the system. This means that some components will have less capacity than they would have if they were designed in isolation.

Sometimes it seems that people like to design an activity with more capacity than necessary for present needs. They plan for growth so they design in extra capacity. Planning for growth is good, but when people see extra capacity just sitting around not being used, they tend to want to use it. That's okay if it's needed, or if it's put to some other use, but it's not good when the capacity is increased in a nonconstraining resource. If you have a lot of extra space in your warehouse, you don't need to fill it up just so your space utilization looks good. If the warehouse isn't a constraint in the supply chain, that extra inventory is just waste and more than likely will end up causing problems.

Examine your supply chain. Evaluate and analyze the system as a whole and the individual components. Generate a value stream map and look for non-value added activities. Then work to improve the system by eliminating non-value added activities and improving the performance of the constraints in the system. For those companies that have extra capacity or higher levels of performance that they can't or won't cut back on, work with them to develop new markets so the capacity can be offloaded to another supply chain.

10.2.3 Variation in the Supply Chain

Variation, although not quite an evil, is something you need to understand and control. We discussed variation in some detail in Chapters 5 and 7, primarily as it affects quality, and we briefly discussed variation and its effect on inventory in Chapters 2 and 4. Variation also has a significant effect on the supply chain. The variation we are concerned with in this context is the variation in demand as the demand flows up through the supply chain.

Demand variation amplifies as it travels upstream from the retailer, or closest point to the consumer, to the raw materials suppliers. This means that the variation increases as it moves up the chain. For a typical product, demand at the retail or consumer level doesn't vary much. There will be small variations, of course, but daily or weekly demand is pretty regular. At the next step up the chain, the wholesale or distributor level, the variation increases. This is because the demand at this level is determined by the orders placed by the retailers, rather than the actual demand at the consumer level. The orders placed by the retailers are determined by their replenishment systems and how well those systems are managed. Although technological changes have been taking place that improve the flow of demand information, these improvements are far from universal. If the retailers place orders at regular time intervals, the size of the orders will likely vary from order to order. If the orders are placed for regular quantities, the time between orders is likely to vary. If there are any ordering glitches — if the system doesn't work as planned — more variation is introduced. One possible glitch is a key employee out of the office, so orders aren't placed when they're supposed to be. Inventory sometimes gets lost or damaged or expires. These conditions might not be discovered in a timely manner, and even if they are they might not be resolved immediately. This increases variation.

Every step up the chain increases the variation, because the same problems and issues occur at every step. Demand is based on forecasts of what you think your customers will order, not on the actual demand at the consumer level. Replenishment systems don't always work as they are designed to and inventory isn't always managed as closely as it could be. The more levels, or links, in the supply chain, the more variation will be introduced, because the variation increases at every step. Figure 10.4 illustrates this bullwhip effect.

As you've learned, variation or uncertainty has a direct effect on inventory levels. Inventory is used to protect against uncertainty, so the more uncertainty, the more inventory is held as protection. As we've discussed throughout this book, inventory is both necessary and useful but can also be wasteful and the source of many problems. There will always be some level of variation in demand and supply that needs to be protected against. There are constraints that need to be protected, and there are other strategic uses for inventory. On the other hand, excess inventory — inventory above what is needed for immediate needs — is waste. It adds cost without benefit. It gets lost, it must be tracked, it is often moved around, and it just generally gets in the way.

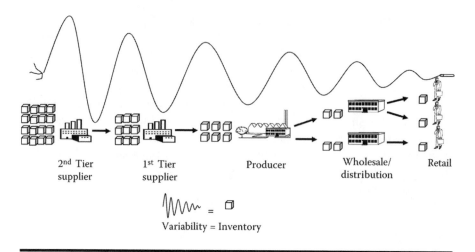

2nd Tier
supplier

1st Tier
supplier

Producer

Wholesale/
distribution

Retail

Variability = Inventory

Figure 10.4   Bullwhip effect.

One way to reduce inventory is to reduce variation. In the supply chain, variation (in demand) can be reduced through closer collaboration, partnership, and information sharing among supply chain partners. Just as materials requirements planning calculates need of materials based on the production of higher level items, demand throughout the supply chain should be calculated based on demand from the consumer. Forecasts have inherent error, or variation. If every link in the supply chain is forecasting demand, more error will be introduced. If consumer demand were the basis for all planning throughout the chain, much of the error and variation would be removed. Demand up the chain would be calculated instead of forecast, reducing error.

This variation is important because it affects the performance of the entire supply chain. If your goal is to improve performance, or throughput, of the supply chain, you need to look at the system and not just your small piece of it. This is a lofty goal, and might be difficult for a small company to take control of, but you should examine it, discuss it, and work toward it nonetheless.

10.2.4 Working with Suppliers and Customers

We've talked a lot about collaborating and partnering within the supply chain, but how do you go about doing it? You have to start with your own suppliers and customers, as obvious as that sounds. Start with the suppliers and customers with which you have the best relationships. If you have a good relationship to start with, taking it to a higher level should be easier than starting from scratch. Technology, such as collaborative planning, forecasting, and replenishment (CPFR) systems, is a tool, but it does not build the relationships that are necessary. There needs to be a high level of trust between partners. You will share a lot of information, some of it proprietary, and performance is important. If you cannot keep confidential information private, or cannot perform as promised, you will lose any trust you had and you won't be able to build trust.

I think it is best to begin with the goal in mind. The goal is a more effective supply chain. You want to increase throughput and profitability. Every individual organization in the supply chain needs to be profitable, and they all want to increase profits. Consumers want lower prices for products. All of these considerations need to be taken into account, and they should be stated openly, up front, and put on the table. As I've said, you have to know where you're going, so you might as well make it clear (see Figure 10.5).

Once you know what you're trying to achieve, you can determine what you need to get there. What information do you need to share (we've already mentioned demand)? What capabilities does each partner have? What processes need to be modified and what new processes need to be created? What functions can be shared, which activities should be performed by each partner, and which activities should be outsourced? What technology is available to manage the supply chain, how will it be deployed, and who will own which part? These are just some of the many questions you must ask. You must also discuss the design of the products, packaging, delivery methods, ordering systems, information flows, and dispute resolution procedures. How payments will be made, payment terms, inventory ownership, and transfer points are other issues. Basically, every aspect of the system will have to be analyzed, designed, improved, or created. Just remember to keep the goal in mind as you work through all these issues.

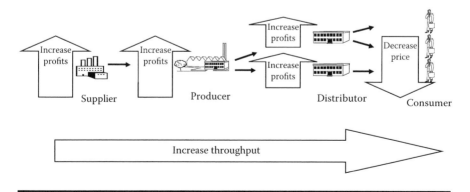

Figure 10.5   Supply chain goals.

These higher level goals of the supply chain will help you in your efforts to improve relations with and work with your suppliers and customers. They provide a context and a framework within which to work. Without this context, your efforts will probably be more negotiation rather than collaboration. You will try to apply more pressure to maximize your own benefits without caring much about your "partners." By working within the supply chain context, you will work more collaboratively and improve the performance of each organization involved.

You can't begin this partnership initiative with all your suppliers and customers at once, and it's not a good idea to try. Start by prioritizing or ranking your suppliers and customers by their importance to your success. The most important will get your focus first. These might not be your largest suppliers or customers, the ones with which you do the most business. They're important, of course, but they might not be number one on the list of those most important to your success. You have to look at your company strategy and your goals. Some of the highest volume suppliers and customers might not be important in the future, depending on your strategic objectives. Keep this in mind.

I want to repeat that you can't, and don't want to, roll out this partnering initiative with all your suppliers and customers at the same time; not just with this activity, but with anything. You have to prioritize. You, and your employees, can only do so many things at the same time. No matter how much you want it, you can't overload your resources, including people. Also, if you're just getting started with this program, you'll be learning a lot along the way. Use what you learn developing your first full-fledged partner to make the next one easier. The first time around, you can say you don't know what you're doing, but you know what you want to achieve. With the next one, and subsequent ones, you can say, "we had this issue, or this obstacle, and this is how we overcame it." Won't that make your future partner more confident and more willing to work with you?

Please note that we're talking about both suppliers and customers here. The concept of prioritizing, rating, and measuring your customers might seem strange, but it is in your best interest. Many companies review their customers and drop or stop supporting those at the bottom of the list. That's only a part of what we're talking about. Not all customers are profitable to you, and not all of them are important. But before you cut them off or threaten them, you need to assess their importance to your success in the context of your strategic goals. Some of those customers will need to be retained and developed. They might not be profitable at present, but they might be of strategic importance. They'll need to be profitable for you in the long run, and you'll need to work with them and develop them to get them there.

After you have prioritized or ranked your partners, you need to determine the attributes or areas that need to be strengthened. What is it about this partner that is most important, strategically? For a supplier, it might be quality (ability to meet specifications). Maybe lead time isn't important with that vendor or for that item, but meeting specs is. Focus on the quality. Later, as your relationship grows and your (plural) success grows, you can work on lead time.

Once you determine the attributes or performance areas that need to be strengthened, you need to develop the performance standards that need to be met. In our example above, this includes the specs, which is the focus area, but it includes other items also. Although lead time isn't the most important element, there need to be lead-time performance standards. Lot sizes, yield rates, pricing, shipment notification, invoice accuracy and timeliness, payment remittance, and dispute resolution are some of the other areas for which you need to develop performance standards. The standards spell out the requirements and responsibilities of both parties — you and your partner. The standards remove any confusion, making the roles and responsibilities clear and understandable.

With the standards in place, you can develop the vendor scorecard or customer scorecard. These scorecards are for record-keeping and serve as records for scoring performance. You will monitor and measure the supplier's and customer's performance against the standards, and you will need a place to record this information. The scorecards record standardized and summarized scores that allow you to easily and quickly review and compare the performance of your partners.

Now it's time to contact the partners at the top of your prioritized list to discuss this new initiative, the reasons for it, the benefits, the work that will need to be undertaken, and all the standards, the scorecard, and the monitoring and reporting processes. During this discussion and when you start working together on the initiative, some of the things you've already developed will change. That's okay, and that's good. That means the process is working. If they just accept everything you throw at them, it doesn't bode well for a successful partnership. That's too one-way. Most likely, your partner will think of things that you haven't or will provide a different perspective. Working together, you can develop even better standards and systems. You may wonder, if that's so, why did I bother to do so much work before I even talked to them about it? For several reasons: it shows you're serious, it prepares you for what comes ahead, and it allows you to do some self-evaluation and make improvements before you begin working with your partners.

You will need to develop the systems, processes, and procedures to capture the information for measuring and monitoring performance. How is the data going to be captured, who will do it, when will it be done, and who is responsible? How is the information going to be reported and communicated and what will be the dispute resolution process? Everything related to capturing, processing, analyzing, and reporting the results must be developed.

Finally, it is time to actually start monitoring the performance and measuring it against the standards. Any new process will have some glitches, and you will have to work through them. But the benefits and rewards of this initiative will soon begin to show and you will be glad you've undertaken the project. As you progress, you should use the results to further strengthen your relationship. Develop a methodology for utilizing the results of the measurements to improve performance. Just undertaking this initiative and working with your partners to develop the systems will strengthen the relationship. By using the results to improve your performance (plural again), it will be strengthened even more. To summarize, here are the steps in supplier partnering:

1. Form team.
2. Prioritize suppliers and customers by importance to success.
3. Determine attributes or areas that need strengthening and monitoring.
4. Develop performance standards.
5. Develop scorecards — vendor scorecard and customer scorecard.
6. Discuss the initiative, scorecard, and reasons behind initiative with partners.
7. Develop metrics and monitoring system.
8. Measure and monitor performance.
9. Use results to further strengthen relationships and improve performance.

We haven't discussed the team you need to put together to start this initiative, to develop the standards and metrics, and to work with your partners. When assembling any team, you need to identify the skills the team will need in addition to basic team skills. Team dynamics, communication, and how to work on a team are some of the skills you need to teach or develop in the team members. A dysfunctional team is a sorry sight. The supplier partnering team members will need skills such as negotiation and problem solving, but they will also need diverse knowledge of your organization and your products.

If you are considering working with a supplier, the obvious choice is to have someone from the purchasing area on the team or leading the team. This is an obvious choice, but who else should be on the team? I can think of a few that will seem almost as obvious, but I'd also want to include someone from the sales department. Sales? On a supplier partnering team? Yes, and here are some reasons.

If you're going to work with your supplier, who do you currently deal with and who do you think will be one of the members of their team who works with you? That's right — someone from their sales team. Who is most accustomed to working with customers? Who is most aware of the issues that usually arise with customers? The salespeople; they are the ones who are closest to the customers, so why wouldn't you include them on your team to contribute their expertise and experience? Another thing is that this team, or teams, you are putting together shouldn't be separated by supplier and customer. They should be partnering teams. You are trying to strengthen the relationships and improve the performance of the supply chain. You're starting with your suppliers and customers, where you have the most contact and influence, but your goal is to improve the supply chain. The team should reflect that goal.

Other members of the team should come from production, engineering, research and development, finance, distribution, quality, customer service, information systems, and any other area that is important to the performance of both your organization and the supply chain. You don't want to overload the team, so some team members might represent more than one area, or some members might work with the team but not be full members of the team.

Getting back to the scorecards, the scorecards will score, or rate, vendors' and customers' performance and track it over time. Without objective records it is difficult, if not impossible, to rate the performance over time. The performance areas to be scored, as well as the rating system, must be communicated to the partners with whom you are working. If any partners consistently rate below acceptable standards, there should be clearly identified actions that will be taken. The action might be required training by your team or verifiable outside training. The action might be that they are refused additional business unless and until their performance is improved. Some possible areas for performance ratings are delivery times (total lead time from time of order to receipt), on-time performance, order accuracy, line-fill rate, invoice accuracy, delivery costs, and item pricing. Other ratings, more in the supply chain context, might be communication of demand forecasts, production plans, delivery schedules, and inventory levels. Supply chain and network distribution costs can be measured and scored, as well as the supplier's or customer's contribution or impact on those costs. Measure what is important to your success, and use the results to make improvements.

A valuable supply chain analysis and performance measurement tool has been developed by the Supply-Chain Council. The supply-chain operations reference model, or SCOR, includes metrics to evaluate the performance of the supply chain. It also includes supply chain best practices, standardized terminology and descriptions, and a relationship framework.

The concept of strategic sourcing is an important topic. With strategic sourcing, the idea is to find, develop, and partner with suppliers that fit in with your strategic objectives, not just find a vendor who can sell you stuff. You might be able to find multiple vendors who could supply you with the materials you need, but which one (or ones) is the best? The suppliers that possess the attributes that are important to you, that will help you reach your objectives, are the best suppliers. Price is an issue, and an important one, but it is not the only issue. The supplier must be able to meet the standards you have developed, but they need to go beyond that. They need to be aligned with your organization and with the direction you are taking. You need to fit in with their strategy too. If the relationship is based on current situations and needs, it might not be aligned with future needs and might not be able to grow the way you want. You need to know what the supplier's strategy is to assess its alignment with yours.

That's strategic sourcing, but what about strategic selling? Most companies don't sell to just anyone, but sometimes it's not too far from that. Some cursory research might be done, probably a credit check, but strategic selling is much like strategic sourcing. Are your customers aligned with your strategy? Will they be able to meet your standards, perform, and grow along with you? Do they fit in with the direction you're taking, you're image, and your markets? What are their capabilities to support you, your products, and the supply chain? You need to analyze both your suppliers and your customers in the context of the supply chain and how they can support and improve its performance.

10.2.5 Network Planning

As I mentioned, the supply chain is more of a web or network, a complex system, than it is a chain or a simple series of suppliers and customers. Each members of the supply chain has its individual goals, and the supply chain has its goal. The design of the network that will allow everyone to achieve their goals and will satisfy all of the customers, especially the final consumer, is quite an undertaking.

There are always tradeoffs. When designing the supply chain network, you must consider numerous tradeoff areas. One tradeoff is multiple suppliers versus a single source. Having multiple suppliers gives you the feeling that you have more protection, whereas a single supplier leaves you more vulnerable. This isn't necessarily true, but it is a persistent perception that must be addressed as a tradeoff.

A common, and important, tradeoff is with the number of production and distribution facilities. Having fewer facilities tends to lower the unit cost of production and warehousing, but having more facilities tends to increase the transportation costs. A single production facility can take advantage of economies of scale, whereas multiple facilities add flexibility. A single warehouse has lower safety stock requirements than the total safety stock held at multiple warehouses. A single production facility must contend with scheduling and resource allocation issues, whereas multiple facilities can be dedicated to particular products or product lines. Other tradeoffs include inventory levels versus customer service levels, lead time versus distribution costs, setup time versus long production runs, and inbound transportation costs versus outbound transportation costs. Some of these issues can be relieved or reduced through many of the tools and techniques discussed throughout this book, but they all need to be addressed and resolved.

The mix of products and their production and distribution points have to be considered in the network plans. Production costs will vary by facility and geographic region. Different areas have different tax rates or may have various trade regulations and customs duties and fees. These issues must be balanced and considered in the design of the network. There might be interplant transfers, third-party production and distribution involvement, and numerous outside service providers involved in the network. Will the network partners operate as make-to-stock or make-to-order, and which partners will operate under which system, and how does that affect the entire network? The support for a make-to-stock producer is different than the support for a make-to-order producer, and the suppliers must be capable of supporting the method of the producer. The cost structure of the network will be different depending on how each partner operates. When planning the network, you must consider the point where each customer is met, and the mix of make-to-stock and make-to-order producers and suppliers should be analyzed for total supply chain cost.

There are advantages and disadvantages to in-house logistics versus third-party logistics providers (3PLs). 3PLs have expertise and resources that many organizations, especially smaller companies, do not have in house. They can often provide logistics services at a lower cost than you can. If that's true, the question then comes down to the strategic fit and the impact on the network. Some items might be better suited for 3PLs, whereas other items are handled in house. The use of brokers and distributors in the network must be analyzed. Again, their strategic fit in the network should be evaluated.

One tool for the supply chain, and a consideration in the design of the network, is foreign trade zones. Many people are unfamiliar with foreign trade zones and how they work. Now's your chance to learn. Foreign trade zones are areas designated by the federal government as being outside of the U.S. Customs' territory. What exactly does that mean? It means that although the area is on U.S. soil and under the control and supervision of the U.S. Customs agency, items entering the zone are not subject to customs duties, taxes, and other requirements. U.S. laws apply within the zone, so don't think you can do anything illegal there. Use of the foreign trade zones offers many advantages.

Goods that are imported into the foreign trade zone do not incur duties and taxes until they leave the zone and enter the United States; they can remain in the zone indefinitely. This allows you to control the timing of the payments of the customs duties. Also, items that are imported into the zone and then reexported are not subject to customs duties. Combined with the fact that items imported into the zone can be manufactured or otherwise processed, this can be an important money-saving process. Domestic goods can be exported to the foreign trade zone and subsequently reimported without customs duties being imposed. Items manufactured in a foreign trade zone can qualify as "Made in the U.S.A." if other requirements are met.

These features are important because different items have different duty rates. If an item can be imported into the zone, then processed into another item with a lower duty rate or no duty, duties can be lowered, avoided, or delayed. This is great for product cost and cash flow management. Many items imported to the United States are restricted by quotas. Items imported into the foreign trade zones are not restricted by these quotas.

Foreign trade zones are cost effective as warehousing and storage facilities. Bonds are not required in the zones, items can be held indefinitely, and fees are assessed only on the actual space utilized. Goods can be unpacked, processed, labeled, sorted, repacked, and displayed for sale. Goods can even be sold from the zone. There are many advantages to operating in a foreign trade zone; anyone involved with importing and exporting should look into working there.

10.2.6 Distribution Requirements Planning

Distribution requirements planning is similar to materials requirements planning, as the names imply. Materials requirements planning calculates raw materials and manufactured components needs, quantity, and timing based on planned production from the master production schedule. This is a forward-looking, proactive approach to planning the materials needs. Instead of forecasting the materials needs, which introduces error into the system and disconnects the materials needs from the finished goods production, materials requirements planning looks ahead at the requirements needed for finished goods production and the expected inventory balances.

Distribution requirements planning does the same thing in the distribution system. If each distribution center and warehouse in the system were to forecast their own demand, error would be introduced at each facility and the needs of each facility would be disconnected from the demand at the end closest to the final consumer. Forecasts must be developed for the facilities closest to the final consumer. The distribution center or warehouse that supports those facilities doesn't rely on forecasts; it uses information provided by those facilities. Let's walk through the example in Figure 10.6.

In this example we have two distribution centers supported by a regional warehouse. The two distribution centers use forecasts to deter-

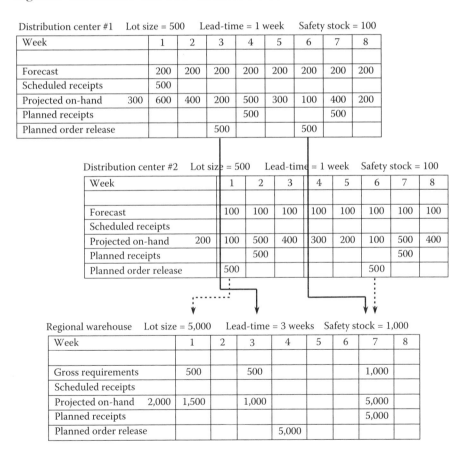

Distribution center #1   Lot size = 500   Lead-time = 1 week   Safety stock = 100

Week		1	2	3	4	5	6	7	8
Forecast		200	200	200	200	200	200	200	200
Scheduled receipts		500							
Projected on-hand	300	600	400	200	500	300	100	400	200
Planned receipts					500			500	
Planned order release				500			500		

Distribution center #2   Lot size = 500   Lead-time = 1 week   Safety stock = 100

Week		1	2	3	4	5	6	7	8
Forecast		100	100	100	100	100	100	100	100
Scheduled receipts									
Projected on-hand	200	100	500	400	300	200	100	500	400
Planned receipts			500					500	
Planned order release		500					500		

Regional warehouse   Lot size = 5,000   Lead-time = 3 weeks   Safety stock = 1,000

Week		1	2	3	4	5	6	7	8
Gross requirements		500		500				1,000	
Scheduled receipts									
Projected on-hand	2,000	1,500		1,000				5,000	
Planned receipts								5,000	
Planned order release					5,000				

Figure 10.6   Distribution requirements planning.

mine their demand. That's the forecast row. In this example we're using weeks as the time buckets. Distribution Center #1 has a forecast of 200 units per week and Distribution Center #2 has a forecast of 100 units per week. The lot sizes of each shipment, lead time to receive shipments, and safety stock requirements are shown in the diagram. Scheduled receipts are incoming shipments that are already released by the system. Projected on-hand is the calculated inventory balance at the end of each time period. The projected on-hand is calculated by subtracting the forecast quantity from the beginning balance (which is the prior period's ending balance) and adding any expected receipts. The forecast quantity is what is expected to go out, and the expected receipts is the quantity that is expected to come in. Receipts are either scheduled or planned. Planned receipts are quantities that are planned in the system (based on these calculations) but not yet released.

The planned order release is the quantity you are planning to order. The order is placed (released) in the time period that is offset from the receipt date by the length of the shipment lead time. For the distribution centers, the lead time is one week, so an order that is to be received in Week 7 must be released in Week 6.

As you know, forecasts have a certain level of error in them. They're not wrong, as many people like to say; there's just an inherent amount of variation in the accuracy of the forecasts. Each forecast has this error in it, so the more forecasts you have, the more error you have. In the supply chain, you can reduce the level of error in the forecasts by forecasting less. Instead of forecasting at each point in the distribution system or each point in the supply chain, requirements should be calculated from downstream needs. The demand at the consumer level should be forecast, but it should be calculated at all the points upstream.

10.2.7 Reverse Logistics

Reverse logistics is the planning and management required to move back up the supply chain for returned items and shipping and packaging material. A well-planned and managed system offers many benefits, including reduced total supply chain costs. Let's start with returned products. If the consumer receives a defective product, it breaks down within the warranty period, they realize they don't like it, or they just don't want it, they'll return it to the place they purchased it. Depending on the condition of the item, it will be restocked and resold, refurbished and resold, or sent back to the producer. The producer might remanufacture the item, scrap the item, or recycle the materials. The producer might perform these activities or contract them out to third parties. Remanufactured items will then reenter the supply chain.

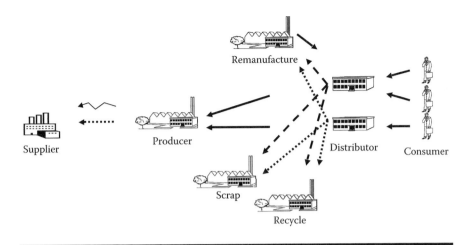

Figure 10.7   Reverse logistics.

If parts or materials can be recovered, they might be returned to the suppliers. If the item was defective because of defective materials, the supplier might just be notified of this and the materials scrapped or recycled. The logistics of returning, remanufacturing, scrapping, and recycling items and materials can be as complex as the original supply chain, maybe even more so because of the additional players (see Figure 10.7).

If the reverse logistics system is not well designed and well planned, a complex and costly mess can develop. All these activities increase the total supply chain costs, although some of them can be recovered by selling scrap, remanufacturing, and recycling. To minimize the costs, eliminate frustration, and improve overall customer service, the reverse flow must be smooth and well managed. All the activities involved in the reverse flow must have clearly defined processes in place. You must plan the flow of the materials — how they'll be moved, whether they'll be consolidated before movement, and who is responsible for the movement. Information must flow too. Some information will flow with the materials, but some will flow without materials. The distributor might send the materials to recycling and the status of the materials to the producer. Financial flows are part of the system too. Who pays, who recovers, and how and when payments are made must be included in the process. Any activities related to the reverse flow up the supply chain must be included in the design of the system.

Because of environmental concerns and cost reduction, packaging is frequently sent back through the supply chain. Reusable materials are often used for packing and shipping, and pallets and shipping containers are returned to the point of origin. These items have to be planned just like the materials discussed above. Again, a well-planned system will reduce frustration and will lower total costs.

10.2.8 Design for Logistics

In the context of the supply chain, the product development process includes not only design for customer value and manufacturability, but also design for logistics. Product development and design need to consider the requirements of the supply chain, including the reverse logistics we just discussed. The packaging of products has an enormous effect on the transportation costs of those products. The more items you can fit in a given amount of space, generally the cheaper it is to ship them. Depending on weight, if you can fit 2,000 units in a container instead of 1,500, or 50 units on a pallet instead of 40, the transportation cost per unit will be lower.

The tighter you pack items, the less storage space they will take up. Not only transportation, but warehousing costs will be lower if less space is used. Retail display space is also at a premium, so items and packaging should be designed to minimize the amount of retail display space needed. Instead of shipping finished goods, you can design items so that final assembly or packaging takes place closer to the consumer. When I was a kid I assembled lawn mowers and wheelbarrows at my family's hardware store. Shipping all the parts in compact boxes is a lot cheaper and easier than shipping fully assembled lawnmowers.

Cross-docking, which has been perfected by Wal-Mart, depends in part on the packing of items. If items are shipped in pallet loads to the warehouse, then broken down to individual items for shipment to customers, the packing and packaging play a key role. The quantities that are packaged together into various selling units need to be carefully designed to facilitate the breakdown of the pallets and the movement of the product. If different items can be easily and tightly stacked together on the outbound trucks, that's even better.

Modularity and standardization in product design is another important and valuable consideration. When parts are interchangeable between products, total production, product, and supply chain costs are reduced. Instead of producing and shipping separate parts for each product, one part that can be used in multiple products is a better strategy. You can reduce total costs by designing a product that can be assembled quickly from parts that fit together. Instead of producing products in an assembly-line format, where each part is added in stages, assemble products in a modular format. The parts can be produced separately or be sourced from different suppliers, then assembled closer to the consumer. This decouples the production of the components from the production of the final product and lowers total production, product, and supply chain costs. Dell computer has perfected modular design and assembly of products.

10.2.9 Outsourcing

Outsourcing is a much-debated and much-discussed topic. Outsourcing to overseas suppliers and service providers is a controversial topic. People tend to forget that outsourcing has occurred since the dawn of civilization, and with good reason. If there were no outsourcing there would be no supply chain. A single entity would perform all the activities from mineral extraction to processing, to distribution to the final consumer. How likely is that to happen in today's world? Not very.

Materials and products are not distributed evenly around the world. Ancient trade routes were established to secure materials and items that weren't available locally and to sell or trade local items that weren't found in other areas. Different geographic areas and different cultures have different capabilities. This is just the natural way of things.

The controversy arises from the changes that occur when work that is currently performed locally or in house moves to another company, area, or country. People lose jobs, their standard of living declines, their work changes — their whole lives change when work is outsourced. That is often true for the organization that is doing the outsourcing. For the people who take over the work, the opposite is true. Those people gain jobs, their standard of living increases, and the entire local economy improves. These are generalizations, of course, and there are cases where the consequences and benefits don't occur, but sometimes reality and perception are hard to separate.

The debate surrounding outsourcing should center on how to extend the benefits to everyone and eliminate the consequences. But that's beyond our scope here. We will discuss some of the things you should consider when you outsource activities or functions. The main reason companies outsource is cost. Someone else can perform the work cheaper than you can do it yourself. There are several reasons for this. Different geographic areas have different wage and tax structures. Materials availability and materials transportation costs vary by region and distance from suppliers. Cost is an important factor to consider.

Capabilities and expertise is another consideration. Companies, even countries, specialize in different areas. If another organization has the expertise and specialized equipment, facilities, and personnel, it can be advantageous to outsource the work to that organization. Capacity is often a reason to outsource. Organizations often outsource work temporarily to increase capacity or relieve capacity constraints. It is often advantageous to outsource work in one area so that capacity can be freed up in another.

The first question you should ask yourself is, why am I outsourcing? Then really examine and analyze your answers. Outsourcing has to fit in with your strategic goals. If you're outsourcing because of cost, is cost the overriding consideration? What about quality, lead time, and control? What is the total cost of outsourcing in relation to the reduced cost per unit? Do you have to keep more inventory on hand and in the pipeline because of increased lead time, increased lead-time variability, or quality issues? Does the lower unit cost make up for the increased transportation, warehousing, and inventory control and management costs? You have to consider the total picture, but the most important thing is to analyze the outsourcing decision in relation to your strategic plans and strategic goals.

10.2.10 Technology

In the book, *Necessary But Not Sufficient*,* the authors discuss the need for technology. In our increasingly complex and globalized world, the use of technology, especially information systems, is a requirement for conducting business. However, technology is a tool. Technology alone won't solve your problems. The technology has to be backed up by effective management systems — effective and high-quality processes and procedures. This is so true, but sadly not always heeded.

Supply chain information systems are similar to enterprise information systems in that they attempt to integrate the supply chain, just as enterprise resource planning (ERP) systems attempt to integrate the enterprise. Most systems tend to be stand-alone systems that are bolted together as needs arise or when somebody gets the great idea to add a new component to the system. These systems are not integrated and often have a difficult time talking with each other and sharing information. They often end up creating more problems than they solve. The problem isn't with the technology itself. As stand-alone systems, they're very useful and helpful. The problem is with the integration, or lack thereof, and the management systems that back up the technology.

There are systems on the market and more being developed that can integrate the supply chain, at least through a couple of layers. Some industries are making progress in integrating the supply chain, and that progress will encourage information system providers to develop improved and more robust systems. When supply chain management becomes more accepted and supply chains become more integrated, technology will be used to support the supply chain systems. Technology is available, and

---

* Eliyahu M. Goldratt with Eli Schragenheim and Carol A. Ptak, North River Press, 2000.

is necessary, but technology alone will not improve supply chain integration and communication.

ERP and supply chain systems comprise many different components or modules. You're probably already familiar with many of these. Standard ERP modules include the financial systems — general ledger, accounts receivable and accounts payable, subsidiary ledgers, asset management, and other components of the financial system. Production and inventory components include inventory master files, bills of materials, routings, master production scheduling, materials requirements planning, capacity planning, purchasing, and all the other necessary functions for production and inventory management. Warehouse management systems are added as modules to ERP systems or act almost as stand-alone systems in distribution companies. Components of warehouse management systems include receiving and shipping, space management, picking and packing, resource utilization and monitoring, and shipping and billing document generation and management.

When you move beyond the walls of your organization into the supply chain, one of the more common applications is customer relationship management (CRM). CRM systems not only collect data on your customers, such as buying history and contact information, but also help you better communicate with your customers. As the name implies, it helps you manage your relationships with your customers by collecting and analyzing customer information. Supplier relationship management (SRM) systems work in much the same way with your suppliers.

Advanced planning and scheduling (APS) systems are not new, but they are not widespread, especially in smaller organizations. APS systems analyze resource needs, capacity availability and constraints, product cost and profitability, and other factors to develop a plan and schedule for production based on actual, planned, and possible customer orders. APS systems are used not only for scheduling, but also for order promising and price quotes. The company can analyze the effects of a potential customer order before a quote and promise date are given to the customer. The system can also present different scenarios and recommend additional resources, including outsourcing of production. All of this depends, of course, on the data and the accuracy of the data you feed into the system and on your data and system management. Use the tools that will help you achieve your goals and make sure you have the management systems in place to back them up.

Chapter 11

E-Commerce: Entering the 21st Century

I'm not much of a techie. Maybe you aren't either. But we are already in the 21st century (by Western calendars). Just to function at the most basic level in business today you have to be somewhat technology savvy. You don't need to be an expert, but you do need to be familiar with the tools available to you. You don't need to be an E-commerce expert, but you do need to understand what it is and how you can use it to improve your organization's performance.

So, what exactly is E-commerce? This is not easy to define. E-commerce means different things to different people. But, being the brave soul that I am, I'll come up with my own definition. E-commerce is simply using any electronic means to conduct business. The most obvious method is the Internet and using a Web site to sell products. Think Amazon.com. While this is probably the first example that comes to mind, there are many other means of conducting business electronically. And you need to keep an eye on the new technologies that arrive almost daily. You can't just think in terms of existing customers and existing channels; you need to undertake the daunting task of keeping up with the times. This is much easier said than done.

The terms *E-commerce* and *E-business* are used interchangeably by most people. Technically, though, the terms have distinct definitions. Take the "E-" out and what do you have? Commerce and business. What's the difference between commerce and business? Commerce encompasses all of the activities of the economy: production of goods, trade, services — everything. Business is a subset of commerce. Business is the individual units and the trade and activity that goes on between them. Your company is a business, and your business is a part of commerce. Now that I've cleared that up, I'll be tossing around both terms pretty freely in our discussion. Let's get started.

11.1 E-Commerce Is Not Just Creating a Web Site

This is as good a place as any to start. If you have a Web site, you're participating in E-commerce, but there's much more to conducting business electronically than just creating a Web site (see Figure 11.1). A Web site is just as much a marketing tool as it is a method to conduct business. It is becoming increasingly difficult to compete effectively without having some sort of Web presence, so let's look at it in more detail.

The first question is, "why?" Why do you have a Web site and what do you expect to accomplish with it? Because you "have to have one," "everybody else has one," or you "want one" are not good reasons to develop a Web site. You've heard this before (unless you jumped to this chapter first): a Web site has to fit in with the strategic goals of your company. There needs to be a justification for the site, a purpose, and an alignment with your strategy. All of the functions and features of the site have to fit in with the strategy.

Steve's manufacturing

Welcome to our Web site. Learn lots of wonderful information about our company and our products.

Buy lots of stuff.

Figure 11.1 Web site.

Question the purpose of the site. Where does it fit in with your current systems? Is it an information-sharing and marketing tool? Is it an extension and enhancement of your "physical" business, or is it more of a stand-alone or separate business? Will the site offer options to current methods of operation, or will it replace them? Will it be available to the public or just to your existing supply chain partners, or both? You need to define what you expect to accomplish with the site. You need to determine its purpose, its goals, and how it will work to achieve those goals. You'll need to develop measurements that will help you determine whether you are reaching those goals.

A Web-based business is a misnomer. Unless everything related to your business is contracted out and your Web site is just an automatic portal or conduit, you have to have physical facilities somewhere. Your primary interface with your customers might be Web based, but the business itself is a physical entity consisting of people and equipment and maybe products.

A Web site is important. I have one and you probably have one too; but it should be an investment, not a cost. To develop and maintain a site requires a bit of money. The more features, the more money. But you should receive a return on your investment; either increased revenues or decreased operating costs. If not, you should probably reexamine and analyze the site; the reasons for it and how it works.

11.2 EDI and Sons

E-commerce is not new; it has been around for many years. It just hasn't been called E-commerce until recently. EDI, or electronic data interchange, has been used since late in the 1960s. Even in today's Internet-dominated environment, EDI is still used by many companies. Basically, EDI allows for computer-to-computer exchange of data in a standardized format. Probably the most common transactions that take place via EDI are the transmission of purchase orders and invoices.

The electronic transmission of the information helps to reduce operating costs. Time is saved in the clerical areas at both ends, especially because most, if not all, of the information being exchanged is already in electronic format. Transcription errors are reduced. Even in electronic environments, many organizations manually enter data from one system or format to another, introducing errors in the process. Time is also saved in the transit of the information. Even overnight delivery takes longer than overnight. By the time the information in a physical form is prepared, sent, delivered, received, and reformatted for use, at least a couple of days have passed. Transmitting the information electronically may take minutes (or hours, depending on any manual activities in your process).

EDI does have some drawbacks. The cost of setting up the system and the lack of standardization in formats are the biggest drawbacks. Standard formats have been attempted, but sadly they have never taken hold. The result has often been that smaller organizations have been forced (bullied) to use the format dictated to them by their larger suppliers and customers. It's not uncommon for smaller companies to be forced to use multiple formats. The costs to implement a single format to communicate with a single partner are substantial. When multiple formats are required, the costs are compounded.

Implementing EDI is not a simple process. The benefits can be substantial, but for many smaller companies the results simply don't live up to the promises. Although EDI is still used, technology has advanced considerably and continues to evolve and improve. Internet-based applications are the norm today, and there are many advantages to this platform.

11.3 World Wide Web: Two Sides of the Coin

Everything in the following discussion can be classified as either business-to-consumer or business-to-business. Sure, there are some overlaps and integration areas, but we can break it down into these two areas. It will help clear up any confusion when we talk about E-commerce too.

11.3.1 Business-to-Consumer (B2C)

Let's start with business-to-consumer, because that is probably what you're more familiar with and what comes to mind first (see Figure 11.2). Take yourself out of business mode and get into personal mode. Think about your other life; the one that involves friends and family and home. Need to buy gifts? Need something for the home? Need to plan your next vacation, check the surf conditions on the North Shore of Oahu, or visit

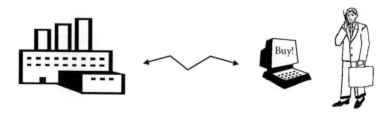

Figure 11.2 Business-to-consumer connection.

the naked mole rat cam at the National Zoo? You're engaging in E-commerce.

Now back into business mode. Let's look at some of this from the other side. How and why does your business interface with your customers, potential customers, and the general public? Three primary reasons for establishing an Internet presence are to provide information, as a marketing tool, and as a sales channel. Using a Web site as a vehicle to provide information is a passive form of communication. Simple information about the company and its products is posted for interested parties to read or see. It is relatively cheap and easy to put up a simple information-sharing Web site. In many cases it is sort of a starter site. The information is usually static, in that it doesn't change often. A brief summary of the business, history of the company, and contact information are usually included. A listing of the products and services offered and their descriptions might be shown, often in summary or family grouping form.

The next level is when the Web site moves beyond simply providing information and acts a dynamic marketing tool. The site still only provides one-way communication, but it is more dynamic. Instead of a passive, information provider, the site elicits some type of response — hopefully excitement. Beyond just providing data about the company and its products, the site tells the customer why they should buy the product or service. It tries to provoke an emotional response; it weaves a story, describing how the customer will feel after they use the product. With a marketing site, news and information will be updated and changed frequently; colors, logos, pictures, music, and video might be used. The site might allow visitors to provide their contact information, or it might allow visitors to provide feedback and comments about the site, the company, and the company's products.

The highest evolution provides information and acts as a marketing tool, but then increases customer participation by allowing them to purchase the company's products. For some businesses this might be their only sales channel; for others it is one of many. For some companies, Internet sales are an insignificant part of their business, but they find it necessary to provide the option. Before you begin working on a Web site, and spending money on it, you need to determine what you expect to accomplish with it. Where does it fit in the business? Why are you building it? How are you going to maintain it? What resources do you need to build and maintain it? What impact will it have? Don't just put up a Web site because everyone else has one. It needs to be well thought out and justified from a business perspective. Technology being where it is today, the functionality that your site offers is almost only limited by your imagination.

11.3.2 Business-to-Business (B2B)

Utilizing the Internet as a business-to-business tool is much the same as using it for business-to-consumer purposes (see Figure 11.3). The difference is in the functions provided at the highest level: the sales and transaction processing level. A business-to-business site can be an information provider, a marketing tool, or a sales channel. The action is at the sales channel level.

A host of functions and transactions can take place between trading partners and businesses. Some are so common they are practically a requirement of any business; others are still fairly rare; and there's always something new on the horizon. Probably the most common activity between businesses is procurement. E-procurement, electronic purchasing, whatever you want to call it; more and more businesses are encouraging or requiring this method of purchasing. In some cases, it's not much more than the purchasing consumers do on a business-to-consumer site, but in other cases it's a very robust system of purchasing, transaction processing, and monitoring.

Some companies have their entire product catalogs online along with real-time inventory status. They might even have advanced availability and order promising capabilities. Many purchasing transactions fall under regular or repetitive ordering. In these situations much of the process can be automated. Human interaction takes place in the negotiations of the parameters (such as price, order quantities, and order triggers), but after that, systems can be integrated so that orders are placed automatically. Orders can be transmitted automatically and electronically, a confirmation can be sent back, and advanced shipping notices as well as invoices can be automatically sent electronically. The payment can even be sent the same way — electronically and automatically. Human intervention is needed only on an exception basis. See Figure 11.4.

Order tracking is another common type of business-to-business function. It is becoming more common in business-to-consumer arrangements too, but in business-to-business transactions, it tends to go deeper. Beyond just shipment tracking, order tracking can get all the way into the pro-

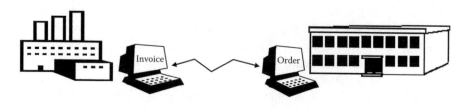

Figure 11.3   Business-to-business connection.

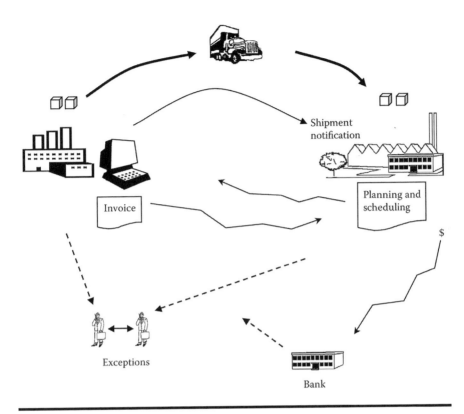

Shipment
notification

Invoice

Planning and
scheduling

$

Exceptions

Bank

Figure 11.4 Automatic purchasing.

duction, or even the design, process. Customers' ability to monitor their orders, from materials availability to its place in the schedule and status in production, can be disturbing to think about. You have a certain measure of power when you control access to information. If a customer calls to complain about a late order or just calls to check on an order, you have the power position. You can assure the customer that everything is fine, that there was a problem, but the order is the top priority and will be arriving shortly. Then you can run out on the floor and expedite the order that somehow got pushed aside and forgotten. You can tell your customer that the order is right on schedule, then check to see if there is even an order in the system for that customer. After all, you can't keep track of everything, can you?

You actually have no power in that situation. It is a false sense of security. You don't have the control you think you have. You are left to react to every problem and question that comes along. The real power is in having a system that is open and available to everyone who needs it. The real power is in a system that runs the way it is supposed to and the way you want it to; a system that only rarely requires intervention; a system that you have enough confidence in that you can open it for all to see.

Beyond order tracking, there is electronic collaboration. With greater access to each other's systems, real-time transaction processing, and open communications, electronic collaboration is almost like sitting down in a room together. Production and delivery schedules can be shared, analyzed, and discussed. Product development can be improved and the time compressed. Engineering drawings can be generated, modified, and approved in an online, collaborative environment. The savings in time to market alone quickly provides the payback for the hardware and software investment.

This sounds great, but there is a lot of work to do to prepare organizations for these systems. It's a new paradigm! That's a worn-out phrase, but it's true. It's a new way of doing business that requires a new way of thinking. Just like we talked about tearing down the walls of the silos in Lean Manufacturing and discussed collaborative engineering, electronic collaboration tears down the walls between companies. Instead of the designers producing a design drawing and handing it off to the manufacturing engineers, the groups can get together and work through the designs and drawings together. The result will probably be a better product; one with fewer design defects, developed in a shorter amount of time.

Internet auctions are well known to just about everyone these days. Even if you have never bid on or sold something on EBay, you've at least heard of it. Auctions are fairly common in the business-to-business arena also. The difference is that consumer auctions generally involve rare or unique items, whereas business auctions involve more commodity-type items. Very similar to auctions are inventory exchanges. In fact, there might be little difference between the two. If there is a difference, it is probably in the services provided. An exchange might act as more than simply a broker by providing logistics services, fund transfer services, qualifying the participants, and offering other services that increase the value of utilizing the exchange.

I don't know how popular they are, but application service provider (ASP) systems are an interesting development that have a certain appeal. Typically, when you implement an enterprise resource planning (ERP) system (or some other type of system), you purchase the hardware and the software. Then, when a software revision or new version is released, you pay more to upgrade the system and convert any data that needs to change with the new system. Sometimes, a new version of the software requires that you buy and install a new or different operating system or new hardware that will accommodate the new operating system. It seems like a never-ending cycle and that you are never fully implemented. ASPs attempt to alleviate some of these problems by hosting the database and the programs and providing you with an interface and portal to the system. The ASP takes responsibility for maintaining the system and performing all upgrades, conversions, and any other systems activities. You lease the system and pay fees for services.

This approach has advantages and disadvantages. The big advantages are with costs and cash management. The price of an ERP system is substantial; even the systems for small companies. The cost of hardware and software alone runs from tens of thousands of dollars for very small companies, to many millions for the big guys. And that doesn't even include the costs of analysis and planning required to select the system, or the costs to find, hire, contract, and train the people who will implement the system. Using an ASP eliminates most of these costs. Selecting the system and finding and training the people who will operate and use the system will incur costs, and you will have to pay for using the system, but the up-front purchase costs and the maintenance costs will be eliminated.

There are some disadvantages and risks to these systems, however. These disadvantages and risks are inherent in any outsourcing situation, but there are even greater risks with ASPs. You're giving up a lot of control of your data and information. Sure, there will be backups and safety built into the system, but you won't have direct control and possession of your data. If something happens to your service provider, you might have an extremely difficult time reclaiming possession of your data. Even if you do, the time delay might have severe consequences to your business. It's something to think about if cash flow and in-house expertise are issues; but you have to weigh the risks involved.

11.3.3 Security

Security is always a relevant topic when discussing electronic systems. For some reason, people tend to not be as vigilant with security when it comes to electronic systems as opposed to physical systems. You secure the stockroom because you are concerned with who has access to the materials. You keep the room locked and you restrict access. You keep detailed records of who has accessed the area and all movement of the materials. You train everyone in the protocols surrounding the area and the materials. In short, you take it pretty seriously.

On the other hand, have you ever overheard someone asking another person for their password so they could get into the system? 'Nuff said.

Computer security is a field unto itself. There are many levels of security, and all of them are important. The hardware needs to be physically secure from accidental or intentional damage. Environmental conditions are a concern. And you don't want anybody to be able to come in and unplug the machine (or turn the power off). Computer components are notoriously sensitive to dirt and dust, moisture, and heat, among other things. Physically secure, environmentally controlled rooms are important for housing your primary systems.

The software and operating systems are where the real vulnerabilities are found. Computer viruses and other malicious computer codes are a constant threat. Many of these viruses and other threats find their way into your systems because of a lack of vigilance, lack of knowledge, or poor maintenance. Firewalls and virus scanners aren't installed, aren't installed correctly, or aren't maintained. Everyone who has access to the company's computers and systems has to be trained in security procedures, and security has to be constantly reinforced. The systems administrators have the daunting task of keeping up with security. They need to keep the security systems updated, keep everyone trained and vigilant, and keep their knowledge of emerging threats up to date.

Threats from outside are important, but threats from inside are just as important. They might also be more difficult to guard against. Threats from outside are in many ways easier to see and monitor, because that's what you're looking for and more people are looking for them. Security guards at a bank are watching out for someone who shows up at the front door with a mask on. They're not watching the tellers, and they're definitely not watching the people in the back office who are working behind the scenes. The same goes for computer security.

You're watching out for attacks on the system from hackers and other nefarious characters. You're not wandering around the cubicle farm making sure people don't have their passwords written down on a sticky note taped to their monitor. Security is a serious concern that you need to take seriously and you need to reinforce regularly. Access to the system needs to be controlled, and access within the system needs to be controlled and monitored. Don't take shortcuts with security.

11.4 The Back End Is Just as Important as the Front End

E-commerce is one of many tools in your toolkit. It is almost a requirement for business today, but some companies remain successful without undertaking any sort of E-commerce initiative. It's likely that you've either entered the E-commerce arena or are planning to do so soon. Many organizations take the plunge in stages. First you post an information-only site, just to have a presence on the Web; then the site evolves into a marketing tool. You grab the viewer's attention and generate interest in your products and company. Then you take the final step to developing the site as another sales channel.

That's when you might to start having difficulties. Like too many organizations these days, you will probably try to tack this new electronic customer interface and sales channel onto your existing systems. You may try to do things the way you've always done them, meaning using a lot of manual procedures. I'll use an example of a company that uses their Web site to sell directly to the consumer, because this should be pretty familiar to anyone who has bought anything online (see Figure 11.5).

You have a great new Web site and you're ready to start taking orders from customers. The customer places an order and makes payment through the Web site. Then, although you have a state-of-the-art Web site and a state-of-the-art ERP system, you print the customer's order and manually enter it into your ERP system. You do this because the two systems don't talk to each other. Then, because you're not utilizing the full capabilities of the ERP system, or you have a separate warehouse management system that doesn't talk to the ERP system, you print a pick ticket and packing slip and send that to the warehouse. Somewhere along the way, you manually record the payment in the ERP system.

You had better hope that there are no problems and the customer doesn't call you to ask about their order, or worse, change it. The customer has placed the order over the Internet, but from then on, they must communicate with you via phone calls. You can look up the order, but

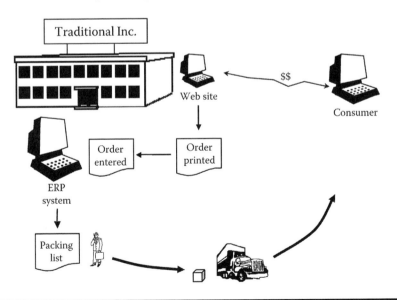

Figure 11.5   Nonintegrated system.

you will have to look in multiple computer systems and find hard copies of some transaction documents.

There is a better way. The problem is that many organizations don't do it, because either they haven't put enough thought into it or they don't know how. You have to integrate your Web site with your internal systems, and all your internal systems should be integrated as well. Let's look at what would happen if your systems were integrated (see Figure 11.6).

The customer would place the order and make payment, just as before. However, this time the customer would instantly receive an order confirmation. The confirmation would verify the order and payment and provide an order confirmation number. Inventory status would have been verified before the customer even placed the order, and the order confirmation would include shipment due dates.

The Web site would be integrated with the company's ERP system. Not only would inventory status be verified in real time, but the customer order would automatically be entered into the ERP system. That would allocate the inventory to that customer and update the inventory status immediately. The inventory and sales transactions would be automatically entered into the accounting system, and the transactions might even be posted automatically to the general ledger.

The order would automatically flow to the warehouse system, whether it is a separate warehouse management system that's integrated with the ERP system or a module of the ERP system. The order would be staged for picking and entered into the pick schedule. Once the order is picked, packed, and shipped, the order status would be updated and a shipping

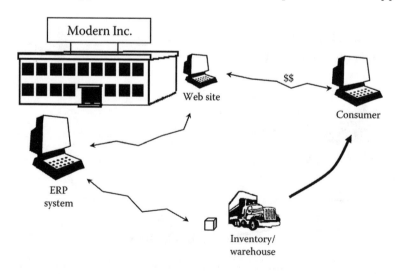

Figure 11.6 Integrated system.

notice might be sent electronically to the customer. The customer could receive shipment tracking information to keep tabs on the order until it arrives.

Think about it. Within seconds the customer would know the order has been received and payment posted. Within hours the customer would know the status of the order and know that it has been picked and shipped and is on its way. The customer will know when to expect the order and who is delivering it. Your customer will be tickled to death with this level of service.

From this example you can see the benefits of integration. You might wonder why every company that enters the E-commerce arena doesn't integrate from the beginning. The two primary reasons they don't integrate are lack of technical expertise and lack of understanding of the need to integrate.

As wonderful as computers are, systems are not always compatible. Integration between hardware and software is still a significant issue. Anyone who has ever tried to install software on their own knows the problems that can arise. Sometimes it won't install at all; sometimes it will seem to install but won't run; and sometimes it will bring your whole computer crashing down. When it comes to large and complex business applications, operating systems, and databases, problems are just compounded. Many organizations have their own IT (information technology) staff, whereas others rely on contracted outside experts. Regardless, you are probably still going to have integration issues.

The level of integration I'm talking about requires systems to not only talk to each other but meld together seamlessly. Many people (meaning the managers who initiate the entry into E-commerce) don't understand the issues involved with integration. They expect things to just work together. They often expect this of people too, with much the same result; but that's a different story. Systems integration needs to be addressed at the beginning stages of E-commerce development. Potential integration issues need to be identified and potential solutions presented. Known incompatibilities may influence the systems that are purchased or developed for the Web site. Integration is a serious concern. Give it the level of attention it deserves.

11.5 Inventory and Operations Planning

One of the biggest problems associated with E-commerce initiatives is the lack of planning that occurs with regard to inventory and operations. Many organizations take the time to develop and design a top-notch Web site, but they either forget about the behind-the-scenes operations or don't consider the potential impacts of this new sales channel. A number of operational and inventory-related issues need to be addressed.

Let's talk about inventory first. You're developing your E-commerce initiative to be a new sales channel. But this channel is different than the channels you're used to. With the Internet, you probably only have one chance to satisfy the customer. If you don't have the item the customer wants in stock and available and can't deliver it immediately, the customer will go somewhere else. You must have the inventory available for immediate delivery. No second chances; no ifs, ands, or buts.

This likely means that you have to increase your inventory levels. You need to support your existing sales channels, with their service level requirements or expectations, and now you have to support this new sales channel that has even higher service level requirements. You can't take away inventory from your existing customers to satisfy the needs of the new customers, but you can't fall short with the new customers. You need to plan ahead. You have to define your service level requirements and determine how you expect to meet those requirements. You need to forecast the sales from the new channel. This will probably be quite different from your existing forecasts. The forecasting method will probably be different, at least until this channel matures, and the forecast will probably look quite a bit different, in terms of volume and timing.

Overall sales volume will hopefully increase. The increase might be dramatic too, at least for a short time. Some of the sales might switch channels, but you probably won't see as dramatic a change in your existing channels. That might not always be true, so you need to be careful with your planning and do some research. Increased sales volume can affect all of your operations if the increase is significant enough. Production planning, scheduling, purchasing, and distribution will be affected. If increased inventory levels are required to meet the growth in demand and the requirements of the new channel, space and warehouse facilities can be an issue. It is better to look into these possibilities in the planning stage rather than after the implementation stage. It is easier and cheaper to plan for warehouse space than it is to find it in a hurry.

Sales (demand) and inventory volatility are another important concern. If you're used to relatively stable or predictable demand patterns, you might be shocked with the increased volatility or new patterns you see when you begin Internet sales. Until you get a handle on the new demand patterns, you might need to increase inventory levels to protect yourself. Remember, you have only one chance with these Internet customers, and you need to maintain your relations and service with your existing customers. If you don't protect yourself with inventory, you will have to protect yourself with production flexibility. Either way, you have to protect yourself.

Another change will likely be in order quantities. Your existing customers probably order in larger quantities, unless you are already selling directly to consumers. This can be a very big change for your operations, and unless you're prepared, you might not be able to handle it. Your existing customers might order in defined, fairly large, lot sizes. The Internet customers might order one at a time or just a few items together (see Figure 11.7). This could put quite a strain on your operations. The number of sales orders will increase, maybe exponentially. The sheer number of orders can be overwhelming if you're not ready for it. Even if you think you're ready, the volume will probably expose flaws in the system that will need to be improved.

Keeping track of all those orders is a big task. The tasks that seemed simple when you had fewer orders become much more complex. If you handled a lot of the order fulfillment process manually, you'll need to add more automation. The processes of picking, shipping, and closing orders can take a lot of time and resources if they are not automated. And of course, the changes that will have to be made to transition to new processes must be considered. It is a big shift in thinking that needs to be planned early and be ready when you go live.

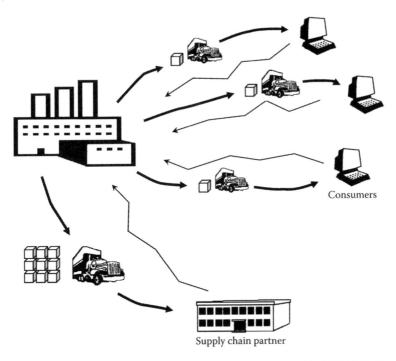

Figure 11.7   Channel differences.

Have I mentioned packaging requirements? If you're used to packaging your products for shipment to business customers, you'll need to start thinking about new requirements. Do you have the packing supplies needed for large numbers of small quantity orders? Are there any other packaging considerations that you need to consider, such as protection from damage, that are different than with your current packaging? Even if you have experience packaging your products for consumer sales, the packaging you use might not be suitable for direct-to-consumer delivery. You'll also need enough packaging supplies on hand to handle the volume you expect. You would hate to miss a shipment because you didn't have the right box available (don't laugh; it has happened). You're used to your current environment; just put some thought into what the new environment will be like and plan accordingly.

Your picking system can be set up in several different ways, but it is set up to accommodate your current customer base; the volume, mix, and timing of your current customers. You might have to make some modifications to your picking and packing system. If you're set up to pick large quantities of items and a mix of items for business customers, the process might not transition easily to large numbers of small quantity orders. In the planning stage, you need to determine what the picking process should look like, how the activities will be performed, and by whom. Don't assume that your existing process, no matter how well it works, will work in the new environment.

After picking and packing, you have the logistics and delivery system to think about. Doesn't this ever end? Can't you just create your Web site and start selling? Well, you could, but it wouldn't be a good idea. Of course, if you set up a site and nobody buys from it, you don't have to worry about the back-end systems. But if you've done your homework and your marketing, you'll hopefully see an uptick in sales. If you're lucky, this could open up whole new markets that you didn't know existed, and sales could explode. That would probably create its own difficulties, but generally pretty good ones to have.

You do have new delivery channels to consider. Instead of shipping pallet loads, truck loads, and containers to warehouses or retail stores, with your own delivery vehicles, or by truck, rail, or ship, you'll be delivering individual items to consumers at their home or office. It's very unlikely that you will be delivering them at all. You will use the U.S. Postal Service or one of the big guys in package delivery (you know who they are). If you do that, you will have to interface with their systems. If you can't integrate with their systems automatically, you will need to develop the systems and find and train the people who will do it. Remember from my integrated view of the future how the customer received an electronic message informing them of the package tracking information? It will be a lot better, especially if your volume is large, if that flows automatically rather than someone having to take it from one system and key it into another.

The final piece of the puzzle is the handling of the financial transactions. Again, if you have a change (or addition) from small numbers of orders with large order quantities to large numbers of orders with small order quantities, this will substantially affect your systems. The finance and accounting system is no different. Even assuming that your accounting system is mostly electronic and highly automated, an increase in transaction volume will have an effect. All the transactions need to be posted to the subsidiary ledgers, then rolled up at some point into the general ledger. The transactions will have to be reconciled and audited. The volume of transactions might affect the method of posting, whether batch processing or real-time posting. You need to determine what your systems are capable of and what method of posting is most appropriate for you.

The reporting of all these transactions, the financial and the operational, will have to be considered. If your number of orders is low and your reports are customer oriented, you might find that with the increase in individual customers your reports are a lot longer than you're used to. More summarized reporting might be in order. Modifying reports is generally fairly easy, but it might be a serious undertaking. How many reports do you have and how many people use them? Modifying all the reports, determining the format the users want them in, and making the actual changes can be a substantial project.

Developing an E-commerce initiative requires you to consider many issues. E-commerce is a different model than many companies are used to operating in. It is a worthwhile and necessary endeavor, but it takes a significant amount of planning to be successful. Don't underestimate the work that needs to go into it. Too often, the Web site gets all the attention even though it is only one piece of the system. The Web site is where the excitement and action are. It's pretty. It's colorful. It has lots of features and functions that you can play around with for hours. The real work, however, is getting the back-end systems prepared to work with, support, and integrate with that Web site. Just like much of manufacturing, sales and marketing gets the glory, but operations and production makes it all work. If it were a rugby team, operations would be the forward pack and sales and marketing would be the backline. Still, it's a team, and one can't work without the other. We forwards make our own glory through the satisfaction of a job well done.

DIAGNOSTICS AND DECISIONS

Chapter 12

Now What?

Wow! We've discussed many different topics and reviewed a lot of different tools that are available to you, the small manufacturer. The big question is, "Now what?" They're all very useful, and they all sound like they would help. But you can't do everything at once. Where do you start? How do you decide which tool to use? You could print them all on a piece of paper and stick it to a dartboard. You probably couldn't go too far wrong with whatever you hit, as long as you were committed to it. But you're not looking for good enough. You want to improve your performance and increase your profitability. This chapter will give you some ideas on how to go about selecting the right tool for *you*.

12.1 Which Tools Are Right for Me?

The selection of tools to use depends on the issue that's being addressed, the desired results, the resources available to the organization, the culture of the company, the timeframe allotted for improvement or for implementation, and other factors. The only place to start is where you are right now.

Where are you right now? This requires an honest, realistic, and possibly harsh assessment. Chances are you fall into, or lean toward, one of two categories: (1) you have deficiencies and you need to make improvements, or (2) your performance is at a high level, but you need or want to take it higher. Although the performance curve flows on a

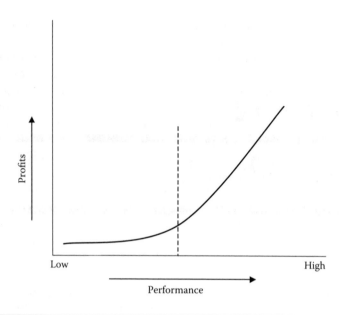

Figure 12.1 Performance continuum.

continuum, you should fall into one category or the other (Figure 12.1). That's one variable in the mix.

In Chapter 2 I said that if you are not doing the basics well you should assess whether you are ready to move on to something new. It takes discipline to perform the basics well on a continuing basis. If you do not have the discipline to do this every day, you should assess whether you have the discipline to move on to more advanced or complex systems. We will use this premise to help determine which tools might be right for you now and in the future.

Knowing where you are is the starting point, but you also have to know where you are going or where you want to go. Without a clear direction, you can end up spending a lot of time and money implementing a system that doesn't give you the benefits you seek. Where do you want to go? You need to have a clearly defined strategy and stated goals. When you determine where you want to go, you will know what resources you need to get there.

Finally, you need commitment. To implement most of the tools we have discussed you need strong and long-term commitment from top management. You cannot implement Lean or an enterprise resource planning (ERP) system in a month. It will take many months, maybe even years. You cannot move from one flavor of the day to another and expect spectacular improvements in performance. And you need committed leaders. Leaders motivate others to action, and that's what you need to make the improvements you want.

The next section uses simple flowchart or decision tree diagrams to help you decide which tools might be appropriate for you. We will start asking questions about where you are now and where you want to go to point you to the tools you might want to use. We will ask questions such as, Do you have inventory problems? Do you have a stable production schedule? Do you want to increase revenue or sales? Do you want to increase market share, decrease lead-time, or increase flexibility? Do you want to increase quality levels or decrease the defect rate?

Some charts will lead you to the top of another chart or another line of questioning. One path may lead you to one set of tools, whereas another path will lead you to a different set. You have to look at the questions, your current status, and where you want to go as a matrix, and put the various pieces together. As you work through the charts, you should begin to get an idea of where you are being pointed.

This diagnostic tool is just a guide and is not comprehensive. Only you know the resources that are available to you right now and which resources you can acquire in a given period of time. Only you know if you are a leader who can stay committed to a particular course of action, or if you have the right people on your team. You might want to create your own diagrams with some questions not posed here. Use the information in Section 12.2 to help you fill in the diagram.

12.1.1 A Selection Tool

There is no particular place to start with these charts (Figures 12.2 through 12.16). Answer the questions as best you can and follow the paths.

12.1.2 Change Management

Do not forget the change management issues that will surround the adoption of any of the tools you choose. Remember that people don't dislike change; they just don't like the way that change is usually presented to them or forced on them. Include change management in any implementation or undertaking you choose. Involve and engage everyone who will be affected. Let them make suggestions, and let them handle many of the changes that directly affect them. Give them the guidelines, resources, training, and education they need.

A formal change management plan should be a part of any significant change or program implementation. If you don't manage the change and

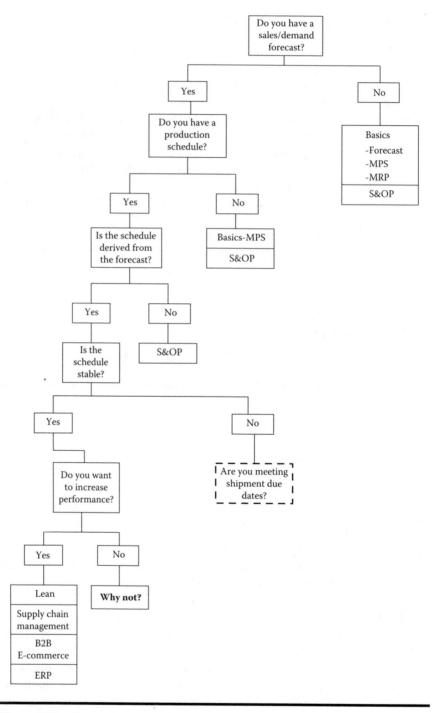

Figure 12.2 Forecast.

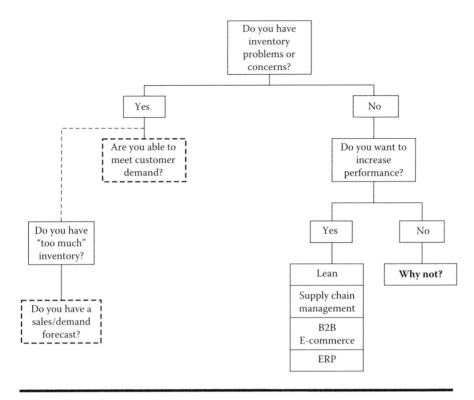

Figure 12.3   Inventory.

prepare the people who will be affected, the program can be derailed quickly and at great expense.

12.2 Application of Tools

Below are brief summaries of the tools that were discussed in this book. Each includes a short description of the primary application of the tool. In conjunction with the decision trees, this will help guide you to the tools that might be most appropriate for your organization at its current stage of development.

12.2.1 The Basics

The basics bring you up to a minimum level of performance, providing the base you need to make improvements. If you can't perform the basics well, or if you haven't already evolved beyond them, you don't have the foundation upon which improvements can be built.

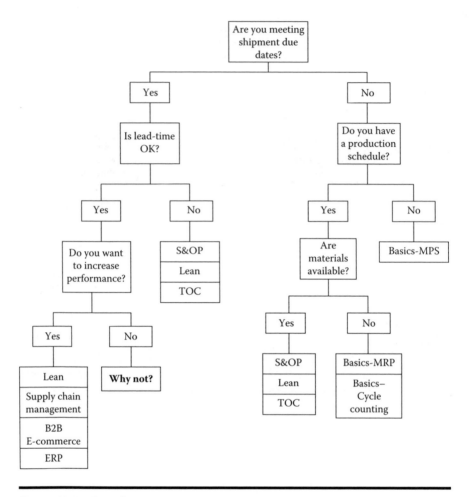

Figure 12.4 Due dates.

A system that is based on the planning hierarchy is a fundamental requirement. The organization needs a direction, a strategy, so that it knows where it's going and what it's trying to achieve. Demand management links strategy to customer and market demand. Demand is forecast and managed for quantity and timing. Production is based on demand and strategy. Components and materials needs are calculated from the production plan and are planned and managed based on the strategy. Plans are checked for validity against available capacity. Performance is measured and compared to the plans, and adjustments are made based on the results.

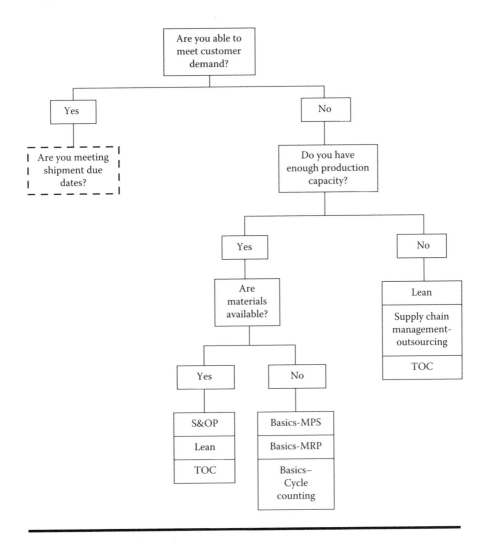

Figure 12.5   Meet demand.

12.2.2 Sales and Operations Planning

Sales and operations planning improves (or adds) the linkages between demand planning and production operations and includes a financial planning component. Sales and operations planning brings the top-level executives into the process and strengthens the link between top management and the operational levels.

Sales and operations planning adds or strengthens the linkages between long-term strategy and short-term planning. It helps top executives make

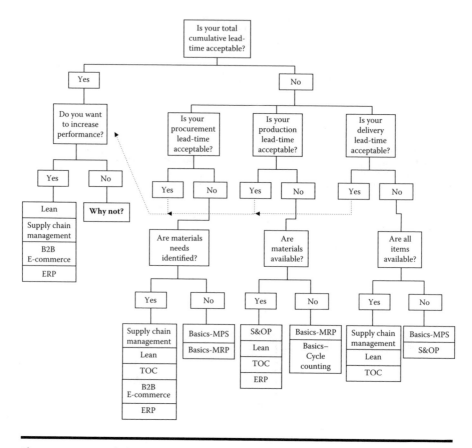

Figure 12.6 Lead time.

better decisions by providing a tool that increases their involvement in short-term planning without needing to get lost in the details.

12.2.3 Lean Manufacturing

Lean Manufacturing principles allow a company to increase customer responsiveness, increase production flexibility, reduce operating costs, increase quality, reduce inventory investment, and increase productivity, efficiency, and effectiveness. Lean is a powerful tool. It is transformational. However, it isn't difficult or onerous to undertake. It just takes commitment, education, and a willingness to change your thinking. Lean tools and techniques can be implemented by any organization, from small to large, from manufacturing to service. Lean should be, and might be in the future, part of the basics or fundamentals upon which improvement is based.

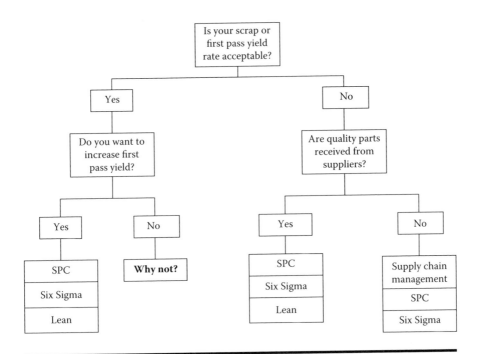

Figure 12.7   Quality yield.

12.2.4 Six Sigma

Six Sigma is a family of tools and techniques for increasing customer satisfaction by meeting customer specifications and customer needs. The primary means of meeting customer specifications is by reducing the variation in the process of producing the product or providing the service.

Six Sigma utilizes a variety of statistical and problem-solving techniques, including teams of specially trained people. Six Sigma is often linked with Lean Manufacturing, in that Six Sigma techniques are useful in implementing many of the components of Lean.

12.2.5 Theory of Constraints

Theory of constraints (TOC) is an interesting, useful, and often misunderstood management system. It is primarily viewed as a production scheduling technique, because that is its origin. As such, it is very good and worthwhile to undertake. But TOC is much more than just a production scheduling tool.

A fairly distinct aspect of TOC is the set of TOC thinking processes. The thinking processes are problem-solving, conflict resolution, and

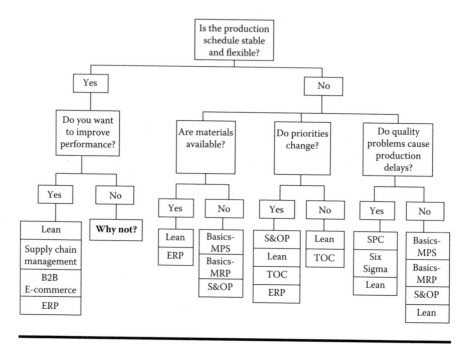

Figure 12.8   Production schedule.

change management tools based on logic and the Socratic method of teaching. Still, TOC is more than that, too.

TOC is a comprehensive system for identifying system constraints, both internal and external to the organization. It then uses the tools from the production scheduling side and the thinking processes side to overcome the constraints and increase the throughput of the entire system.

12.2.6 Statistical Process Control

Statistical process control (SPC) should be used by anyone who manufactures anything. Service providers can and should use it too. SPC is an integral part of Six Sigma and is a well-known and fundamental tool of the quality profession. It is a tool for measuring the output of a process to see whether the output conforms to defined standards or if the process is operating within defined limits.

If you're not using SPC to monitor your processes, you should. If you have any quality problems or if you want to improve the quality of your products or services, you should start using SPC. If you have any variation in the output of a process or if you are having trouble controlling the

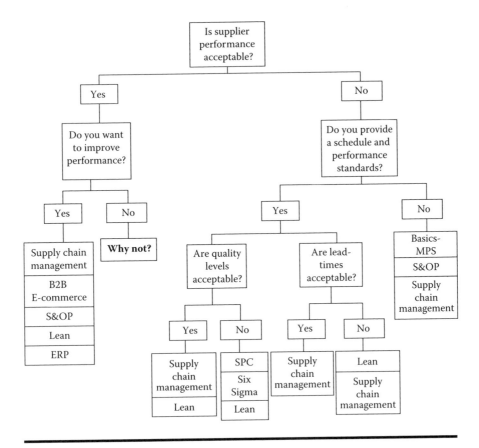

Figure 12.9 Suppliers.

process, you should start using SPC to help you monitor the process so you can begin to get it under control.

12.2.7 ISO 9000

ISO 9000 primarily is undertaken only when it is required by a customer or is a requirement to enter a particular market. If it is required, you will undertake the task of becoming certified to whichever particular standard is called for. Two things: if you begin the certification process because it is required, you should embrace the standards and use them as a basis for developing an improved quality, and management, system; even if ISO certification is not a requirement, you can use the standards as a guide to develop an improved quality management system.

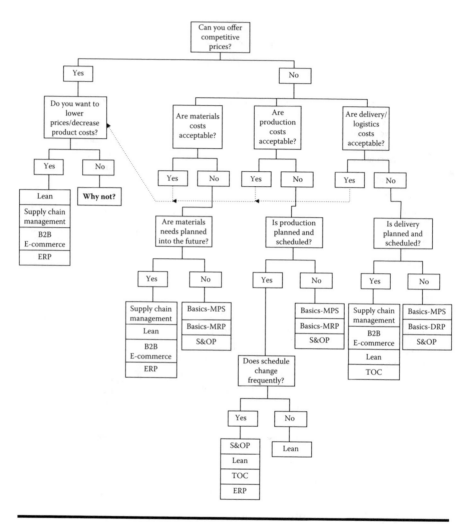

Figure 12.10 Prices.

12.2.8 Malcolm Baldrige National Quality Award Criteria

The Baldrige Award criteria are a framework for a comprehensive management system. The award part is secondary. The use of the criteria as the framework and basis of a comprehensive business management system is the primary benefit. The criteria do not proscribe any particular method for meeting the requirements. This allows each organization to develop its own operating systems and methods for meeting the requirements of the criteria. Many of the other tools described in this book can be used to meet the requirements. The criteria are proven and effective, thus

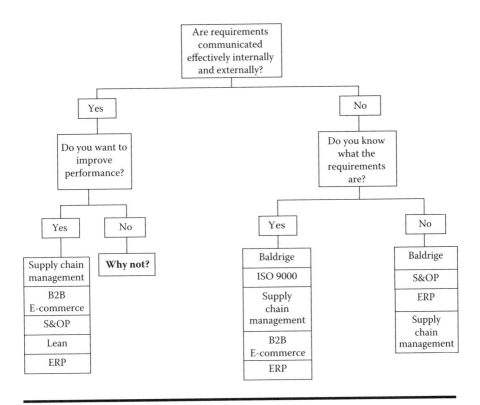

Figure 12.11   Communications.

freeing the organization from having to reinvent the wheel when it comes
to developing an overall management system to run the business.

12.2.9 Project Management

You're already managing projects, maybe even on a daily basis. Why not
manage them effectively so you're more likely to meet the stated goals
of the project? By using proven methods for planning and managing a
project, you're more likely to meet your goals and to complete the project
on time and on budget. Use a project management methodology developed and used by project management professionals. Learn about critical
chain project management as an alternative to standard project management time and resource planning.

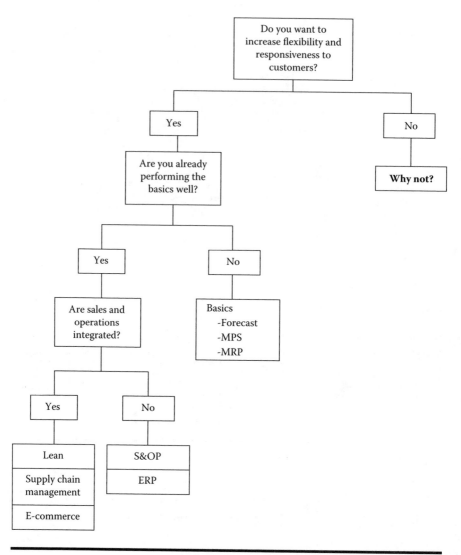

Figure 12.12   Flexibility.

12.2.10 Enterprise Resource Planning

You are almost certainly using a computerized system to manage at least some part of your business. If you haven't already taken the leap to an enterprisewide, integrated system, you're probably looking into it or considering it. Or you could be looking for a new system, an upgrade, or some enhancements or add-ons. An ERP system is a tool to help you improve your performance. It is not a tool that will make a poorly run

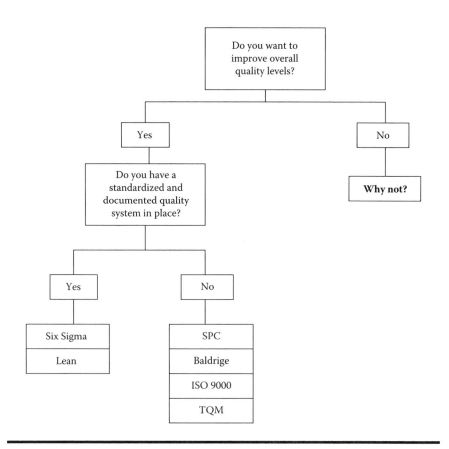

Figure 12.13   Overall quality.

system work better. Communications, number crunching, and data storage and manipulation are the main inner workings of an ERP system. The benefit comes from using these functions to make improvements and raise the level of performance. If you're not communicating effectively, don't have well-developed plans, and don't have a high level of performance already, an ERP system won't help you much.

If you're going to implement an ERP system, or any other technology for that matter, you need an effective methodology to do it. Analysis, justification, team selection, training and education, and system selection and installation must be performed well to achieve the desired results.

12.2.11 Supply Chain Management

You're already part of the supply chain, the global supply chain. A formal supply chain management system designed to improve communications

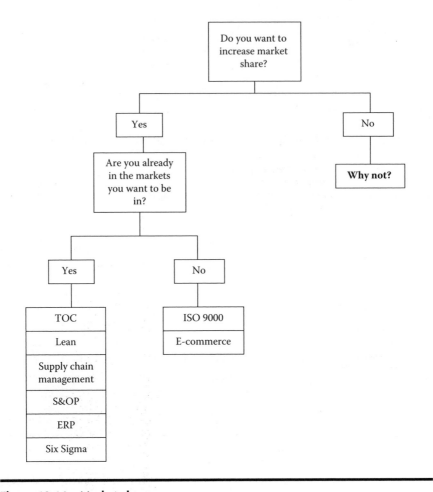

Figure 12.14 Market share.

and the flow of products and information among trading partners in order to reduce total supply chain costs and increase customer satisfaction is a worthy goal. It's not easy to link systems, coordinate schedules, align goals, and increase cooperation among a complex web of suppliers, customers, and service providers, but it should be a desire and it should be attempted.

Supply chain management formalizes and enhances many of the processes that already exist among trading partners. It adds mechanisms to improve relationships and includes performance measures that help improve the system.

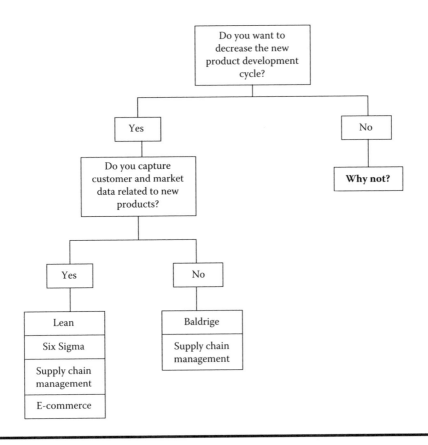

Figure 12.15   New product development.

12.3 E-Commerce

E-commerce is simply a natural evolution of the normal activities of an economy. Just as the industrial revolution transformed the way business was conducted, modern technology in the form of computers and other electronic systems has again transformed the way business is conducted. E-commerce leverages technology, which allows an organization to reach broader goals with fewer resources. The smallest of companies — one person working part time out of their home — can reach a global market cheaply and easily. Technology allows partner organizations to communicate and conduct trading activities with fewer resources than in the past. Transactions ranging from solicitation of bids and engineering collaborations to the transfer of funds can be performed in real time.

Entry into E-commerce takes a significant amount of planning and investment in time, money, and equipment. The rewards can be great,

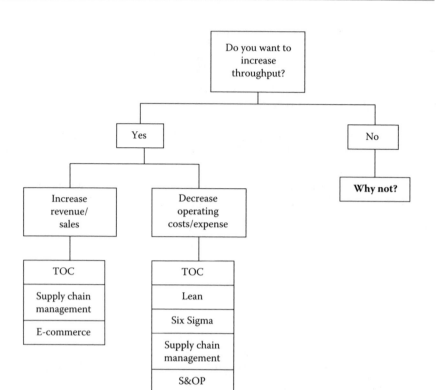

Figure 12.16   Throughput.

however. Reaching new markets, shortening the business cycle, and leveraging resources are valuable benefits.

12.4 Service — Your Competitive Weapon (But I'm a Manufacturer!)

U.S. manufacturers, including small manufacturers, *can* compete in the global marketplace. It isn't easy, but you have a lot of tools at your disposal to help you. We've discussed a few in this book. We talked about how you can select the tools, and we looked at how and when you can use them.

It is becoming ever more difficult, if not impossible, for U.S.-based manufacturers to compete on price. Even if you can't compete on price, you still have to have reasonable prices, which means you have to have a competitive cost structure. That means you must have the best performing operations possible. This book has focused primarily in that operations area. Flawless performance and superior quality are no longer market differentiators; they're market qualifiers. In other words, you're not going to differentiate yourself from your competitors based on your performance and your quality; those aspects are expected and are a requirement just to enter the market. So where does that leave you?

Many organizations still need to improve their performance and quality, so use the tools to help yourself do that. But once you do that you still need to differentiate yourself. That's where service comes in. You need to provide such outstanding service that customers search you out rather than you looking for them. People are willing to pay for outstanding service, especially because so much substandard service still exists out there.

People won't spend $100 for an hour's massage at a spa when they can get the same massage for a fraction of that price at a massage school or from an independent masseuse, unless they get something extra at the spa. That something extra is service. They pay for the experience of the spa and the service. You need to offer the service and the experience to all of your supply chain partners and the final consumer that makes them want to come back to you day after day. Service isn't a discipline that many operations types throw themselves into, but you should be thinking about service in everything you do. How does the production scheduling process impact your customer service? How does cycle counting impact service? How do warehouse operations affect the customer's experience? Incorporate the service perspective in everything you do, in every operation you perform, and every process you design or improve, and you'll reap the rewards in the marketplace. Good luck.

12.5 The Last Word

There is no quick and simple way to determine which tool or tools you should choose to help improve your business. I do hope this book has given you some insight and knowledge into some of the various tools at your disposal and helps you on your journey to increased profitability and success.

Appendix A provides a short list of books, magazines, and organizations where you can find more information. There are more consultants out there than you can shake a stick at who are more than willing to help you (for a fee, of course). Use all the resources at your disposal, but don't forget to look inside your own organization first. Talk to, listen to, and engage people throughout your organization, at all levels. You just might be surprised at what (or who) you find. Encourage your people to contribute. Encourage radical thinking. Encourage involvement in outside organizations and activities. Find the gem that's just looking for the right setting.

Appendix A

References and Resources

A.1 Organizations

APICS

The Association for Operations Management
5301 Shawnee Road
Alexandria, VA 22312-2317
1-800-444-2742
www.apics.org

ASQ

American Society for Quality
600 N. Plankinton Avenue
Milwaukee, WI 53203
417-272-8575
www.asq.org

SME

Society of Manufacturing Engineers
One SME Drive, P.O. Box 930
Dearborn, MI 48121
313-271-1500
www.sme.org

IIE

Institute of Industrial Engineers
3577 Parkway Lane, Suite 200
Norcross, GA 30092
800-494-0460 / 770-449-0460
www.iienet.org

PMI

Project Management Institute
Four Campus Boulevard
Newtown Square, PA 19073-3299
610-356-4600
www.pmi.org

The Supply Chain Council

150 Freeport Road
Pittsburgh, PA 15238
412-781-4101
www.supply-chain.org

Warehousing Education and Research Council (WERC)

1100 Jorie Boulevard, Suite 170
Oak Brook, IL 60523-4413
630-990-0001
www.werc.org

Council of Logistics Management/Council of Supply Chain Management Professionals

2805 Butterfield Road, Suite 200
Oak Brook, IL 60523-1170
630-574-0985
www.clm1.org

ISM

Institute for Supply Management

P.O. Box 22160
Tempe, AZ 85285-2160
480-752-6276 / 800-888-6276
www.ism.ws

Lean Enterprise Institute

P.O. Box 9
Brookline, MA 02446
617-713-2900
www.lean.org

The Goldratt Institute

442 Orange Street
New Haven, CT 06511
203-624-9026
www.goldratt.com

National Association of Manufacturers (NAM)

1331 Pennsylvania Avenue, NW
Washington, DC 20004-1790
202-637-3000
www.nam.org

Manufacturing Extension Partnership

NIST MEP
100 Bureau Drive, Stop 4800
Gaithersburg, MD 20899-3460
301-975-5020
www.mep.nist.gov

IMA

Institute of Management Accountants
10 Paragon Drive
Montvale, NJ 07645-1718
800-638-4427 / 201-573-9000
www.imanet.org

A.2 Books

APICS Dictionary

Eleventh Edition, APICS, 2005

5-Phase Project Management: A Practical Planning and Implementation Guide

Joseph W. Weiss and Robert K. Kysocki
Perseus Books Publishing, L.L.C., 1992

All I Need to Know about Manufacturing I Learned in Joe's Garage: World Class Manufacturing Made Simple

William B. Miller and Vicki L. Schenk
Bayrock Press, 1996

Back to Basics: Your Guide to Manufacturing Excellence

Steven A. Melnyk and R. T. "Chris" Christensen
St. Lucie Press/APICS Series on Resource Management
CRC Press LLC, 2000

Bills of Material: Structured for Excellence

Dave Garwood
Dogwood Publishing Company, 1988

Capacity Management

John H. Blackstone, Jr.
South-Western College Publishing, 1989

Critical Chain

Eliyahu M. Goldratt
North River Press, 1997

Distribution Planning and Control

David Frederick Ross
Kluwer Academic Publishers, 2000

ERP: Tools, Techniques, and Applications for Integrating the Supply Chain

Carol A. Ptak, CFPIM, CIRM, Jonah, PMP, with Eli Schragenheim
St. Lucie Press/APICS Series on Resource Management
CRC Press LLC, 2000

The Goal

Second revised edition
Eliyahu M. Goldratt and Robert E. Fox
North River Press, 1992

Integrated Learning for ERP Success: A Learning Requirements Planning Approach

Karl M. Kapp with William F. Latham and Hester N. Ford-Latham
St. Lucie Press, 2001

Integrating Your E-Business Enterprise

Andre Yee and Atul Apte
Sams Publishing, 2001

Introduction to ISO 9000:2000 Handbook

Roderick S. W. Goult
The Victoria Group, 2001

Introduction to Materials Management, Third Edition

J. R. Tony Arnold, CFPIM, CIRM
Prentice Hall, 1998

Inventory Record Accuracy: Unleashing the Power of Cycle Counting

Roger B. Brooks and Larry W. Wilson
John Wiley & Sons, 1995

The ISO 9000 Book: A Global Competitor's Guide to Compliance & Certification, Second Edition

John T. Rabbitt and Peter A. Bergh
Quality Resources, 1994

Just-In-Time: Making It Happen: Unleashing the Power of Continuous Improvement

William A. Sandras, Jr.
John Wiley & Sons, 1989

Lean Thinking: Banish Waste and Create Wealth in Your Corporation

James P. Womack and Daniel T. Jones
Simon & Schuster, 1996

Management Dilemmas: The Theory of Constraints Approach to Problem Identification and Solutions

Eli Schragenheim
St. Lucie Press/APICS Series on Constraints Management
St. Lucie Press, 1999

Managing the Supply Chain: The Definitive Guide for the Business Professional

David Simchi-Levi, Philip Kaminsky, and Edith Simchi-Levi
McGraw-Hill, 2004

Manufacturer's Guide to Implementing the Theory of Constraints

Mark J. Woeppel
St. Lucie Press/APICS Series on Constraints Management
St. Lucie Press, 2001

Manufacturing Planning and Control Systems, Third Edition

Thomas E. Vollmann, William L. Berry, and D. Clay Whybark
IRWIN/APICS Series in Production Management
IRWIN Professional Publishing, 1992

Master Scheduling: A Practical Guide to Competitive Manufacturing

John F. Proud
John Wiley & Sons, 1994

Master Scheduling in the 21st Century: For Simplicity, Speed and Success — Up and Down the Supply Chain

Thomas F. Wallace, and Robert A. Stahl
T. F. Wallace & Company, 2003

MRP and Beyond: A Toolbox for Integrating People and Systems

Carol A. Ptak, CFPIM, CIRM
Irwin Professional Publishing, 1997

MRP II: Making It Happen — The Implementers' Guide to Success with Manufacturing Resource Planning

Second edition
Thomas F. Wallace
John Wiley & Sons, 1990

Necessary But Not Sufficient

Eliyahu M. Goldratt with Eli Schragenheim and Carol A. Ptak
North River Press, 2000

Orchestrating Success: Improve Control of the Business with Sales and Operations Planning

Richard C. Ling and Walter E. Goddard
John Wiley & Sons, 1988

Orlicky's Material Requirements Planning, Second Edition

George W. Plossl
McGraw-Hill, 1994

Production Activity Control

Steven A. Melnyk and Phillip L. Carter
McGraw-Hill, 1987

A Guide to the Project Management Body of Knowledge: PMBOK® Guide — 2000 Edition

Project Management Institute, 2000

The Race

Eliyahu M. Goldratt and Robert E. Fox
North River Press, 1986

Sales & Operations Planning: The How-To Handbook

Thomas F. Wallace
T. F. Wallace & Company, 1999

Sales Forecasting: A New Approach

Thomas F. Wallace and Robert A. Stahl
T. F. Wallace & Company, 2002

The Business of E-Commerce: From Corporate Strategy to Technology

Paul May
Cambridge University Press, 2000

The Six Sigma Way: How GE, Motorola, and Other Top Companies are Honing Their Performance

Peter S. Pande, Robert P. Neuman, and Roland R. Cavanagh
McGraw-Hill, 2000

SPC Simplified: Practical Steps to Quality

Robert T. Amsden, Howard E. Butler, and Davida M. Amsden
Quality Resources, 1998

TPM Development Program: Implementing Total Productive Maintenance

Edited by Seiichi Nakajima
Productivity Press, 1989

What Is This Thing Called Theory of Constraints and How Should It Be Implemented?

Eliyahu M. Goldratt
North River Press, 1990

What Is Total Quality Control? The Japanese Way

Kaoru Ishikawa, translated by David J. Lu
Prentice Hall, 1985

World-Class Warehousing and Material Handling

Edward H. Frazelle
McGraw-Hill, 2002

A.3 Magazines

APICS Magazine

APICS — The Association for Operations Management
5301 Shawnee Road
Alexandria, VA 22312-2317
703-354-8851
www.apics.org

Quality Progress

American Society for Quality
600 N. Plankinton Avenue
Milwaukee, WI 53203
417-272-8575

A.4 Articles

"Blitzing the 'Information Factory'"

Frank Kieffer, CIRM, Jonah, and Lynn Gates
APICS 2000 International Conference Proceedings

"A Little Lean"

Rebecca Morgan
APICS — The Performance Advantage
July/August 2002

Appendix B

Normal Distribution versus Sigma Conversion

These tables show the difference between the normal distribution and the Sigma level of performance, or the Sigma Conversion. If the normal distribution was used, a yield rate of 68.3% would be a 1 Sigma level of performance (meaning that only one standard deviation fits within the specification range). By comparison, the accepted definition of a 1 Sigma level of performance is equal to a yield rate of 30.9%.

Normal Distribution

Standard Deviation = "Un-shifted" Sigma Level	% of Values = Yield Rate	Theoretical quantity of holes that meet specifications	Theoretical quantity that meet specifications if 1,000,000 holes drilled	Defect Rate = 1 - Yield Rate	Theoretical Defect Quantity	Theoretical Defects per million holes drilled
1	68.3%	10,928	683,000	31.7%	5,072	317,000
2	95.4%	15,264	954,000	4.6%	736	46,000
3	99.7%	15,952	997,000	0.3%	48	3,000
4	99.994%	15,999	999,940	0.006%	0.96	60
5	99.9999%	15,999.98	999,999.00	0.0001%	0.016	1.0
6	99.9999998%	15,999.99997	999,999.99800	0.0000002%	0.000032	0.0020

Sigma Conversion Table

Sigma Level	Yield Rate	Theoretical quantity of holes that meet specifications	Theoretical quantity that meet specifications if 1,000,000 holes drilled	Defect Rate = 1 - Yield Rate	Theoretical Defect Quantity	Theoretical Defects per million holes drilled (DPMO)
1	30.9%	4,936	308,500	69.2%	11,064	691,500
2	69.2%	11,064	691,500	30.9%	4,936	308,500
3	93.3%	14,931	933,200	6.7%	1,069	66,800
4	99.380%	15,901	993,800	0.620%	99.20	6,200
5	99.9770%	15,996.32	999,770.00	0.0230%	3.680	230.0
6	99.9996600%	15,999.94560	999,996.60000	0.0003400%	0.054400	3.4000

The difference between the normal distribution and the defined Sigma Level is because of what is known as the "1.5 sigma shift." The originators of Six Sigma at Motorola discovered that processes are not stable over time and that relatively small amounts of data, such as a week's or month's worth are representative of data over the long term, such as a year or five years. Because of this, a conversion from the normal distribution to the "shifted" Sigma Level was developed. The normal distribution and the Sigma Conversion are shown here for comparison.

Index

Q